Global Change – The IGBP Series

Springer

Berlin
Heidelberg
New York
Barcelona
Hong Kong
London
Milan
Paris
Tokyo

Peter Tyson · Roland Fuchs · Congbin Fu · Louis Lebel · A. P. Mitra
Eric Odada · John Perry · Will Steffen · Hassan Virji (Eds.)

Global-Regional Linkages in the Earth System

With 154 Figures and 43 Tables

Springer

Editors

Peter Tyson
Climatology Research Group
University of the Witwatersrand
PO Wits, Johannesburg, 2050
SOUTH AFRICA

Congbin Fu
START TEA RC
Institute of Atmospheric Physics
Chinese Academy of Sciences, PO Box 2718
Qijia Huo Zi, De Wai Street, Chao Yang District
Beijing 100029
CHINA

Roland Fuchs
International START Secretariat
2000 Florida Ave. NW Suite 200
Washington, DC 20009
USA

Louis Lebel
Department of Social Science
Chiang Mai University
THAILAND

A.P. Mitra
National Physical Laboratory, New Delhi
K. S. Krishnan Road
New Delhi 110012
INDIA

Eric Odada
Department of Geology
University of Nairobi
Box 30197
Nairobi
KENYA

John Perry
6205 Tellyho Lane
Alexandria, VA 22307
USA

Will Steffen
IGBP Secretariat
The Royal Swedish Academy of Sciences
Box 50005
S-104 05 Stockholm
SWEDEN

Hassan Virji
International START Secretariat
200 Florida Ave. NW Suite 200
Washington, DC 20009
USA

ISSN 1619-2435
ISBN 3-540-42403-2 Springer-Verlag Berlin Heidelberg New York

Die Deutsche Biliothek - CIP-Einheitsaufnahme

Global regional linkages in the earth system : with 43 tables / Peter Tyson
... (ed.). - Berlin ; Heidelberg ; New York ; Barcelona ; Hong Kong ; London ;
Milan ; Paris ; Tokyo : Springer, 2002
(Global change)
ISBN 3-540-42403-2

Springer-Verlag Berlin Heidelberg New York
a member of the BertelsmannSpringer Science+Business Media GmbH

Printed in Germany

Cover Design: International START Secretariat and Erich Kirchner, Heidelberg
Dataconversion: Büro Stasch, Bayreuth

SPIN: 10847577 3130 – 5 4 3 2 1 0 – Printed on acid-free paper

Dedicated to
Tom Malone, who inspired so many,
and
developing country scientists who strive against the odds.

Preface

Global environmental change occupies a central niche in the pantheon of modern sciences. There is an urgent need to know and understand the way in which global biogeochemical cycles have changed over different time scales in the past and are likely to do so in the future. Equally important, it is necessary to determine the extent to which natural variability and that induce by anthropogenic activities are bringing about change. A number of international co-operative scientific programmes address these issues. Chief among them are the International Geosphere-Biosphere Programme (IGBP), the World Climate Research Programme (WCRP) and the International Human Dimensions Programme (IHDP) for global change. This book is one of a series of IGBP syntheses drawing together findings in global environmental change over the past decade or so.

One focus of IGBP activities is the System for Analysis, Research and Training (START). Co-sponsored by the WCRP and IHDP, START establishes regional research networks for global change science in developing countries, stimulates and carries out global change research in developing regions of the world, and builds capacity to undertake such research at personal, institutional and regional levels. Several regional global change networks have been established, and much regional research has been accomplished in the last five years or so. In this book, work relating to four of the older START regions, Southern Africa, South Asia, Southeast Asia and East Asia, will be used as case studies to illustrate regional-global linkages in Earth System Science.

The results of START regional research form a major component of the synthesis and reflect in part the outcome of research-led START capacity-building efforts. In addition, all other relevant and accessible global change research has been considered. As far as has been possible, only published research has been considered. No claim is made that all the research integrated into the syntheses has been sponsored or fostered by START. The synthesis itself is, however, that of the developing country scientists involved in START. While every effort has been made to include as much relevant material as possible, undoubtedly some will have been missed depending on the willingness of individuals to become involved, and on a variety of other normal human oversights.

Inevitably, coverage within and between regions is uneven, being a function of the interests and support given to regional scientists, the degree of IGBP, WCRP and IHDP core-programme regional involvement and the degree of national global-change-research carried out by national institutions (particularly universities) and individuals in regions. Case studies, and in some instances transect work, provide the foundation of the integration. Extensive use is made of modelling to interpolate between local findings and to gain maximum regional coverage. In global studies, the interests of individual scientists become subsumed in the greater whole. In contrast, the nature of the work undertaken in regional studies often depends upon the interests of individuals and research groups, however cohesive and comprehensive regional science planning may have been. This diversity is reflected in the four regional syntheses offered in this book. Two examples make the point. In the case of Southern Africa, it is argued that regional change is mainly the consequence of natural driving forces of global change, modulated substantially by anthropogenic influences. In the instance of Southeast Asia, a case is made that the primary control of regional change is the human impact of economic globalisation, modulated by natural forces of change. Such contrasting approaches are a great strength and result in fascinatingly different regional insights into the way in which global change may be impinging on regions and the way in which regional change may be contributing to the global system as a whole.

Undoubtedly a regional approach is going to add substantially to understanding the future of the global system in all its complexity. It is hoped that in a small way this book on regional-global linkages in the Earth System will point the way to a sound alternative way of studying the global system as a whole.

It is appropriate to acknowledge with gratitude the considerable assistance given by many in the preparation of the book. At the Climatology Research Group, University of the Witwatersrand, Johannesburg, Wendy Job prepared the figures; Kristy Ross checked all the references. At the International START Secretariat, Washington, DC, Amy Freise helped edit the East Asia chapter; Mayuri Sobti assisted with compilation of the South Asia chapter and together with Ching Wang helped with cover design, layout, references, tables and the seeking of permissions to use copyright material. Regional maps were prepared by the US State Department. Referees of individual chapters and readers of the book as a whole are thanked for their time, effort and constructive comments.

Roland Fuchs, Executive Director of START
Peter Tyson, Chairman, START Scientific Steering Committee

Washington, DC, January 2002

Contents

Contributors

Y. P. Abrol
Division of Plant Physiology, Indian Agricultural Research Institute
New Delhi 110 012, INDIA

Congbin Fu
START TEA RC
Institute of Atmospheric Physics, Chinese Academy of Sciences
PO Box 2718, Qijia Huo Zi, De Wai Street, Chao Yang District
Beijing 100029, CHINA

Hideo Harasawa
Environmental Planning Section, Social & Environment System Division
National Institute for Environmental Studies
16-2 Onogawa, Tsukuba, Ibaraki 305, JAPAN

Naveen Kalra
Divsion of Environmental Sciences, Indian Agricultural Research Institute
New Delhi 110 012, INDIA

Vladimir Kasyanov
Marine Biology Institute, Far East Branch, Russian Academy of Sciences
17 Palchevskogo UL, Vladivostok 690 032, RUSSIA

Jeong-Woo Kim
Department of Atmospheric Sciences, Yonsei University
134 Shinchon-Dong, Seodaemun-ku, Seoul 120-749, REPUBLIC OF KOREA

M. Dileep Kumar
National Institute of Oceanography
Dona Paula, Goa-403004, INDIA

K. Rupa Kumar
Indian Institute of Tropical Meteorology
Homi Bhabha Road, Pune-411008, INDIA

Louis Lebel
Department of Social Science, Chiang Mai University
THAILAND

A. P. Mitra
National Physical Laboratory, New Delhi
K. S. Krishnan Road, New Delhi 110012, INDIA

S. W. A. Naqvi
National Institute of Oceanography
Dona Paula, Goa 403 004, INDIA

Eric Odada
Department of Geology, University of Nairobi
Box 30197, Nairobi, KENYA

Dennis Ojima
Natural Resources Ecology Lab, Colorado State University
Fort Collins, CO 80523-1499, USA

Roland Schulze
Department of Agricultural Engineering, University of Natal
PO Box 375, Private Bag X01, Pietermaritzburg 3200, SOUTH AFRICA

Peter Tyson
Climatology Research Group, University of the Witwatersrand
PO Wits, Johannesburg, 2050, SOUTH AFRICA

M. Velayutham
National Bureau of Soil Survey & Land Use Planning
Amravati Road, Nagpur 440 010, INDIA

Coleen Vogel
School of Geography, Archaeology and Environmental Studies
University of the Witwatersrand
PO Wits, Johannesburg, 2050, SOUTH AFRICA

Zhibin Wan
START TEA RC, Institute of Atmospheric Physics, Chinese Academy of Sciences
Qi Jia Huo Zi, De Wai Street, Chao Yang District, Beijing 100029, CHINA

Shidong Zhao
Commission for Integrated Survey of Natural Resources, Chinese Academy of Sciences
PO Box 9717 Building 917, 3# Datun Road, Chaoyang District, Beijing 100101, CHINA

Chapter 1

Regional Studies and Global Change

The phrase, 'the future of the past,' is in many ways the epitome of Earth System science and global change. For the natural component of the system, the statement is to a large extent true; for anthropogenically induced change, it is true only for the Anthropocene – the latest era in Earth's history in which the influence of humanity has come to rival that of nature. While humans have modified their environment from earliest times, the scale of that modification only assumed global proportions after the Industrial Revolution. Initially, global change science was concerned only with the global dimensions of natural and human-induced change. Increasingly, however, scientists have come to recognise that regional processes must be taken into account in global change science.

Understanding the Earth System requires that the two-way linkages between regions and the global system be well understood and predictable. Most studies of global change are undertaken thematically across limited disciplinary boundaries; few are attempted holistically across many disciplines within regions. The regional approach offers many advantages. Here several diverse linkages from four regions are presented to illustrate the power of a regional approach to Earth System science.

Global environmental change is manifested in many different ways at sub-global scales. Traditionally, the Earth System and its perturbation by humans are studied by deconvolution of the whole into functional physical and biophysical component parts (the climate subsystem, the oceanic subsystem, the carbon cycle, etc). After research into these, synthesis of the results leads to a better understanding of the fully integrated system.

An alternative approach is offered by regional decomposition. Regions are a natural scale for such assessments, as regions are often defined by shared cultural, political or biogeographical contexts, or by common resource bases and pollution sources and sinks. Regions, moreover, offer the opportunity for holistic, systems-based studies of linkages between regional and global change and so provide significant insights into the functioning of the Earth System itself.

In this book, it is demonstrated that an integrated approach to studying *regional* environmental change in its own right is a powerful tool for enhancing understanding of the Earth System at the *global* scale. Regional-global linkages are ubiquitous, bi-directional and critical for system functioning. Four case studies of individual aspects of integrated regional studies are presented to illustrate how their outcomes may contribute to a better understanding of the Earth System as a whole. The four regions are Southern Africa, South Asia, Southeast Asia and East Asia (Fig. 1.1).

Depending on the interests and backgrounds of the contributors in the different regions, different approaches have been used in the various regional syntheses. They vary significantly. On the one hand, arguments are based on the notion that regional change is mainly the consequence of natural driving forces of global change, modulated substantially by anthropogenic influences. On the other, it is proposed that a primary control of regional change is the human impact of economic globalisation, modulated by natural forces of change. Such contrasting approaches are a great strength. They result in important different regional insights into the way in which global change may be impinging on regions, and the way in which regional change may be contributing to the global system as a whole.

In the Southern African case study, the production, prolonged atmospheric recirculation of trace gases and aerosols within the region before subsequent transport out of the region are shown to have significant effects on far-removed areas of the globe. An example is given of how atmospheric circulation can link particulate nutrient sources with distant biospheric sinks to affect ecosystem functioning at the regional scale and planetary metabolism at the global scale. It is also shown how rapidly evolutionary change in lacustrine biota took place in the last few millennia and how changes in *gradients* of climatic change across the region affect water resources and agro-pastoral activities.

In the South Asian region, the frequency and magnitude of South Asian monsoon failures over the last millennium is demonstrated using high-resolution Himalayan ice-core records. These show that during the twentieth century the increase in anthropogenic activity in the regions and its adjacent neighbours to the west is recorded by a doubling of chloride concentrations and

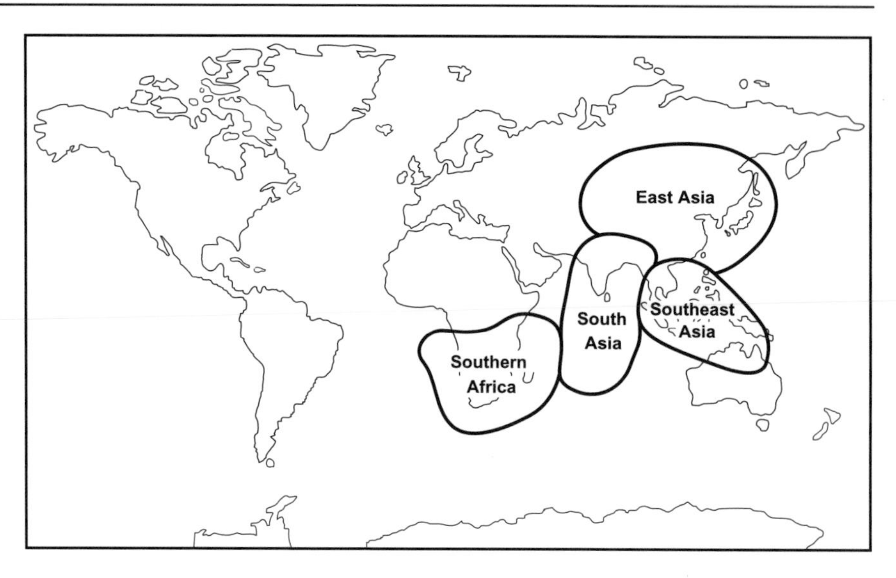

Fig. 1.1. The four regions considered in this book

a fourfold increase in dust. Over the Tibetan Plateau it is shown that surface twentieth-century warming appears to be amplified at higher elevations. To the south, global warming appears to be manifest in an increase in near-surface maximum, rather than minimum, temperatures over large parts of the region. It is argued that changes in trace gas emissions in South Asia may have impacts on the global distribution of tropospheric ozone. Alterations in the biogeochemical balances of the Arabian Sea and Bay of Bengal following atmospheric and land transport of material offshore, and the effect of global change and climate variability on crop production are also considered.

In the East Asian case study, the most important finding to emerge is based on regional modelling. It shows how human modification of land cover over a long period of time may have resulted in changes in albedo, surface roughness, leaf area index and fraction of vegetation cover significant enough to have altered the complex exchanges of water and energy from the surface to the atmosphere. Without any changes due to greenhouse gas forcing, changes in land cover alone are shown to have had the potential to alter both vertical and horizontal moisture transport over the region and to have brought about significant changes in the East Asian monsoon.

In the Southeast Asian chapter, the importance of the human forcing of global change is taken further. Here it will be argued that globalisation of economic activity, especially through its effects on industrialisation and the commercialisation of agriculture and forestry, has been a primary driving force for change in the region. Rapid development in the region, responding to and part of economic globalisation, has lead to a situation where the trace gas and aerosol contents of the atmosphere are increasing much more rapidly than the gross domestic product, in some places at up to ten times the rate. The consequences for global change are examined.

The common message from the contrasting perspectives of the different regional syntheses undertaken is that the regions are being significantly affected by environmental change, with different impacts in different regions. At the same time, changes originating in the regions are having impacts that transcend the regions and may affect the planetary metabolism at a global scale.

Chapter 2

Regional-Global Change Linkages: Southern Africa

Peter Tyson · Eric Odada · Roland Schulze · Coleen Vogel[1]

2.1 Introduction

Unravelling the skein of global change effects in southern Africa is a non-trivial task. It is made all the more interesting since Africa is the birthplace of humanity. Southern Africa preserves an impressive five-million-year record of human-environmental interaction. From the evolutionary cradle onwards, environmental change has profoundly affected the development of the early and later hominids into *Homo sapiens* (Vrba et al. 1995). More recently, over the past two millennia, environment was a major factor affecting migrations of Bantu people into southernmost Africa. Until as late as the nineteenth century, environment continued to be a dominant factor affecting the settlement and survival of the population of the region.

Over the past few centuries the influence of environment on human activities has weakened and in many instances reversed as humans have increasingly been modifying the environment, often irreversibly. The synergies between the effects of environmental change on humans and *vice versa* are subtle, complicated and ever changing. They affect not only the manner and degree of global change taking place, but also the way in which regional manifestations of global change are expressed. This is particularly so in southern Africa, here defined as all countries in Africa south of the equator, including all of Kenya and none of the Congo (Fig. 2.1). Southernmost Africa will be taken to mean South Africa, Lesotho and Swaziland.

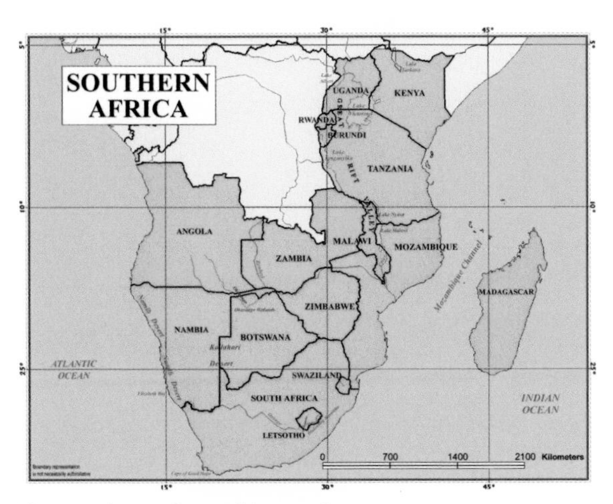

Fig. 2.1. The southern African region

One of the factors that gives regional unity to so much of southern Africa is its climate and the subcontinental circulation that maintains it and transports air, and whatever trace gases and aerosols it carries, over the region. However, regional change arising out of global change is much more than just the effect of climatic change; it is the interaction of a variety of forcings, the human dimensions of which are of fundamental importance. Nonetheless, whichever way one looks at it, a large component of the regional manifestation of global change inevitably turns out to be driven by changing climate, which is in turn driven by a changing atmosphere and land surface. In this regional study, changing atmospheric circulation, and all its consequences, will be a primary focus for the integration and synthesis that will be presented.

A synopsis of what is to be discussed may be useful. In working from the distant past to the present and future, the increasing role of human activity as a major driver of regional change is an underlying theme. After considering changing environmental conditions over the past half million years, present-day regional transport of air over southern Africa will be considered. Fluxes of trace gases into the lower troposphere, transport of these gases and aerosols, their recirculation and consequences over the region and adjacent oceans will be examined. Thereafter the oceans around southern Africa will be considered, as too will be regional changes in the hydrological cycle and terrestrial ecosystems. Human acceleration of regional change and impacts of natural and human-induced change on small-scale and commercial agriculture will be considered before finally presenting scientific surprises and major findings from the synthesis.

The 1998 population of southern Africa was around 220 million and growing at an annual rate of around 3%. This rate of growth represents a doubling about every 25 years. Since 1950 there has been a more than five-fold increase in the population of the region. To a significant degree, it is this increase in the human pressure on the

[1] *Contributing authors:* J. Arntzen, R. Chanda, C. Gatebe, T. Hoffman, J. Lutjeharms, L. Marufu, G. Midgely, P. Monteiro, D. Obura, L. Otter, T. Partridge, L. Perks, S. Ringrose, M. Scholes, R. Scholes.

land that is driving the regional environmental changes that are occurring and are likely to occur in future.

Approximately two-thirds of the regional population is rural, ranging from half in South Africa and Botswana to three-quarters in Tanzania and Lesotho, and 85% in Malawi. Throughout the region, rates of urbanisation are exceeding population growth as migration of the rural poor to cities accelerates. Most of the region's countries are among the world's low-income group, with annual per capita Gross National Product (GNP) being lowest in Malawi, Mozambique and Tanzania at around US$200 (1998 figures). Only South Africa and Botswana can be classified as middle-income countries with per capita GNP exceeding US$2 800.

With its urbanisation and industrialisation, South Africa is the largest user of motor vehicles and consumer of energy in the region. Almost all the energy used in the country derives from fossil fuel burning, mainly in large (3 600 MW) coal-fired power stations. Coal is relatively abundant in South Africa, Zimbabwe and Mozambique, while major oilfields are located offshore of Angola. Mozambique and Namibia have extensive natural gas deposits, presently unexploited. In South Africa, only 4% of electrical energy is derived from hydroelectric, nuclear or renewable sources. In contrast, several thousand MW of installed hydroelectric capacity is available on the Zambezi River at Kariba and Cahora Bassa. The countries of southern Africa share electrical power within a connected grid.

Even where electricity is available, there is a high dependence on biomass fuels (wood, maize cobs and stalks and cattle dung) for domestic energy needs. For example, in South Africa, half the households use wood as a primary fuel; this fraction rises in less-developed countries. The harvesting of fuelwood is a major factor leading to the thinning of woodlands, especially within an economically viable transport distance of urban centres. This radius is extended by the practice of reducing the wood to charcoal in primitive kilns, an activity which is a significant source of carbon monoxide, methane and other hydrocarbons. The resulting deforestation is not only threatening the stability of the environment, but also the sustainability of the biomass supply base.

The level of development and the energy mix over the region means that net carbon emissions from Southern African Development Community (SADC) countries remain low. Per capita carbon emissions are estimated to be below the global average of 1.2 t yr^{-1} (Subak et al. 1993) and are well below those of developed countries. The contribution to total global emissions is less than 2% for all greenhouse gases. With the exception of emissions from the land, South Africa emits vastly more carbon than the rest of the countries of the SADC region combined (Table 2.1). For the region as a whole, net carbon emissions from land use in the region are higher than from energy and SADC contributes about 5% to global carbon emissions from land use (Subak et al. 1993).

Given these apparently low emission figures for the region, it is tempting to minimise the contribution of the region to global change. This would be misleading. The urban and industrial emissions of aerosols and trace gases to the atmosphere, particularly in South Africa, together with those emitted by biomass burning in the tropics, have major implications for regional change. Likewise, the circulation of much natural dust over the subcontinent has the potential to bring about consider-

Table 2.1. Estimated emissions of greenhouse gases for SADC countries. First column is total carbon emissions (energy and biota) in t per capita. All other columns are gross emissions. Data apply roughly to 1988 (after Subak et al. 1993)

	Total C per capita (t)	C emissions (Mt)		CH_4 (kt)	N_2O (kt)	CFC-11 equiv. (kt)
		Energy	Biota			
Angola	1.4	2.0	12.4	334	3	1
Botswana	0.9	0.7	0.5	117	0	0
Lesotho	N/a	0.1	N/a	61	0	0
Malawi	0.5	0.3	5.2	110	2	1
Mozambique	0.5	0.5	6.5	229	2	1
Namibia	0.9	0.7	0.7	185	0	0
Swaziland	1.7	69.0	−1.1	3760	22	14
Tanzania	−0.6	0.2	−0.7	37	0	0
Zambia	0.2	0.5	6.4	1049	3	2
Zimbabwe	1.4	0.8	11.2	231	3	1
South Africa	0.6	3.7	2.3	363	2	1
SADC	1.0	78.4	43.3	6476	37	21
Global	1.2	6431.9	855.9	352398	3783	1369
SADC as % of global	82	1.2	5.1	1.8	1.0	1.5

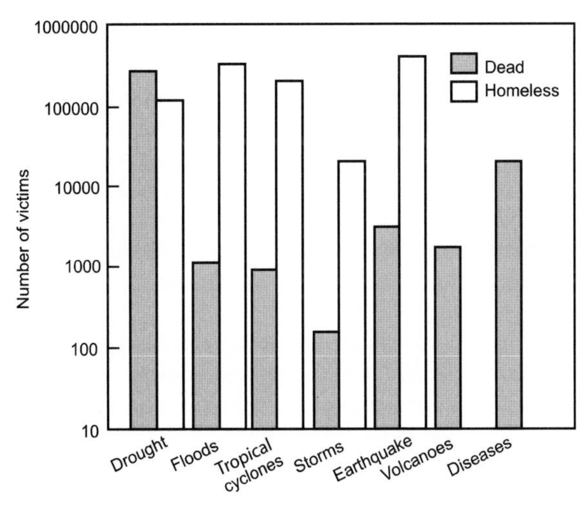

Fig. 2.2. The dead and homeless resulting from recorded disasters in Africa during the 1980s (World Meteorological Organization 1990)

able change in the climate. This is rendered so by unique features of the regional atmospheric circulation affecting the accumulation of aerosols and trace gases and their subsequent transport in the atmosphere. The prevalence of anticyclonic circulations in the subtropics also has implications for other aspects of climate, like the occurrence of extreme droughts and floods.

Droughts are an endemic feature of the climate of southern Africa, always exacerbate prevailing conditions of poverty that compound situations of human stress, and are the single greatest cause of climate-induced deaths in Africa (WMO 1990) (Fig. 2.2). The cost of droughts for all countries of the region is high; they impose the greatest of the climate-induced burdens on national economies. Likewise, floods extol high costs, both in terms of human misery and financially. Droughts and floods represent extreme conditions, and are precisely those that are foreseen to increase in future with global change. Periods of abnormal drought and times when rainfall may have been much above the present-day mean have occurred frequently in the past, and as the palaeo-record indicates, often for centuries at a time.

2.2 The Changing Climate of the Region

The manner in which the climate has changed over southern Africa from late-Quaternary times onwards is best considered in terms of temporal and spatial variability within specific time periods.

2.2.1 Half a Million Years of Change

Data recovered from Antarctic ice cores provides the best high-resolution information on environmental change in the southern hemisphere over the last half-million years. The Vostok core from eastern Antarctica is an excellent example (Petit et al. 1999) (Fig. 2.3). It records changes in atmospheric composition and climate over the last four glacial cycles. The pattern of successional changes, from the rapid onset of warming, through variable declining temperatures to the glacial minimum and abrupt termination of cold conditions, was the same in each cycle. Throughout the last 420 000 years carbon dioxide, methane and other atmospheric and climatic parameters oscillated within regular limits. Two quasi-steady states are apparent. The upper one illustrates the extent to which present-day concentrations of carbon dioxide and methane are without precedent in the past half-million years (Petit et al. 1999). The changes observed in Antarctica were widespread throughout the southern hemisphere and beyond (Imbrie et al. 1984; Partridge et al. 1999; Tyson and Partridge 2000). They are reflected in ocean sediment cores taken off the south-eastern coast of South Africa just north of the subtropical convergence in the Indian Ocean sector of the Southern Ocean (Lemoine 1998) and in cores off Namibia (Kirst et al. 1999) (Fig. 2.4). The correspondence between the Vostok sequence and those off Southern Africa is

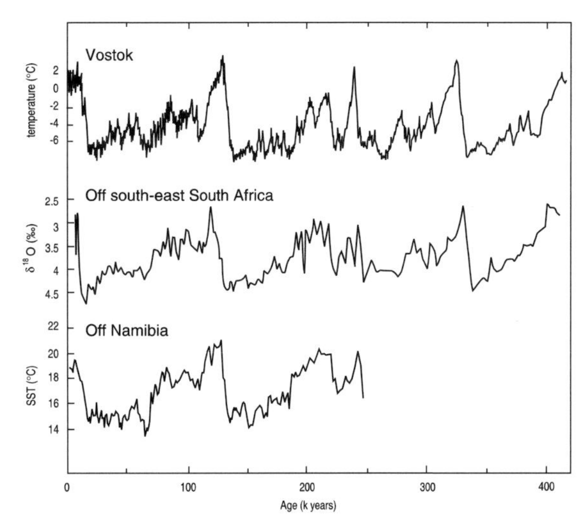

Fig. 2.3. Temperature and oxygen isotope variations during the late Quaternary and Holocene: **a** at Vostok in Antarctica (Petit et al. 1999), **b** off the southeast coast of South Africa (Lemoine 1998) and **c** off Namibia in deep-water conditions (Kirst et al. 1999)

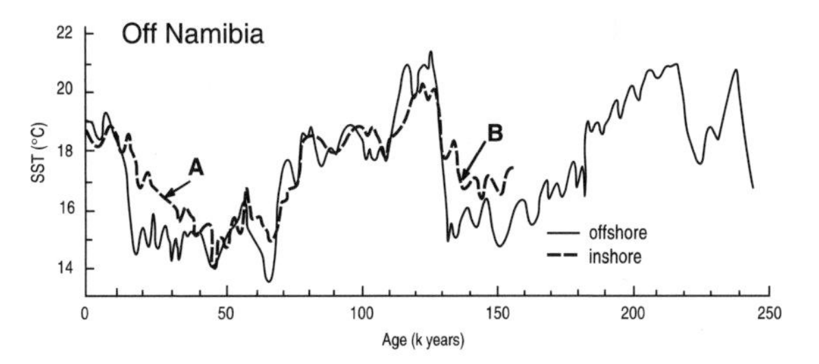

Fig. 2.4. Sea surface temperature variations determined from ocean sediment cores taken along a transect from the coast in offshore and near inshore waters off Namibia (Kirst et al. 1999)

good. The region as a whole was responding to global change. However, this was not invariant and local exceptions occurred.

Alkenone-derived ocean temperatures off Namibia (Kirst et al. 1999) reveal broad similarities in some respects to the Vostok sequence and distinct differences in others (Fig. 2.3). The deep-water profile corresponds to the Vostok record. During cooler (glacial) times the South Atlantic Anticyclone is displaced equatorward, the Angola-Benguela Front (separating the two currents and warm Angolan water from colder Benguela water) is likewise displaced north, the Benguela Current is colder and the coastal upwelling is stronger and more persistent. During warmer (interglacial times) the South Atlantic Anticyclone is displaced poleward, the Benguela Current is warmer and the coastal upwelling weaker and less persistent. This accords with present-day experience within daily and seasonal time scales (Kostianoy and Lutjeharms 1999).

The inshore situation on the continental shelf off Namibia is more complicated. The near-shore profile reflects distinctive regional or local changes occurring in the Benguela upwelling system. It appears that within individual glacial-interglacial cycles, and perhaps within individual upwelling cells, inshore sea surface temperatures begin to rise locally around 20 000 years before the abrupt change is initiated offshore (Kirst et al. 1999) (Fig. 2.4a,b). Conditions became unstable and variable. It is difficult to account for the initiation of near-shore warming so long before the abrupt hemispheric warming occurs. It has been suggested that the process responsible may be similar to that producing present-day spatial variations in the Angola-Benguela Front and Benguela Niño conditions (Kirst et al. 1999). In the modern analogue, perturbations in the tropical easterlies over the western tropical Atlantic Ocean, are thought to weaken the Angola-Benguela Front to allow warm Angola Current water to extend southward along the coast to produce an increase in mean inshore sea surface temperatures in a manner analogous to the occurrence of El Niño along the west coast of South America (Shannon et al. 1986). Though plausible, the explanation presents problems, since extended-period perturbations in the South Atlantic Ocean easterlies are unlikely to have been divorced for as long as 20 000 years from the rest of the general circulation of the atmosphere. Since terrestrial areas of the subcontinent responded to hemispheric atmospheric adjustments during the late Quaternary and Holocene in the same manner as the deep-oceans around southern Africa (Tyson 1986; Cockcroft et al. 1986), the near-coastal changes off Namibia must have been local in origin.

2.2.1.1 *Solar Forcing*

Solar (Milankovitch) forcing of climatic change in the southern African sector of the southern hemisphere has been clearly discernible over the past half million years. The 100 000- and 41 000-year effects, in which astronomical forcing is weaker and the inertia of high latitude ice sheets plays a correspondingly larger role, are evident in the in ocean-core data for the region. Precessional forcing with a period of ~23 000 years has been particularly clear over the past 200 000 years. Currently, perihelion (when the earth is nearest to the sun) occurs in early January in the southern hemisphere summer, and aphelion (when the earth is furthest from the sun) in early July in winter. This causes both present-day southern hemisphere summers and winters to be more severe than their northern hemisphere counterparts. In 11 500 years the opposite will be the case and southern hemisphere summers and winters will be milder than those in the northern hemisphere. After a further 11 500 years the present situation will again prevail, completing the 23 000-year cycle.

Granulometric analysis of the sediments in a 90 m core taken from the Tswaing impact crater lake (formerly the Pretoria Saltpan) in the northern part of South Africa has permitted the derivation of a proxy rainfall series for the region (Partridge et al. 1997). The 23 000-year cycle of covariation between the rainfall and the Milankovitch precessional solar radiational curve for the approximate latitude of the site over the past 200 000 years is striking (Fig. 2.5). At times when perihelion occurs in the middle of the southern hemisphere summer, as at present, the meridional temperature gradient between equator and South Pole strengthens, the southern-hemisphere summer Inter-Tropical Convergence Zone (ITCZ) does likewise and tropical forcing of summer southern Africa climates is enhanced. Wetter conditions prevail. At the same time drier conditions occur over Africa north of the equator. At times when aphelion occurs in the middle of the southern hemisphere summer the opposite conditions prevail. Rainfall changes forced by precessional changes in this manner have been shown to be 180° out of phase in Africa south and north of the equator, as theory requires (Partridge et al. 1997).

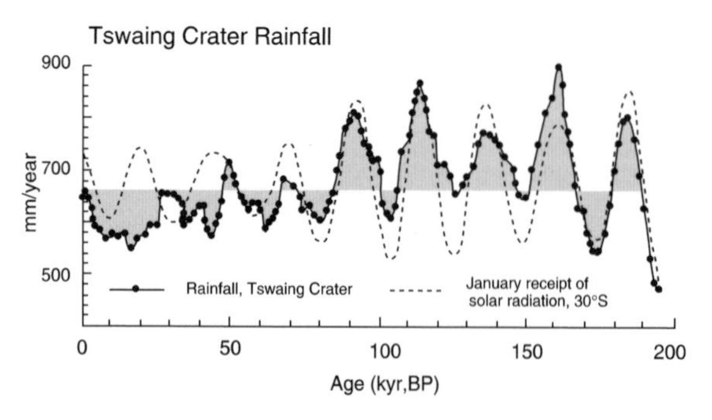

Fig. 2.5. The variation of rainfall at the Tswaing Crater north of Pretoria and changes in solar irradiance at 30° S consequent upon the 23 000-year cycle of the precession of the equinoxes over the last 200 000 years (Partridge et al. 1997)

2.2.2 The Past Twenty Thousand Years

A clear correspondence exists between the abundance of marine molluscs found in different cultural and human occupance stratigraphic units in Klasies River Mouth Cave and the Vostok sequence (Thackeray 1988, 1992) (Fig. 2.6a). The decline to the most severe glacial conditions in Antarctica is replicated in many respects in the mollusc record from the southern coast of South Africa. In each cultural horizon the greatest relative abundance of molluscs corresponds to periods of regional warming.

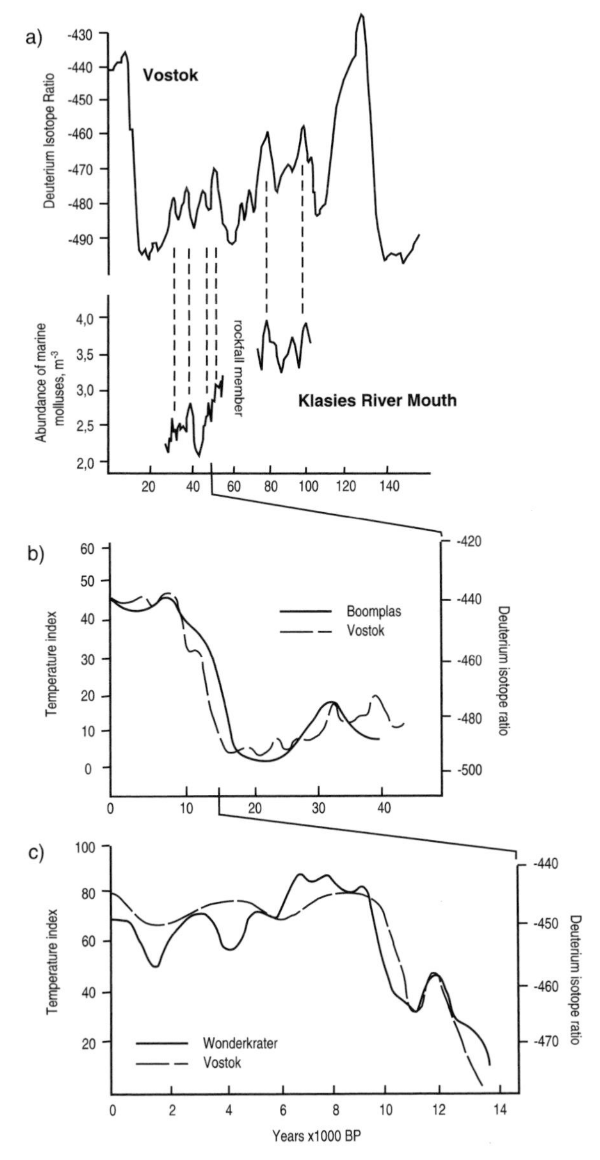

Fig. 2.6. The correspondence between deuterium isotope ratios for the Vostok ice core, Antarctica and: **a** the abundance of marine mollusc samples taken from different cultural stratigraphic horizons in the Klasies River Mouth cave, South Africa (Thackeray 1992), **b** a principal components pollen assemblage temperature index for Boomplaas, South Africa (Thackeray 1987, 1990), **c** a principal components pollen assemblage temperature index for Wonderkrater (Scott and Thackeray 1987; Thackeray 1990)

Regional proxy palaeotemperature series have been derived from micromammalian faunal assemblages (Thackeray 1987; Thackeray and Avery 1990) and pollen sequences (Scott and Thackeray 1987) for Boomplaas in the all-seasons rainfall region of southernmost Africa and for Wonderkrater in the Bushveld of the summer rainfall region of the central plateau interior (Thackeray 1990). They show a reasonable general correspondence with the Vostok record (Fig. 2.6b and c).

Distinctive spatial palaeoclimatic and vegetational gradients prevailed over southern African at the time of the Last Glacial Maximum (Partridge 1997; Partridge et al. 1999) (Fig. 2.7). Rainfall appears to have been lower everywhere, with a minimum of 60% less than that of the present over the Kalahari region. The smallest decrease in rainfall occurred over eastern regions of South Africa. The reconstructed palaeotemperature field shows a clear south-to-north meridional gradient. Analyses of artesian groundwater and stalagmite stable isotopes (Talma and Vogel 1992; Stute and Talma 1998) show that temperatures were depressed by about 5–6 °C at the time of maximum cooling in southernmost Africa. At the same time, lowering of permafrost levels to around 4 000 m on Mts Kilimanjaro and Kenya (Downie 1964; Baker 1967) and extension of mountain glaciers by between 1 400 m and 2 200 m (Hastenrath 1984; Osmaston 1989; Rosqvist 1990) suggest a temperature decrease of up to 10 °C in East Africa at higher altitudes.

Reconstructions of the probable distribution of major biomes during the Last Glacial Maximum (Fig. 2.7) reflect the combined influence of temperature depression and desiccation (Partridge et al. 1999). South of about 13° S, desert evidently occupied almost the entire western half of the subcontinent. Most of the remainder of the area was covered by xerophytic woodland, shrubland and grassland, with rainforest restricted to a small area in the western part of the equatorial belt (Maley 1993). Around the Lesotho highlands and the mountains of East Africa, steppe vegetation apparently occupied significant areas. At the southern tip of the subcontinent the Fynbos (sclerophyllous thicket) Biome of the Western Cape seems to have maintained, or even increased, its areal extent.

The sediments accumulated in the large lakes of the region record a history of environmental change in the Southern African tropics that has a temporal resolution of decades to centuries. In many lake basins the record may be continuous for several million years (Johnson and Odada 1996). Major fluctuations in equatorial and tropical lake levels have occurred over the past 40 000 years (Fig. 2.8a). Many of the large lakes respond to shifts in rainfall and evaporation by switching between open and closed-basin configurations. Lake Turkana, for example, has been closed for the past 6 000 years. Lake level, water

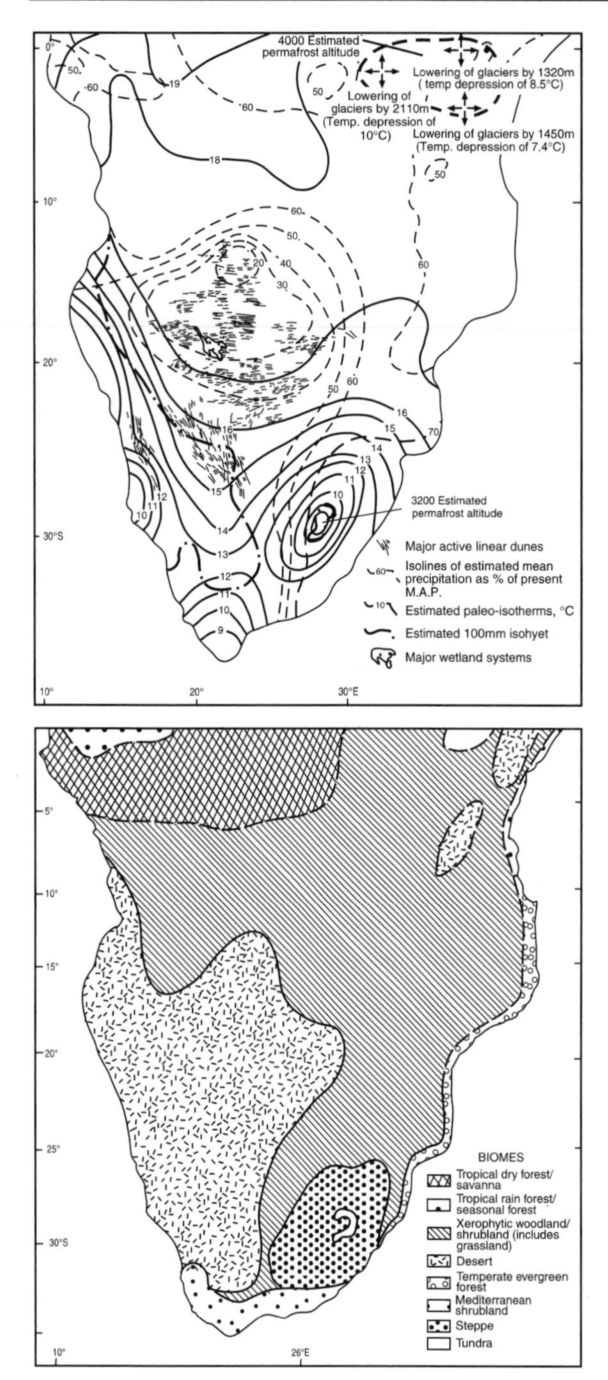

Fig. 2.7. Southern African rainfall, temperature and vegetation distributions at the time of the Last Glacial Maximum, 21000–18000 B.P. (Partridge 1997)

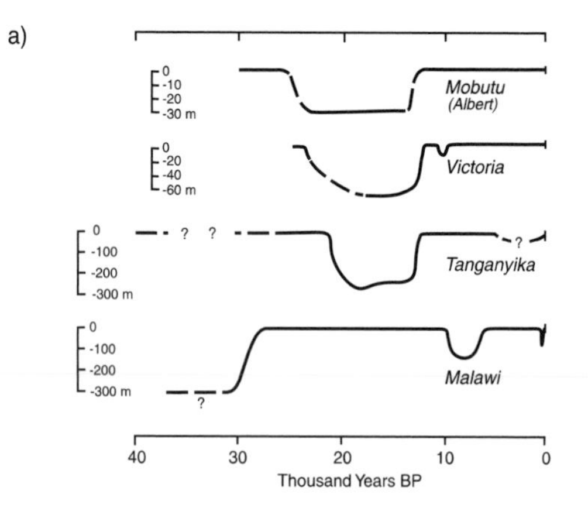

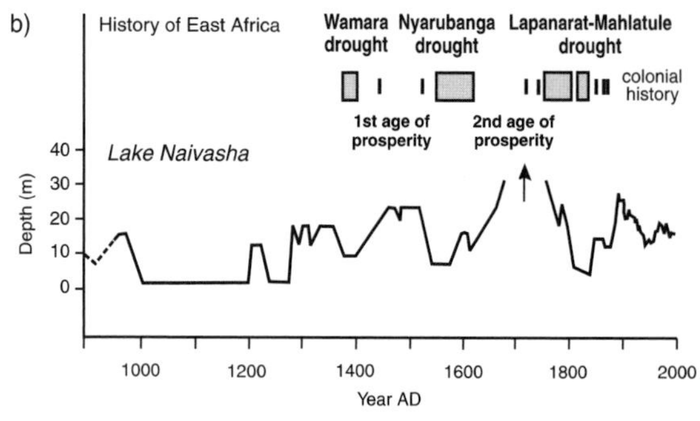

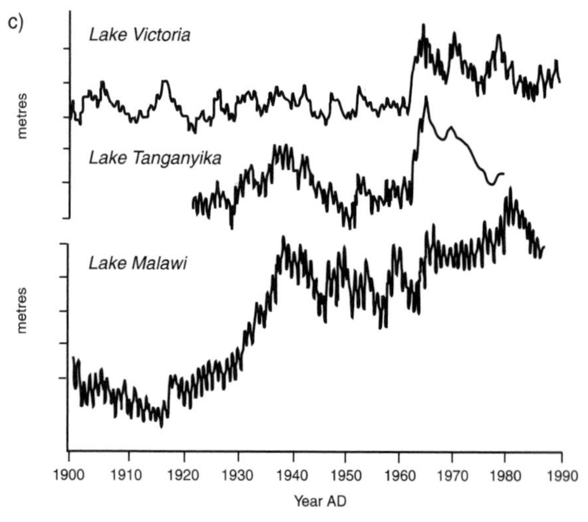

Fig. 2.8. Lake level fluctuations for large East African lakes: **a** millennial-scale variability over the past 40000 years (Johnson et al. 1996), **b** decade-to century-scale variability in the lake level of Lake Naivasha (Verschuren et al. 2000), **c** inter-annual variability during the twentieth century (Grove 1996) (ordinate markings are at 1-m intervals)

chemistry and biota have varied significantly as a consequence of climatic change and have left a clear record in the sediments (Johnson et al. 1991). A decade-scale reconstruction of rainfall and drought is available for a small crater lake on an island within Lake Naivasha, Kenya (Verschuren et al. 2000) (Fig. 2.8b). It shows a significantly drier climate during medieval times (A.D. 1000–1270) and wetter conditions, punctuated with prolonged dry episodes, during the Little Ice Age (1270–1850).

Lake Malawi is an open basin at present with outflow through the Shire River at the southern end of the lake. There was no outflow for the first three decades of the twentieth century as a result of conditions being somewhat drier than the present (Pike and Rimmington 1965).

Sedimentary evidence indicates that the lake was at least 100 m lower than at present for a brief period within the past four centuries (Owen et al. 1990), possibly associated with the desiccation during the Little Ice Age. From 6 000–10 000 B.P. and 33 000–40 000 B.P. Lake Malawi levels were about 150 m below the present datum (Finney and Johnson 1991; Finney et al. 1996). Similar (but not synchronous) records have been obtained from Lake Tanganyika (Livingstone 1980; Gasse et al. 1989) and Lake Victoria (Kendall 1960). During the twentieth century, inter-annual lake-level variability has been considerable (Fig. 2.8c). Although twentieth century temperature variability in the region has not been large compared to that in higher latitudes, the variability in the hydrological budget has been striking.

2.2.3 The Period 18 000 to 10 000 B.P.

Recent investigations during the International Decade for the East African Lakes (IDEAL) have significantly advanced knowledge of the history of Lake Victoria, the largest lake in Africa and the source of the River Nile (Johnson and Odada 1996). The seismic profiles across the lake often show distinct unconformaties, such as the most recent one (Fig. 2.9). Prior to 13 000 B.P. Lake Victoria was a temporarily dry, grass-covered depression (Johnson et al. 1996) (Fig. 2.10). The lake began to refill at about 13 000 B.P. and overflowed at around 7 500 B.P. (Beuning et al. 1998). Immediately prior to the overflowing, diatom productivity dropped to a minimum (Johnson et al. 1998). In the first 500 years after the lake began to fill after 13 000 B.P., primary production was extremely high, nourished by high input of nutrients from the flooded landscape (Johnson et al. 1998). A few species of cichlids and other fish emerged out of their fluvial refugia to colonize the new lake, generating hundreds of new endemic species over the last thirteen millennia (Johnson et al. 1998).

Widespread evidence is available to demonstrate that rapid warming took place in southern Africa after 16 000 B.P. (Partridge et al. 1990; Partridge 1993, 1997), with sea-surface temperatures in the nearby tropical Indian Ocean showing a marked increase about 1 000 years later (Sonzogni et al. 1998). The sudden cooling associated with the Younger Dryas, which at around 11 000 B.P. punctuated the general amelioration of conditions after the Last Glacial Maximum, is evident in the Wonderkrater pollen record (Scott and Thackeray 1987; Thackeray 1990) and from mollusc stable isotope data for the Atlantic Ocean seaboard (Cohen et al. 1992). Thereafter, temperatures rose over much of the subcontinent. From the Boomplaas and Wonderkrater records it would appear that the Holocene altithermal (warm period) was manifest in South Africa between about 7 000 and 5 500 B.P. (Fig. 2.6b and c). At this time temperatures were at their highest values since the Last Interglacial at 125 000 B.P. and higher than those of the present. Rainfall increased in some areas as temperatures rose, with maximum receipt occurring at different times in different regions. Contemporaneously, the proportion of rain falling in summer increased as that in winter diminished (Partridge et al. 1990; Partridge 1993, 1997).

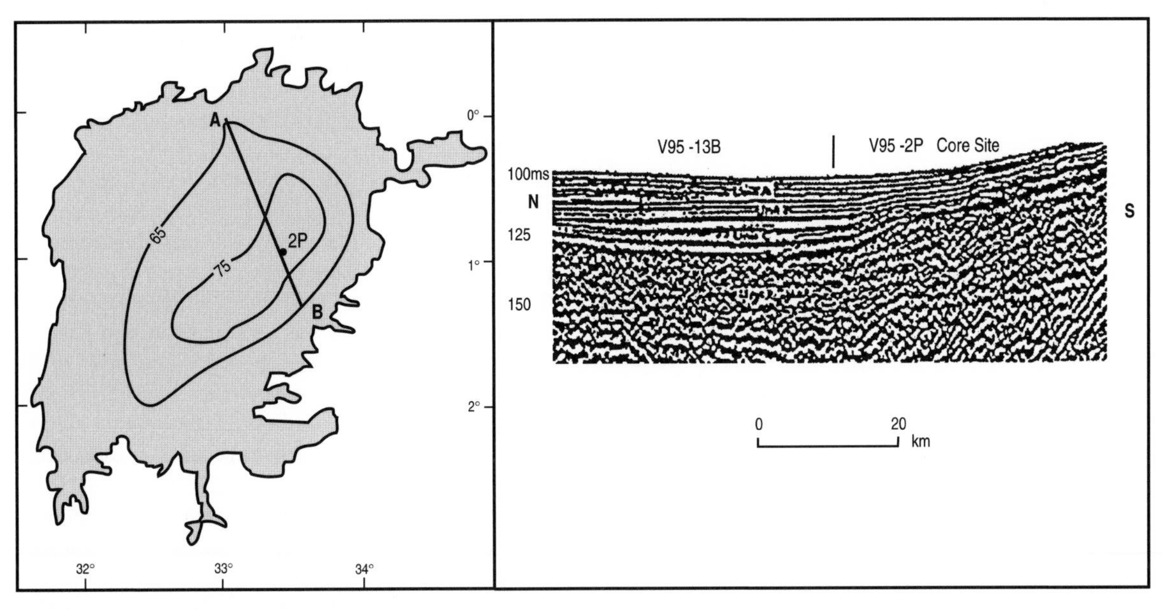

Fig. 2.9. A seismic profile taken along transect A–B across Lake Victoria to illustrate desiccation events over the past 100 000 years, together with more uniform sediment layers associated with flooding of the formerly dry lake after 12 400 B.P. (Johnson et al. 1996). Present-day lake depths are in metres

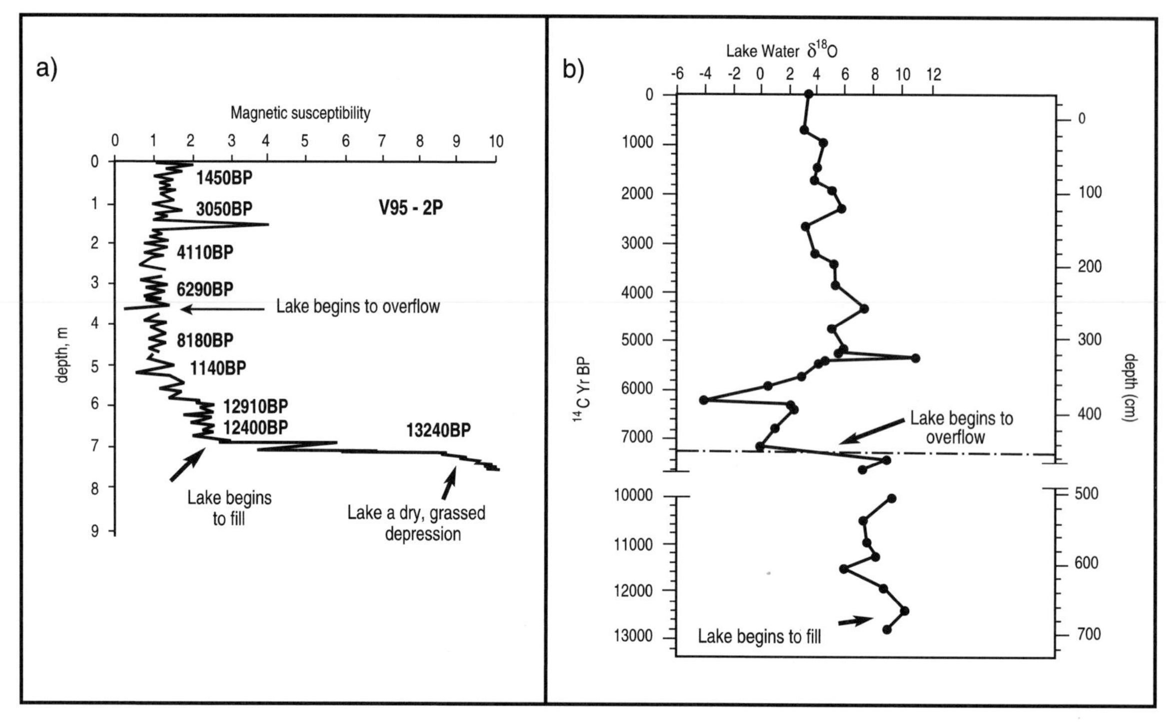

Fig. 2.10. Lake Victoria sediment-core variations in **a** magnetic susceptibility (after Johnson et al. 1996) and **b** lake water $\delta^{18}O$ (Beuning at al. 1998) to illustrate the early Holocene refilling and overflowing of the lake

2.2.4 The Middle to Late Holocene: 6 000 to 1 000 B.P.

The best regional high-resolution record from the middle Holocene to the present is the stable isotope record for the Makapansgat Valley in northern South Africa. The climate change experienced at Makapansgat appears to be widely representative of conditions obtaining over large areas of Southern Africa during the Holocene (Holmgren et al. 1999; Tyson and Preston-Whyte 2000). Two major events are apparent in the $\delta^{18}O$ record, where enriched values indicate warmer, wetter conditions and lower values cooler drier conditions (Holmgren et al. 1999) (Fig. 2.11). The first is the altithermal warming until about 5 500 B.P.; the second is the five centuries of cooling associated with the Little Ice Age from A.D. 1300 to 1800 (Tyson et al. 2001). A high degree of variability on decadal to century scales, accompanied by frequent abrupt changes in climate, has characterized the environmental conditions prevailing over the past six millennia.

It is instructive to compare the Makapansgat record to that of the Cango Cave on the south coast of Africa (Fig. 2.11). All the major cool events recorded at the northern site over the past six millennia are replicated at the southern location, except for that occurring around 2 200 B.P., which had a warm counterpart at Cango Cave. The cool periods appear to have been of longer duration in the south, whereas warm periods appear to have been longer in the north, indicating the manner in which cooling spread equatorward from the south and warming poleward from the north with expansion and contraction of the circumpolar vortex of westerly winds over the region. General circulation models (GCMs) show that the expansion and contraction of the vortex is driven by changing equator-to-pole meridional temperature gradients (Rind and Overpeck 1993; Jousaumme and Taylor 1995; Rind 1998).

A regional reconstruction of the climate and vegetation at the time of the altithermal has been attempted (Partridge et al. 1999) (Fig. 2.12). Considerable spatial variation is evident. The semi-arid western interior of the southern part of the region appears to have received more rainfall, while the area to the east (including the Tswaing Crater) was drier on the evidence of pollen and other proxy evidence. Rainfall over the plateau areas of the northern tropics of the region appears to have increased. Nearer the equator, the high lake levels of the early Holocene declined somewhat (Hamilton 1997), temperatures rose by around 2 °C and mountain glaciers receded by around 500–1 000 m relative to their early-Holocene extent (Mahaney 1990; Rosqvist 1990; Hastenrath 1991). Reconstructed vegetation distributions over southern Africa during the Holocene altithermal are much the same as those of today.

Variations in the $\delta^{13}C$ record in southernmost Africa speleothems yield a continuous record of the proportion of C_4 grasses in the overlying biomass and indicate changes in the proportion of C_3 (woodier) and C_4

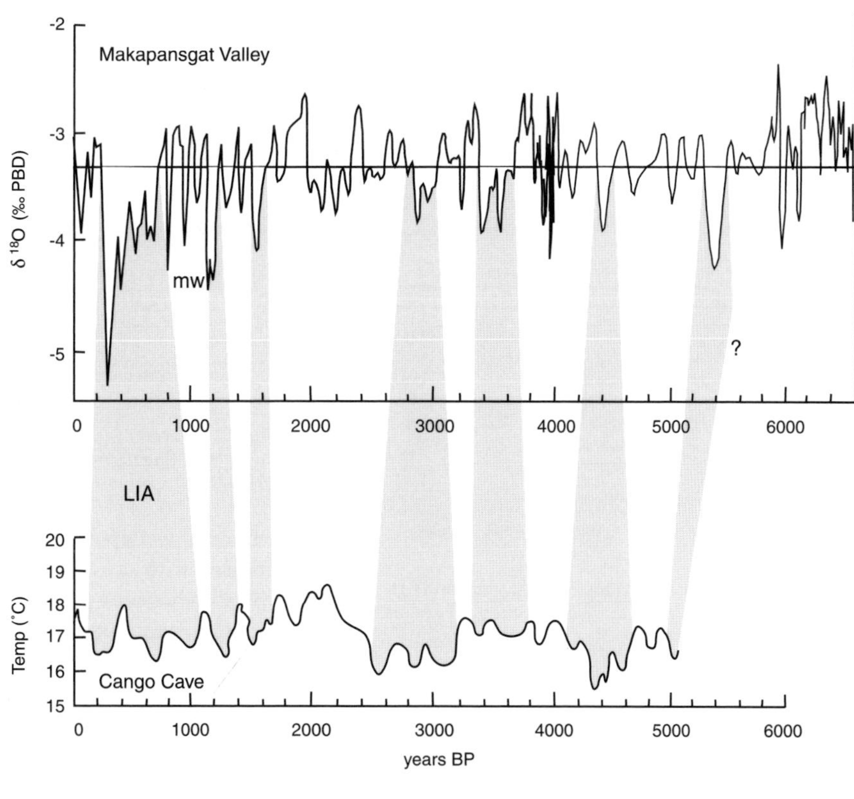

Fig. 2.11. Regional teleconnections between Makapansgat (Holmgren et al. 1999) and Cango Cave (Talma and Vogel 1992), South Africa. The Little Ice Age is denoted by *LIA* and the period of medieval warming by *mw*

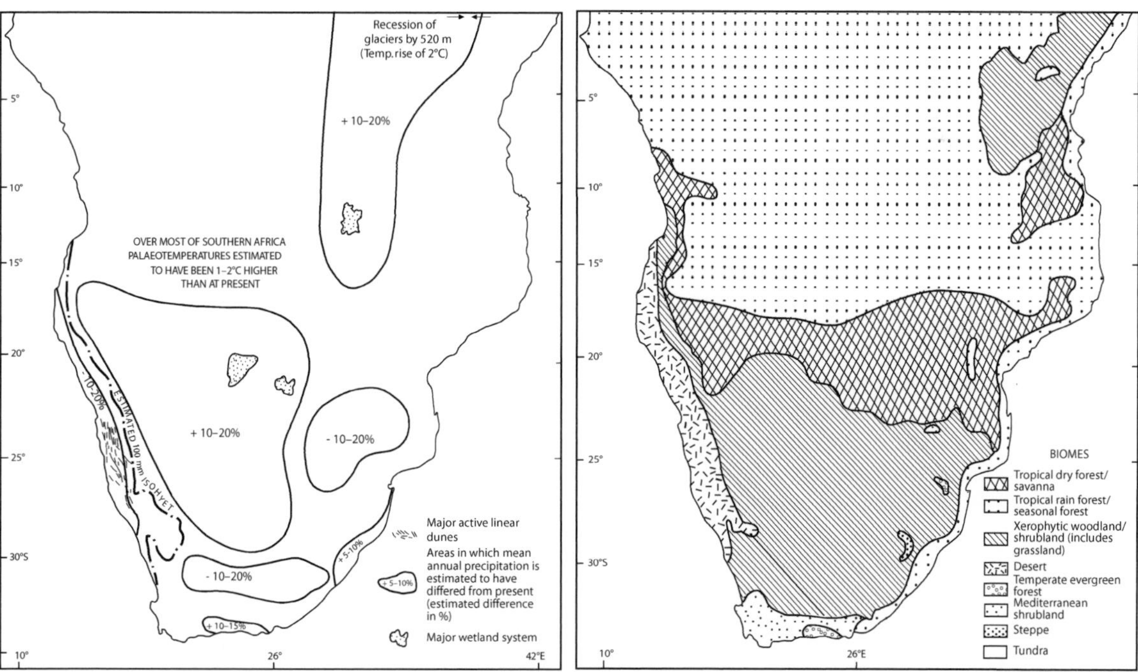

Fig. 2.12. A palaeoclimatic reconstruction of southern African rainfall, temperature and vegetation conditions at the time of the Holocene altithermal at about 7 000 B.P. (Partridge 1997)

(grassier) environments (Holmgren et al. 1995; Lee-Thorp and Talma 2000). At both Makapansgat and Cango Cave the variation in amount of grass cover was similar over the past six millennia (Fig. 2.13). A maximum of grass cover occurred between 2 000 and 2 400 B.P. Over the last ~4 000 years the correlation between colour banding, band width, $\delta^{18}O$ and $\delta^{13}C$ values suggest that periods of highest abundance of grass were warm and moist, whereas those of lowest abundance coincided with cooler, drier periods (Holmgren et al. 1999). Prior

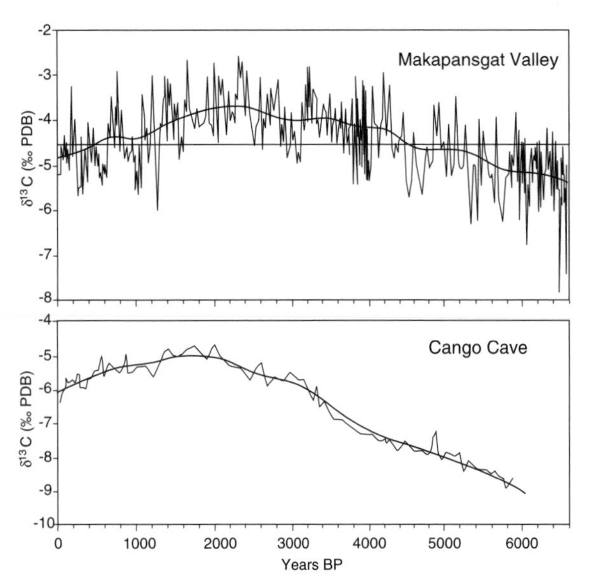

Fig. 2.13. Millennial and inter-decadal variations in the $\delta^{13}C$ records for Makapansgat (Lee-Thorp et al. 2000) and Cango Cave (Talma and Vogel 1992)

to 4000 B.P., lower $\delta^{13}C$ and weaker colour banding of the Makapasgat stalagmite suggest that increased woody or shrub cover may have been coeval with warmer conditions (Lee-Thorp and Talma 2000).

2.2.5 The Past Thousand Years

Two extreme events are manifest in the Makapansgat oxygen isotope record of the past millennia in South Africa. The first is the period of medieval warming that prevailed from A.D. 900 to 1300. The period was characterized by highly variable conditions. Maximum warming occurred at around 1250 when the climate is estimated to have been 3–4 °C warmer than present. The second distinctive extended anomaly was that of the 1300–1800 Little Ice Age. Maximum cooling prevailed at around 1700; another pronounced cool spell occurred about a century or so earlier. Mean annual daily maximum temperatures were depressed by around 1 °C from present-day values at the time of maximum cooling. The 1700 event was coeval with a similar event in the oxygen isotope record derived from a coral in the ocean off southwestern Madagascar (Tyson et al. 2001) (Fig. 2.14). It is also concordant with cool events recorded in a large variety of proxy data from sites all over southern Africa (Tyson et al. 2001).

The degree to which the Makapansgat and Lake Naivasha records have varied inversely over the past millennium is striking (Tyson et al. 2002; Fig. 2.15). The inverse correlation is similar to that which occurs during present-day ENSO events. Whereas the period of medieval warming in South Africa was wet, in Kenya it was dry; the Little Ice Age in South Africa was dry and punctuated with

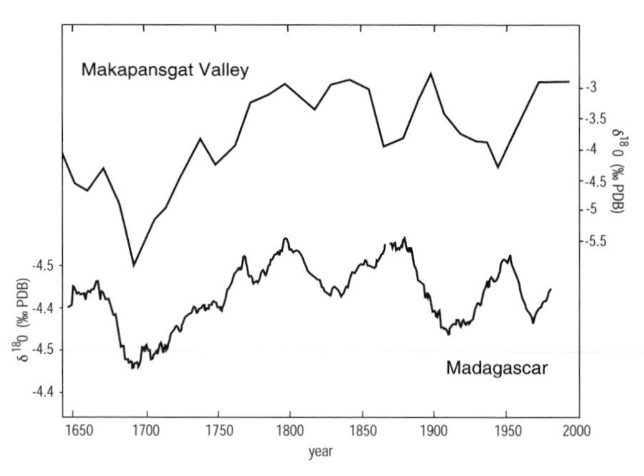

Fig. 2.14. Variabilty in the $\delta^{18}O$ records for a Makapansgat valley stalagmite (Holmgren et al. 1999) and a coral off southwestern Madagascar (Tyson et al. 2001)

extended wet episodes, in Kenya it was wet and interspersed with extended dry intervals. The inverse relationship has long been prevalent and is clearly evident in the meteorological record of the twentieth century, when east and southern Africa have experience out-of-phase rainfall correlations on the scale of decades (Nicholson and Entekhabi 1986), on ENSO scales (Nicholson and Entekhabi 1986; Ogallo 1988; Allan et al. 1996) and inter-annually (Nicholson 1986). The long history of changing gradient of climatic change between equatorial and subtropical latitudes has had significant implications for the people of southern Africa. It also indicates the extent to which the teleconnections evident during ENSO events occur over much longer time scales, and in fact transcend ENSO variability.

The link between changing climate and human activities is clear. The Lake Naivasha proxy data record accords well with the oral history of people living in the region prior to nineteenth century colonization (Verschuren et al. 2000). Periods associated with drought-induced famine, political unrest and large-scale human migrations matched periods of low levels of Lake Naivasha (see Fig. 2.8b). The intervening ages of prosperity, agricultural expansion and population growth (Webster 1980) are coeval with high lake levels. Further south, it has been suggested that the medieval warming and wetter conditions around A.D. 900–1300 may have been responsible for changes in Iron Age settlement patterns occurring over much of South Africa at the time (Hall 1984; Huffman 1996). The deteriorating conditions associated with the onset of the Little Ice Age may have contributed to the collapse of the nascent Mapungubwe state in the presently semi-arid Shashi-Limpopo basin, the area in which the Botswana, Zimbabwe and South Africa boundaries meet (Vogel 1995; Huffman 1996). In the Kuiseb River Delta of Namibia, settlements were likewise abandoned between 1460 and 1640 with the advance of the Little Ice Age (Burgess and Jacobson 1984).

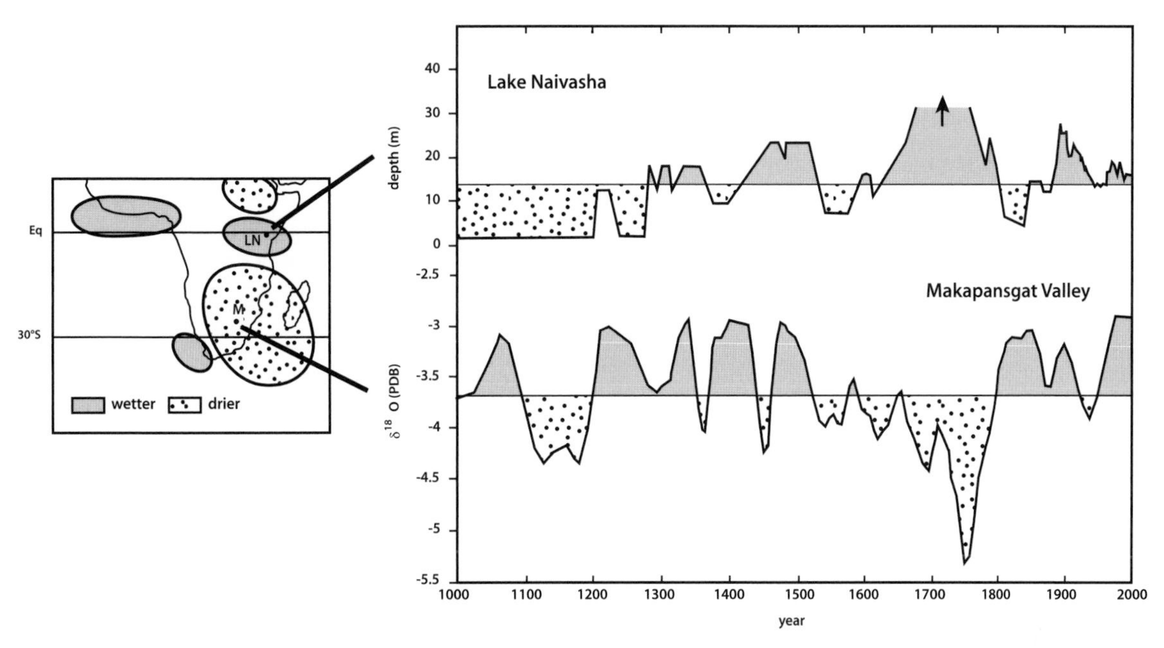

Fig. 2.15. *Left*: areas in Africa south of the equator showing teleconnection patterns associated with El Niño (Allan et al. 1996) (*LN* denotes Lake Naivasha, Kenya and *M* the Makapansgat Valley, South Africa); *right*: Makapansgat Valley stalagmite $\delta^{18}O$ variations, smoothed with a 5-term binomial filter (Holmgren et al. 1999) and the Crescent Island Crater, Lake Naivasha lake level record (Verschuren et al. 2000) over the past millennium. In the Makapansgat record, high values of $\delta^{18}O$ are associated with warmer, wetter conditions; lower values with cooler, drier climates (Holmgren et al. 1999)

An intriguing possibility is suggested by the probable existence of the out-of-phase rainfall gradient between East and southern Africa in the centuries before the beginning of the Naivasha record. Bantu-speaking agropastoralists first penetrated into the coastal regions of southeastern Africa by about A.D. 100–200. Thereafter, they expanded rapidly into northeastern interior regions south of the Limpopo River, where numbers of fairly large settlements were in evidence by A.D. 400–500 (Vogel 1995). Ceramic studies show cultural links between East Africa and the early southeastern coastal settlements and those in the interior (Huffman 1989; Maggs 1984). The Makapansgat climate record indicates almost continuously moister conditions in southern Africa between A.D. 0 and A.D. 500 (Holmgren et al. 1999). Might not the climate gradient between equatorial and subtropical regions, coupled with the need to find additional or new suitable lands for cultivation, have provided an incentive for people to migrate southward to a more supportive environment?

This seems to have been the case at the end of the first millenium when further migrations occurred, and when the Naivasha and Makapansgat records show the gradient of change to the south was particularly pronounced. The appearance of a new ceramic style (Huffman 1989) and linguistic affinities between some of the peoples of southern and east Africa have been argued to reflect the first migration of Sotho-Tswana people into the region (Huffman and Herbert 1994; Iliffe 1995). The climatic gradient towards ameliorating conditions in the south may well have been a significant factor in augmenting the movement of populations in this direction.

The evidence of the oxygen isotope record shows the termination of the Little Ice Age at Makapansgat in South Africa was abrupt. Such abrupt changes are a feature of the record, particularly in the earlier period of medieval warming, when oscillations in annual mean daily maximum temperature of 2–3 °C in a few decades were observed. Abrupt climate changes of this order in periods as short as 3–5 years have been observed in Greenland ice cores (Alley et al. 1993, 1997) and in tree ring chronologies from Fennoscandia (Briffa et al. 1992).

2.2.6 The Present

Over much of South Africa, annual mean maximum temperatures have been rising steadily over the past few decades (Tyson et al. 1998) (Fig. 2.16a). The same is true of mean annual temperatures over the wider Southern Africa region (Hulme et al. 1996, 2001). The increase is associated with a concomitant increase in South African surface rock temperatures derived from borehole temperature profiles (Tyson et al. 1998).

Rainfall over the southern parts of the region as a whole has shown no large systematic linear trends during the twentieth century (Tyson et al. 1975; Tyson 1986; Hulme et al. 2001) (Fig. 2.16b). However, local areas have shown weak upward and downward trends of up to 10% per century, except over the central tropical area of the

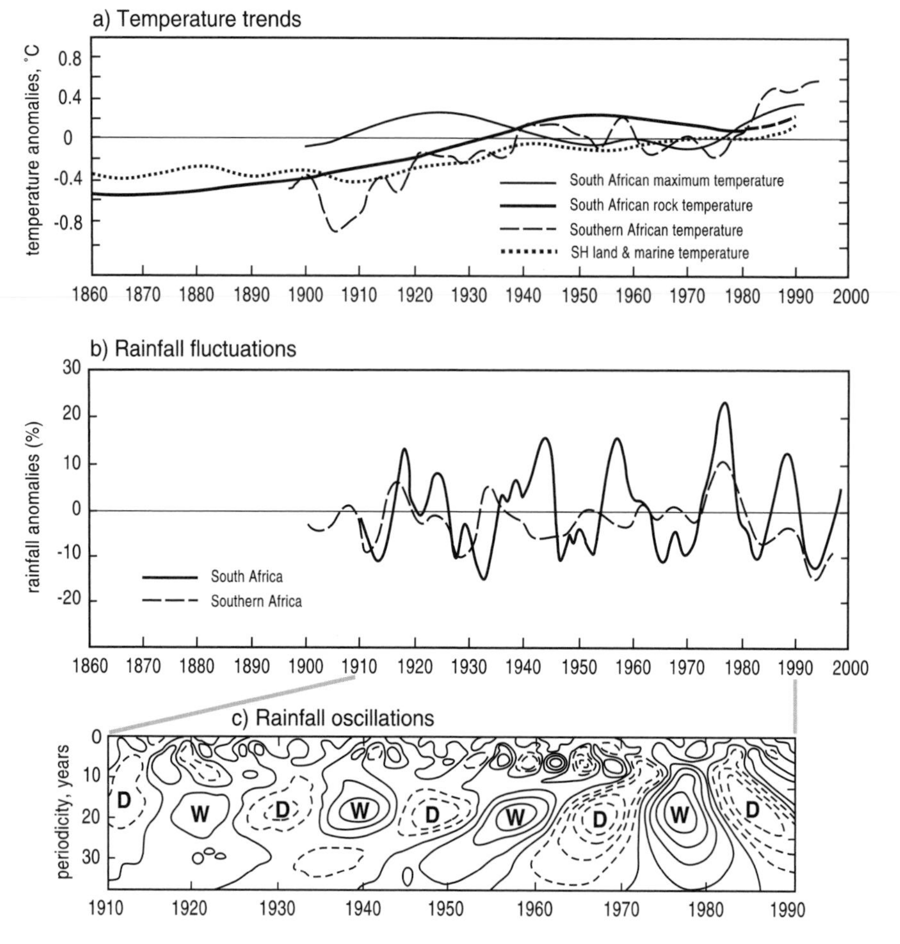

Fig. 2.16.
a Decadal southern hemisphere combined land-sea temperatures (IPCC 1996), regional South African air temperatures and South African borehole-derived surface rock-temperature anomalies (°C) (Tyson et al. 1998), **b** smoothed rainfall variations for South Africa (Mason and Tyson 2000) and southern Africa (Hulme 1996), **c** wavelet analysis to show the ~18-year oscillation in southern African rainfall during the twentieth century

region and western Angola where increases appears to have been greater (Hulme et al. 2001). Over shorter time periods, local trends may have been more pronounced, e.g., in the Lowveld of South Africa, Zimbabwe and Mozambique in recent years (Mason 1996). A high degree of inter-annual and inter-decadal rainfall variability characterises the whole region (Tyson 1986; Nicholson 1986, 1993) and is inversely correlated with changes occurring in east Africa, especially on ENSO time scales (Nicholson and Entekhabi 1986; Ogallo 1988; Allen et al. 1996). The most pronounced variability in subtropical southern Africa as associated with the multi-decadal oscillation centred at around 18 years (Fig. 2.16c).

Evidence exists to suggest that variability and extremes in the southernmost Africa may be increasing (Mason 1996), especially in the drier western parts (McClelland 1996). Between 1931 and 1990, the intensity of extreme events increased significantly over South Africa (Mason et al. 1999; Poolman 1999).

Understanding biogeochemical changes occurring at present and in the future depends on appreciating the roles played by the subsidence of air associated with dominant anticyclonic flow fields, the longevity of thermodynamically stable layers in the atmosphere for so much of the year, the persistence of haze layers that blanket vast areas and the transport of aerosols and trace gases within these haze layers. Southern Africa is dominated by the subsidence limb of the southern Hadley cell of the general circulation of the atmosphere, particularly in winter (Newell et al. 1972) (Fig. 2.17a). The mean circulation over the subcontinent is anticyclonic throughout the troposphere for most of the year (Jackson 1961; Taljaard 1981). The July circulation at the 850 hPa level in the atmosphere (at an altitude of ~1.5 km, approximately the mean height of the inland southern African plateau) is a case in point (Fig. 2.17b). Circulation changes associated with the ITCZ affect northern tropical and equatorial areas, but even at such times and in such regions, anticyclonic curvature in the windfield is common (Thompson 1965; Hastenrath 1985; Garstang and Fitzgarrald 1999). The annual variation in anticylonic airflow in the subtropics is illustrated by conditions obtaining over northern South Africa (Garstang et al. 1996) (Fig. 2.17c). The subsidence associated with such flow is ubiquitous and blankets the whole of southern Africa in winter (Fig. 2.17d) (Tyson 1997).

Anticyclonic circulation has long been dominant in the region and small changes in the position of the sub-

a) June - August meridional circulation

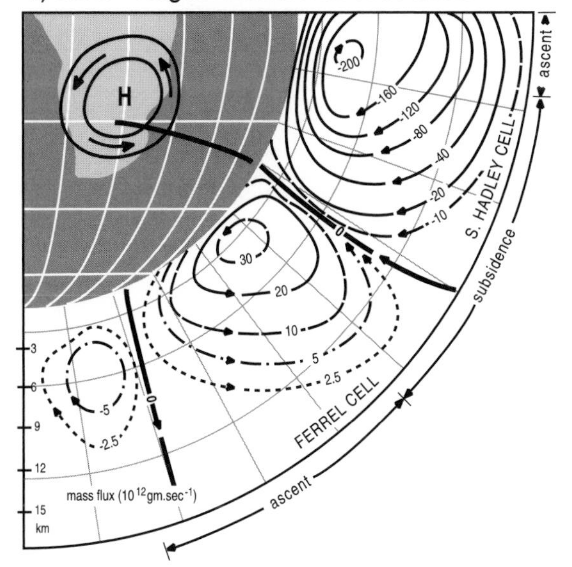

b) July 850hPa circulation

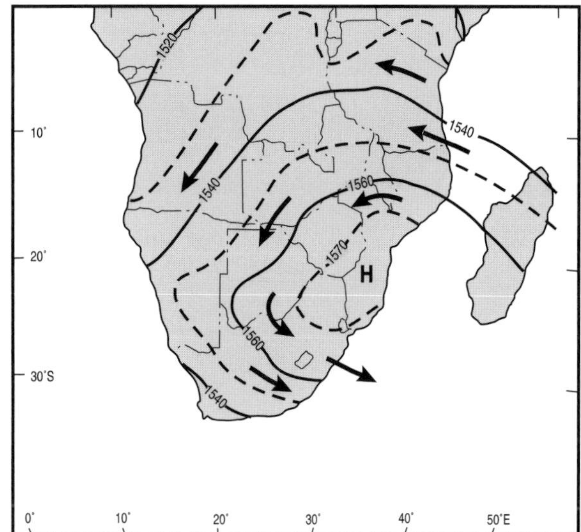

c) Anticyclonic Frequency

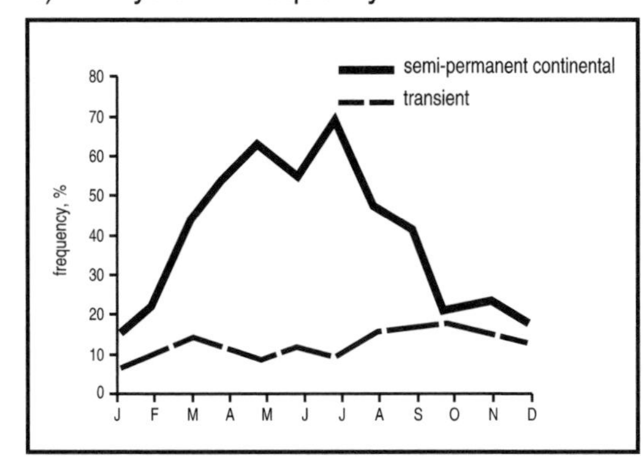

d) July 500-800 hPa subsidence, hPa h^{-1}

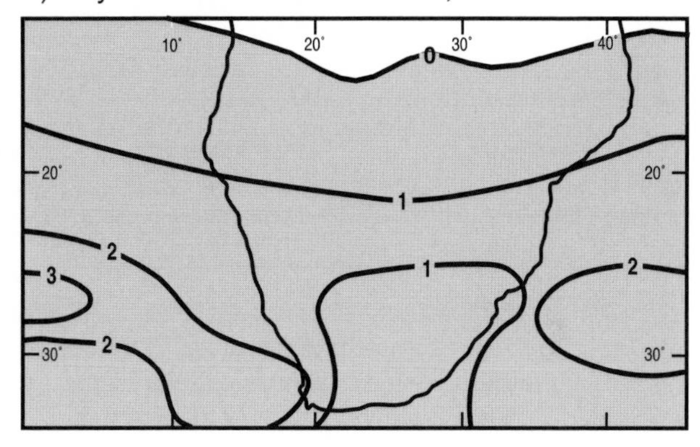

Fig. 2.17. **a** South Hadley cell subsidence over southern Africa (modified after Newell et al. 1972), **b** winter 850 hPa circulation over southern Africa (Jackson 1961), **c** anticyclonic frequency over South Africa (Tyson et al. 1996), **d** July subsidence rates over southern Africa (Tyson 1997)

tropical high-pressure system have been associated with major changes in climate over time. The alignment of Kalahari dunes of different morphologies (Lancaster 1981, 2000; Stokes et al. 1997, 1998) illustrates the point. The dune directions are directly linked to the locus of the system prevailing during extended spells of aridity. From 50 000–40 000 B.P. northern and eastern Kalahari dunes, formed in earlier times, were reworked extensively when the present continental anticyclone over Southern Africa was displaced about 2° equatorward of its present position (Lancaster 1981) (Fig. 2.18). In another period of extensive reworking from 17 000 to 10 000 B.P., the southern dunes acquired the trends seen at present and resulted from sand transport similar to that of today. During the drier periods of dune activity, the southern circumpolar vortex and semi-permanent high pressure belts, together with the Antarctic and Subtropical Ocean Convergences (CLIMAP 1981; Howard 1985; Jousaumme and Taylor 1995), were compressed equatorward and sea surface temperatures decreased off Namibia (Stokes et al. 1997; Stokes et al. 1998).

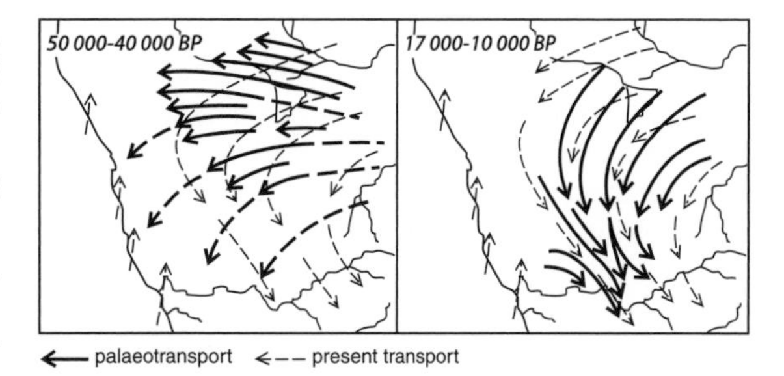

Fig. 2.18. Palaeowinds inferred from the alignment of presently fixed sand dunes in the Kalahari compared to present conditions (Lancaster 2000)

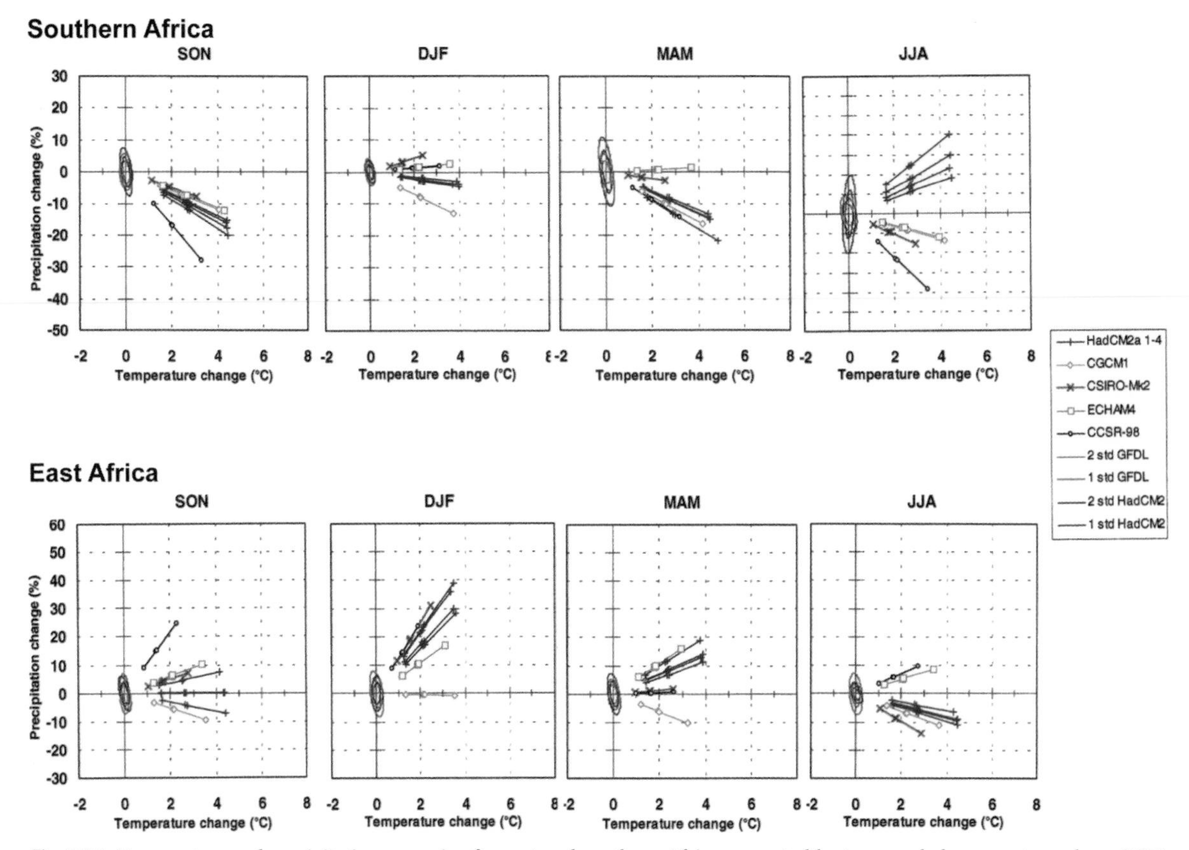

Fig. 2.19. Temperature and precipitation scenarios for east and southern Africa suggested by ten coupled ocean-atmosphere GCMs using four different emission scenarios (becoming less constrained towards the right of the panels) (Carter et al. 2000)

2.2.7 Future Climates

It is not yet possible to predict the future climate, but it is feasible to present scenarios that may approximate what might happen. Ten different coupled climate-ocean circulation models have been run for four different emission scenarios (Carter et al. 2000). The results suggest that by 2050 for subtropical southern Africa temperatures throughout the year will be 2–4 °C higher, with model predictions being more consistent for spring and autumn months (Fig. 2.19). Midsummer rainfall in the summer rainfall region may decrease by ~10%, whereas in spring and autumn the decrease may exceed 20% (winter changes are not meaningful since winter is by and large rainless). For east Africa temperatures may be 2–4 °C higher and December to February rainfall may increase by ~30%. During the rest of the year, different models give different projections of future conditions, diminishing the significance of the interpretations that may be placed on them.

One of the models shown to perform best for subtropical southern Africa is the HaDCM2 (Joubert and Kohler 1996). Using this model some idea of the spatial variability of mid-twenty-first-century conditions may be obtained (Joubert and Kohler 1996) (Fig. 2.20). Greatest increases are likely to be in winter minimum temperatures. Uncertainty still attaches to such simulations and is greatest in the case of precipitation estimates (IPCC 1996).

Downscaling to small regions or local areas is possible using regional, limited-area models nested in and forced by a GCM. Resulting precipitation fields resemble observed conditions more closely than those of the forcing GCM (Joubert et al. 1999). However, the models remain beset with uncertainties and do not always produce entirely realistic conclusions. Downscaling using neural networks that link rainfall to circulation types derived from the forcing GCM provides an alternative approach (Hewitson 1999).

Given that the palaeoclimate record is consistent in associating wetter conditions with warmer periods in the past, it is surprising that GCMs predict that future warming will lead to drier conditions in much of subtropical southern Africa. Either the interpretations of past conditions are in error or the models may be based on invalid assumptions about regional climate processes. Alternatively, it needs to be recognised that past warming is not analogous to present warming, which contains an unprecedented component of anthropogenic forcing. The current and future levels of atmospheric CO_2 and other greenhouse gases are unprecedented in the past

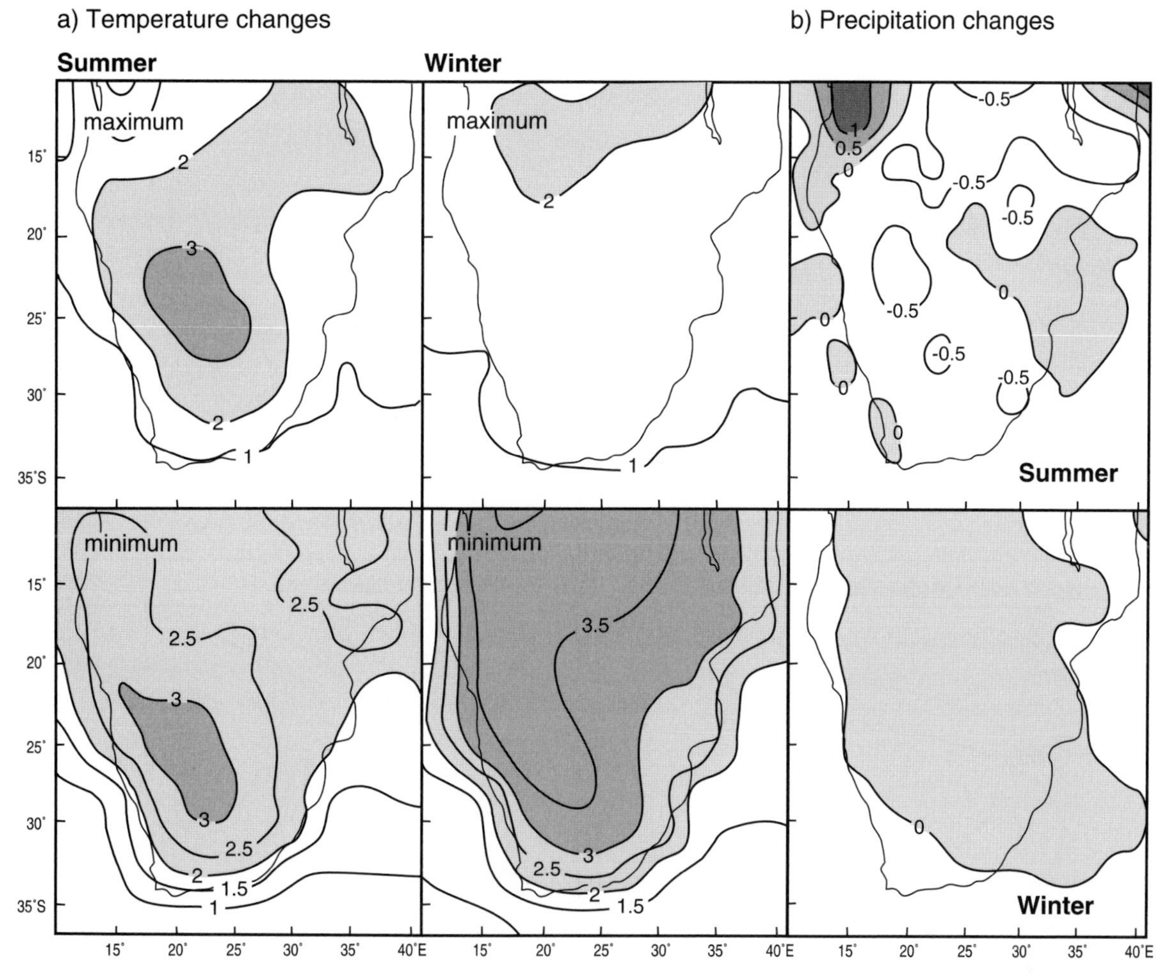

Fig. 2.20. HadCM2 model simulations for Southern Africa assuming combined greenhouse gas and sulphate aerosol forcing with a transient increase in carbon dioxide until doubling in 2030–2059: **a** summer and winter maximum and minimum temperature increases (°C) (Joubert and Kohler 1996). Shaded areas indicate degree of warming; **b** summer and winter daily precipitation changes (mm d^{-1}) over southern Africa (Joubert 1995). Shading indicates drier conditions

half million years. The linked ocean-atmosphere system is strongly non-linear and it is possible for the system to be producing regional anomalies with the present global warming.

Over the past half-million years, spatial and temporal variability of climate over southern Africa have been distinctive. Small changes in the dominant atmospheric circulation types have produced major effects, particularly changes in the subtropical high-pressure systems. Today, it is these semi-permanent systems that are especially important in respect of the transport of air, trace gases and aerosols over the region.

2.3 Regional Transport of Air over Southern Africa

The dominance of anticyclonic curvature in the windfield and its controlling effect on horizontal transport patterns is evident throughout the region, except in the near-equatorial tropics where ITCZ effects are more important. This is illustrated by the degree to which in March transport of air to Kenya is a function of both northern and southern hemisphere circulation forcing (Gatebe et al. 1999) (Fig. 2.21a). Transport to Kenya from the southern hemisphere (via the Mozambique Channel region) reaches a maximum in May (67%) in contrast to January when it rarely occurs. At all times of the year maximum transport from Kenya takes place in the tropical easterlies along an equatorial duct towards the Atlantic Ocean. In contrast, south of the Congo transport is dominated by the semi-permanent subtropical high-pressure systems associated with the southern Hadley cell over Africa. This is illustrated by regional transport patterns originating from the Highveld of South Africa (Piketh et al. 1999) (Fig. 2.21b) and from Mozambique, Malawi, Zambia and Zimbabwe (Swap 1996; Tyson et al. 1996a) (Fig. 2.21c). Given the anticyclonic forcing of transport pathways over southern Africa, it is not surprising that the major outflow duct for aerosols and trace gases from the subcontinent south of

a) Transport to and from Kenya

37%
21%
41%
61%
20%
March

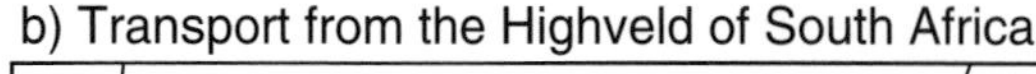

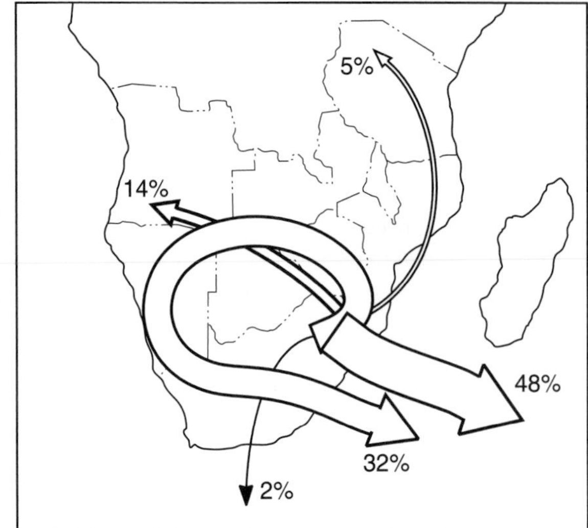

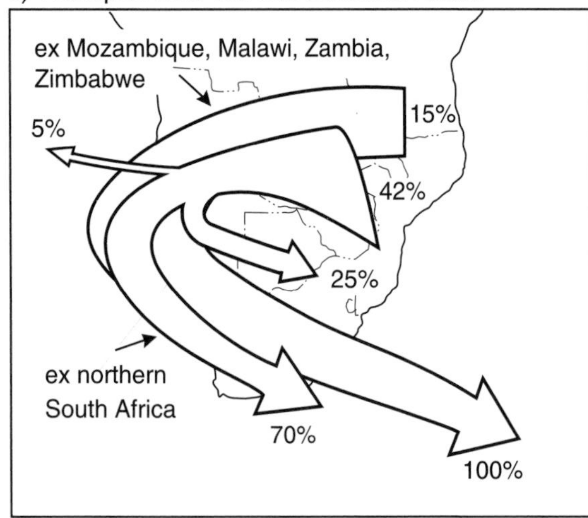

d) The Indian Ocean Plume

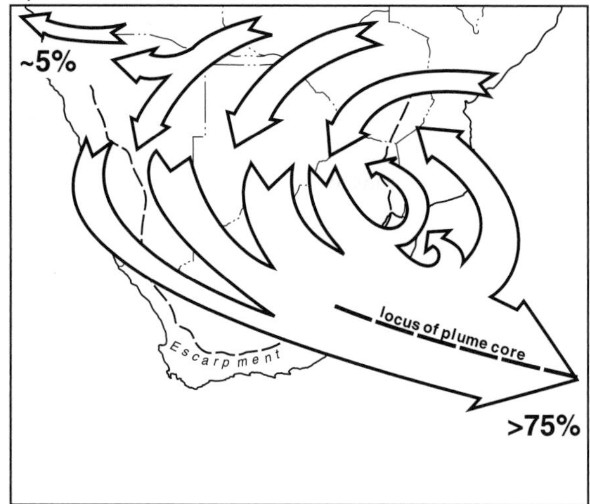

Fig. 2.21. Aerosol and trace gas horizontal transport patterns over southern Africa: **a** integrated March 700–500 hPa transport to and from Mount Kenya for the period 1991–1993 (Gatebe et al. 1999); **b** transport from the South African Highveld, 1990–1994 (Piketh et al. 1999); **c** transport from Mozambique, Malawi, Zambia and Zimbabwe (Swap 1996); **d** schematic depiction of mean annual transport over the subcontinent (after Tyson 1997). Percentages at the start of arrows indicate transport to a point of origin; those at the end of arrows transport from the origin

northern Zambia is to the Indian Ocean over South Africa (Fig. 2.21d) (Tyson 1997). The locus of the mean annual plume is at about 31° S over southern Lesotho. More than 75% of all air circulating over South Africa and countries to the north exits the subcontinent at this point (Tyson et al. 1996a,b; Piketh et al. 1999). What is carried in that air is a function of a combination of natural and anthropogenic processes transferring material from the surface to the atmosphere.

2.4 Trace Gas Fluxes from the Surface

Tropical and subtropical regions in southern Africa, particularly the savannas that occupy two thirds of the land surface, have the potential to emit large amounts of trace gases and aerosols into the atmosphere owing to their high biological activity and dryness for much of the year. These emissions can take the form of trace gas fluxes from the soil and vegetation, where gaseous species are produced and consumed by living organisms, by smoke emissions from burning vegetation and by the transfer of mineral soil dust from the surface to the atmosphere by turbulent wind action. In addition, industrial emissions and the burning of fossil fuels (including atmospheric pollution from urban areas) and the production and consumption of charcoal and domestic use of biofuels, contribute significantly to the total gas and aerosol loading of the troposphere.

2.4.1 Biogenic Production of Trace Gases

Soil respiration rates of CO_2 from tropical and subtropical savannas typically are of the order 0.4–0.8 g C $m^{-2} d^{-1}$ (Morris et al. 1982; Delmas et al. 1991; Scholes and Walker 1993; Zepp et al. 1996). They are increased substantially (by at least a factor of 20) from the dry to the wet season and are less influenced by temperature than soil moisture (Scholes and Walker 1993; Zepp et al. 1996).

Concentrations of the hydroxyl (OH) radical in the atmosphere (the major chemical scavanger of the atmosphere) are strongly influenced by its reaction with CO, which in turn can lead to the production or consumption of tropospheric ozone, depending on the NO_x mixing ratio (Crutzen 1995). In the biogenic production of CO the net flux from the surface depends on a kinetic competition between chemical oxidation of soil organic matter to produce the gas (Conrad and Seiler 1985a), which is favoured by dry soil conditions (Conrad and Seiler 1982), and biological oxidation of CO by soil microorganisms (Seiler and Conrad 1987). Phenolic structures in soils are the primary source of the CO production (Conrad and Seiler 1985b). Emission of CO from degrading leaf and grass litter are much the same in South African savannas as elsewhere, but soil fluxes are substantially higher than in the savannas of South America (Fig. 2.22). In southern Africa the photochemical CO source is active primarily during the dry season, when the savannas and grasslands become a net source to the atmosphere (Schade et al. 1999).

It is estimated that vegetation is the source of >90% of all non-methane hydrocarbons (NMHCs) in the atmosphere (Guenther et al. 1995). In southern Africa the savannas are most likely the major biogenic source. The major hydrocarbons emitted from vegetation are isoprene and monoterpene. The emission of isoprene is influenced by light intensity, leaf temperature and nutrient availability, that of monoterpene by leaf temperature, vapour pressure and relative humidity. Estimates for isoprene production range from 0.6 to 9.0 mg $m^{-2} h^{-1}$ and for monoterpenes from about 0.05 to 3.0 mg $m^{-2} h^{-1}$ (Guenther et al. 1996). Both gases show a clear seasonal dependence with a winter minimum and summer maximum that is a reflection of foliar density, light intensity and temperature. Southern African savannas are estimated to emit between 44 and 88 Tg C yr^{-1} as isoprene and monoterpene (Otter et al. 2001).

Methane is an important greenhouse gas 21 times more effective as such than CO_2 on an equal mass basis. During winter and in periodic droughts, uncultivated dry soils of the southern Africa savannas are net sinks for atmospheric CH_4 (Zepp et al. 1996). Once wet they become sources. This is illustrated by values of −1.5 mg CH_4-C $m^{-2} h^{-1}$ being absorbed during the peak of an August dry season, only to be followed by an emission of 1.8 mg CH_4-C $m^{-2} h^{-1}$ at the start of the rainy season in the following October. The source effect lasts only for a few months into the wet season. Termites are often thought to be a primary source of CH_4 from savannas soils (Poth et al. 1995; Anderson and Poth 1998). This appears not to be the case in southern Africa (Otter et al. 2001). Instead microbial activity in aerobic soils is the dominant source. Soil moisture is an important factor in this process and hence there is a relationship between the elevation of the land above valley-floor level and CH_4 production. The shallow valleys of many savanna landscapes in southern Africa are occupied by hydromorphic clay soils that periodically become saturated and are thus treeless. These transient wetlands, called *vleis* in South Africa and *dambos* further to the north become CH_4 sources once saturated in the rainy season. Under such conditions emissions of 9.0 mg CH_4-C $m^{-2} h^{-1}$ have been recorded (Scholes and Andreae 2000). Maximum CH_4 emissions are observed with saturated soils. Once water rises to create a wetland lake, emissions drop, but do not cease altogether. Once the vleis dry out they behave identically in respect of CH_4 fluxes in soils of adjacent non-vlei savanna regions. It is estimated that up to 10 Tg yr^{-1} of CH_4 could be

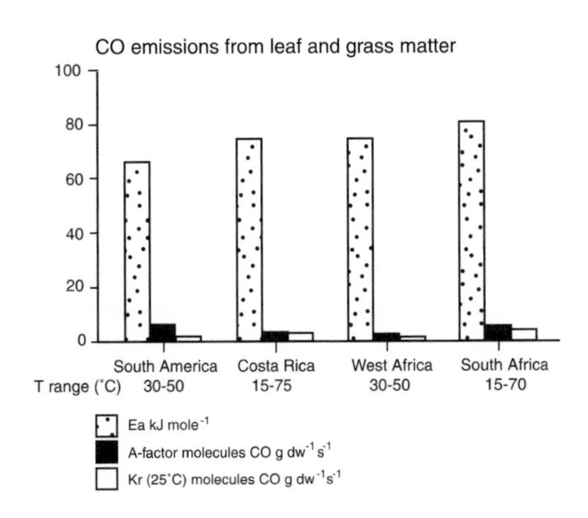

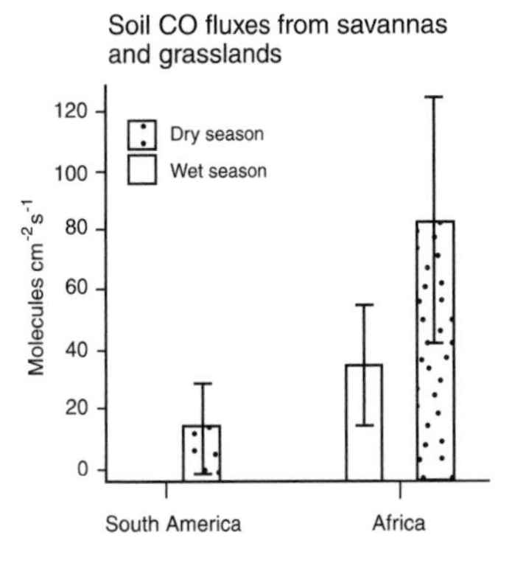

Fig. 2.22. Evaluated Arrhenius parameters and measured CO emissions at 25 °C of some degrading tropical deciduous leaf and grass matter (Schade et al. 2000) and soil CO fluxes for African and South American tropical and subtropical savannas and grasslands (Zepp et al. 1996)

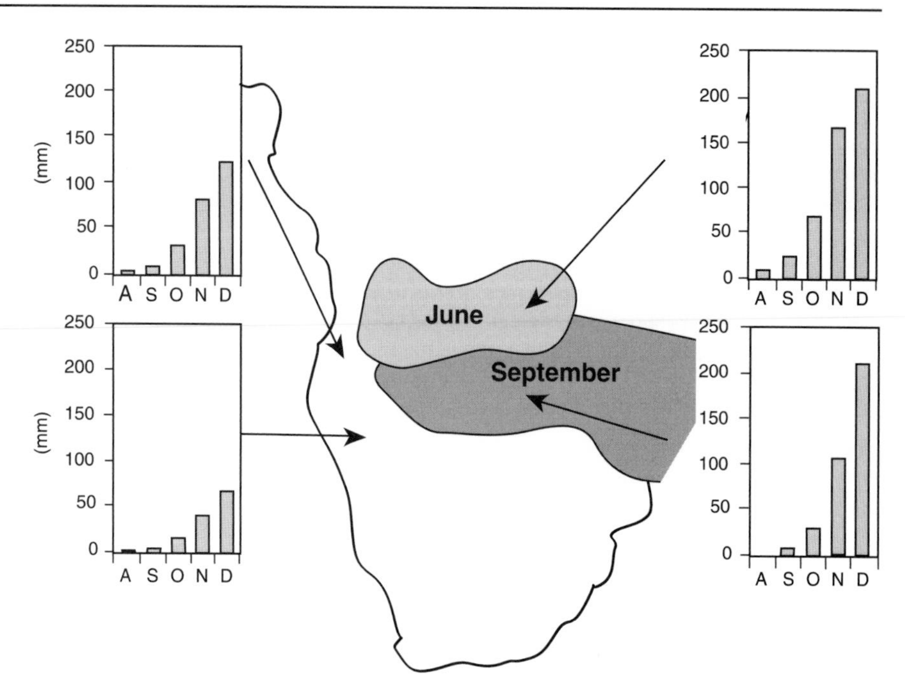

Fig. 2.23.
The onset of summer rains and peak biomass-burning seasons (Swap 1996)

added to the atmosphere in this way in the region. Few data are available for methanogenesis and oxidation in African soils. However, estimates suggest that the net CH_4 soil sink in the subcontinent could be about 0.04 Tg yr^{-1} (a net sink since dry periods in winter and summer are longer than wet periods in summer) (Scholes and Walker 1993; Otter et al. 2001). Emissions from enteric fermentation in large herbivores in Africa south of the equator are estimated to be around 0.32 Tg yr^{-1} (Scholes et al. 1996a).

Nitric oxide (NO) and nitrous oxide (N_2O) are key species involved in the chemistry of the troposphere and stratosphere and in global change. Nitric oxide is involved in the photochemical production of ozone and chemical production of nitric acid, the fastest growing component of acid rain, as well in the chemistry of the OH radical. The gas is an important greenhouse gas that has a global warming potential 200 times that of CO_2 on a per molecule basis. Microbial soil emissions are a significant source for both NO and N_2O in the atmosphere. Global modelling suggests that savannas are a major source of soil biogenic NO (Yienger and Levy 1995; Potter et al. 1996) as a result of their high temperatures and large geographic extent. In southern Africa the range of NO emissions reported is from 0.05–100 ng N m^{-2} s^{-1} depending on different conditions of soil disturbance, soil moisture and soil nutrient availability (especially N) (Harris et al. 1996; Levine et al. 1996; Parsons et al. 1996; Serça et al. 1998). Mean values for savanna emissions appear to be around the order of 10–15 ng NO-N m^{-2} s^{-1} (Levine et al. 1997).

The role of the savannas in the global budget of N_2O is uncertain. Where measurements have been made in South Africa, Zimbabwe and Namibia it appears that NO is the dominant nitrogen species emitted. The overriding reason for this is the long dry winter season, followed by rain interspersed with many dry episodes. Consequently, water filled pore space seldom reaches the threshold needed for N_2O emissions. During the transition to the wet season strong pulses of NO flux are observed, particularly after the first rains in September or October (Fig. 2.23), however light these may be. NO fluxes up to 60 times the preceding dry period values may be experienced for period of a few days (Scholes et al. 1997; Meixner et al. 1997). Soil moisture is the dominant factor controlling NO fluxes. The rain-induced biogenic flux is at least as important in producing NO as the immediately preceding biomass burning; it may even exceed the burning effect.

2.4.2 Biomass Burning Emissions

The burning of biofuels occurs in both rural and urban situations. It contributes a gross quantity of 138 Tg yr^{-1} to the CO_2 loading of the atmosphere in southern Africa, but the net loading is substantially less, since part of the biomass regrows. The fluxes of other pyrogenic trace gases produced by biofuel consumption, such as 5.5 Tg yr^{-1} of CO, 0.16 Tg yr^{-1} of NO_x and 0.1 Tg yr^{-1} of NMHCs and CH_4 (Marufu et al. 1999, 2000) are not taken up again, and are therefore net fluxes.

Savanna fires are by far the largest single source of pyrogenic emissions to the atmosphere over southern Africa. Latest estimates suggest that the amount of biomass consumed annually in southern Africa (excluding forest clearing, agricultural waste burning and biofuel burning) is around 177 Tg DM yr^{-1} (Scholes et al. 1996b).

The emission of CO_2, CH_4, CO, NO_x and N_2O from this biomass burning has been modelled to give distributions by country over most of southern Africa (Scholes et al. 1996a) (Fig. 2.24). Small counties have been ignored since the coarse simulation renders their estimates unreliable. The amount of pyrogenic CO_2 that is exchanged with the atmosphere is large, around 20% of the net primary production. This is not a net flux of CO_2 to the atmosphere at the scale of the continent, since over the period of a few years all of the carbon is taken up again by the vegetation as it regrows after the fire. Most pyrogenic CO_2 is produced in Angola, Zambia, Tanzania and Mozambique where the majority of the fires occur in the dry winter season (Fig. 2.25). The pyrogenic emissions of CH_4 are of the same order of magnitude to those produced by enteric fermentation in large mammal herbivores (Scholes and Andreae 2000), with the same countries again providing the largest emissions. The pyrogenic CO produced over the whole region is about 1% of the global CO budget and is approximately equal to that of the annual industrial emission from South Africa, the most industrialized country of southern Africa. Regional pyrogenic emissions of NO_x are a significant portion of the global NO_x budget and are an order of magnitude higher in Angola, Zambia and Tanzania than in countries to the south. The same applies to N_2O emissions and again reflects closely the distribution of biomass burning fires.

Owing to the relatively clean, flaming-dominated combustion typical of savanna fires reduced trace gases

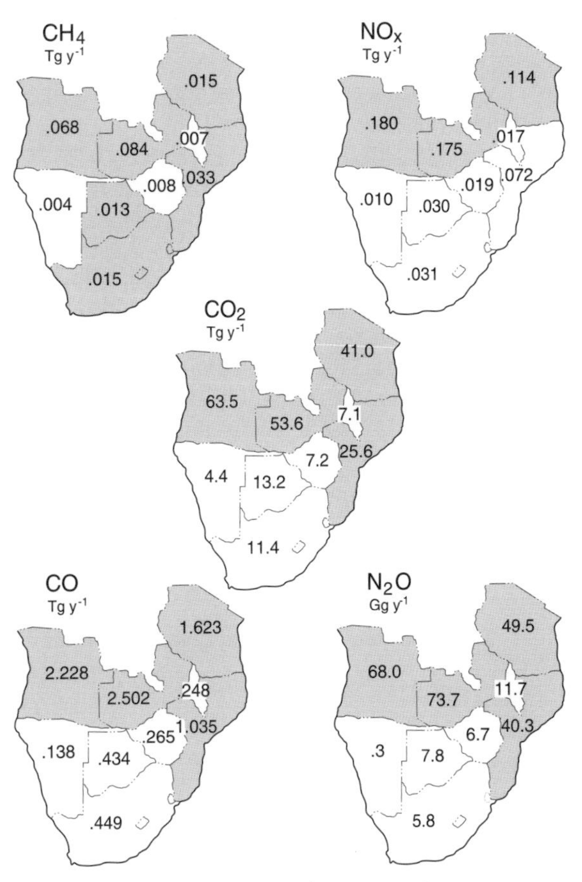

Fig. 2.24. Pyrogenic emissions of trace gases by country over southern Africa (Scholes et al. 1996)

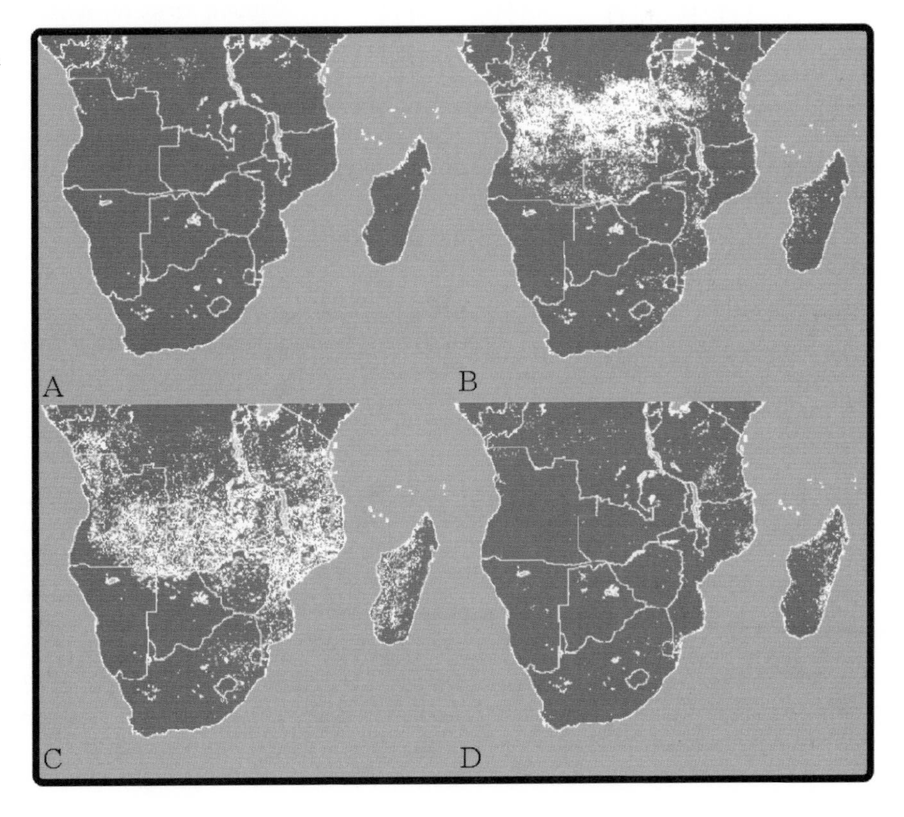

Fig. 2.25. Biomass burning over southern Africa during 1993: **a** March, **b** June, **c** September and **d** December (European Space Agency 1999)

(CO, CH_4, NMHCs) are emitted somewhat less than expected. For example, in case of CO, savanna fires produce only 35% of the global pyrogenic emission. In contrast, species that are connected to flaming combustion are favoured in savanna fires and in fact 52% of global pyrogenic NO_x emissions originate from savannas (Andreae 1997). A consequence of this flaming-dominated emission profile of savanna fires is that the ratio of NO_x to NMHC in smoke plumes is relatively high. Hence higher specific ozone production can take place in savanna fire plumes than in forest fire emissions (Andreae et al. 1994). Hydrogenated species such as CH_4, NMHCs, CH_3Cl and CH_3Br, as well as other important trace gases such as CO are produced predominantly in the smoldering phase of a fire characterized by insufficient O_2 supply. Savanna fires, together with the burning of agricultural wastes provide the greatest source of pyrogenic methyl halide species (Scholes and Andreae 2000) (Fig. 2.26).

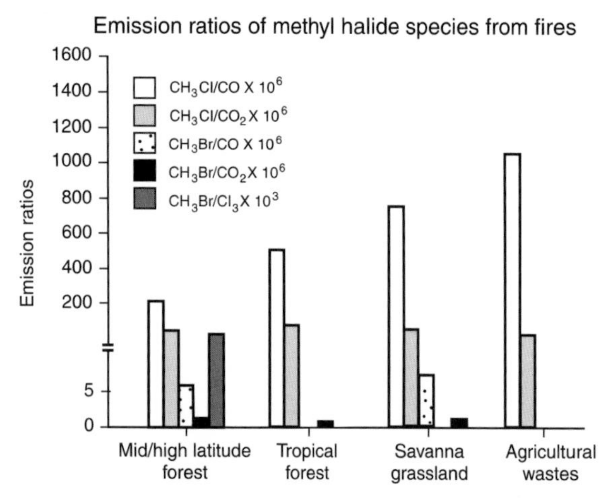

Fig. 2.26. Emission ratios of methyl halide species from fires in various ecosystems (Scholes and Andreae 2000)

Trace gases are produced in large quantities in industrial processes, as well as in the many activities that contribute to urbanization and urban living. It is useful to compare estimates of industrial trace gas production in southern Africa to biogenic, pyrogenic and biofuel burning production of the same gases (Otter et al. 2001) (Table 2.2). The single largest source of trace gases in southern Africa is by industrial emission of CO_2. Biogenic processes produce most NMHCs and CH_4. Ignoring CO_2 production, biomass burning is responsible for most CO. More NO_x results from industrial activities than from any other activity on its own.

2.4.3 Atmospheric Aerosols

Aerosols are a product of biomass burning. Smoke aerosols contain both inorganic compounds and black carbon in addition to many organic substances (Cachier et al. 1991; Cachier et al. 1996; Andreae et al. 1998). However, they are not the only source of aerosols in southern Africa. Generation of aeolian mineral (soil) dust from bare or sparsely vegetated surfaces by wind action, industrial emissions of particulate matter into the atmosphere and marine aerosols injected into the air from the oceans constitute other important ways in which the atmosphere over the region becomes laden with aerosols to produce the dense haze layers which are such a feature of southern African climate on all days without rain, even in summer.

The prevalence of subsidence in the atmosphere over southern Africa on most rainless days results in the formation of stable layers, which, though are not as strong as inversions of temperature, are sufficiently stable to inhibit vertical transport of aerosols and trace gases. The stable layers form preferentially at levels around ~700 hPa (~3.5 km altitude), ~500 hPa (4–6 km) and ~300 hPa (~8–9 km) over the interior plateau (Cosijn

Table 2.2. Comparison of trace gas emissions (Tg yr^{-1}) from biogenic, pyrogenic, biofuel burning and industrial sources in southern Africa

Trace gas	Biogenic	Savanna burning[a]	Biofuel burning[b]	Industrial
CH_4 soils	−0.23	0		0
CH_4 wetlands	0.2 – 10			
CH_4 vegetation	0	0.42	0.1	2.59
CH_4 animals	3.2	0		0
NMHC	44.2 – 87.8	0.55	0.1	0.61
NO_x	1.0	0.55	0.16	1.75
N_2O	?	0.45		?
CO soil	?	0		5.6
CO vegetation	1.5 – 65	11.51	5.48	0
CO_2	?	290.28	138.26	360.0

[a] Calculated with savanna fire activity level estimates of Scholes et al. (1996) and best guess savanna fire emission factor of Andreae (1997).
[b] Based on biofuel comsumption rates and emission factor estimates by Marufu et al. (1999) and Marufu et al. (2000).

and Tyson 1996). Over the region between the coast and escarpment in southernmost Africa a further layer is present at around the 850 hPa level (~1.5 km) (Fig. 2.27a). All three lower-tropospheric layers trap material below the level at which they occur. The ~500 hPa layer is the most persistent. On occasions it may prevail without disruption, while oscillating in height about its mean level, for more than 40 days over South Africa in winter and early spring (Garstang et al. 1996). The ~700 hPa and ~850 hPa layers tend to be less persistent and are disrupted approximately weekly by the passage of frontal disturbances (Preston-Whyte and Tyson 1973) over the central and southern areas of the region. This promotes vertical mixing of aerosols and trace gases to the ~500 hPa level. It is the stable layer at this level that exerts the most visible effect, since it usually marks the top of the southern African haze layer at altitudes between 4 and 6 km. The haze layer may cover the entire southern African region and extend over more than 30° of latitude (Garstang et al. 1996; Tyson 1997). An example of such a layer extending from South Africa to northern Zambia (Browell 1993) is given in Fig. 2.27b, together with measures of the aerosol loading within the haze layer (Anderson et al. 1996). Hydrocarbon trace gases

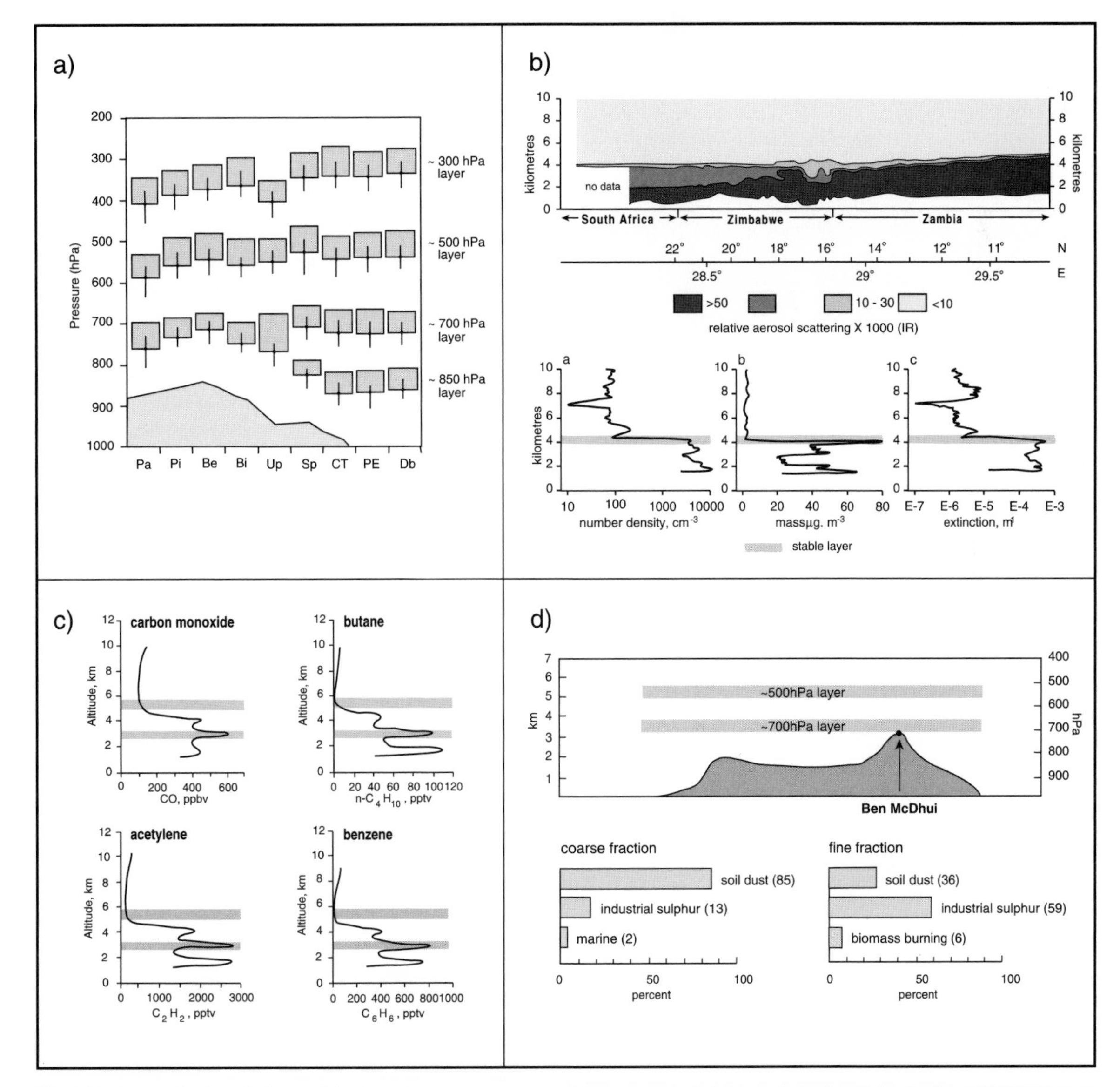

Fig. 2.27. Atmospheric stability and the haze layer over southern Africa: **a** mean heights and depths of absolutely stable layers over South Africa (Cosijn and Tyson 1996): **b** aerosol concentrations in the haze layer observed in a transect from Johannesburg, South Africa to northern Zambia (Browell 1993), vertical profiles of aerosol number density, mass and extinction in and above the layer (Anderson et al. 1996) capped by an absolutely stable discontinuity; **c** vertical profiles of hydrocarbon trace gases over southern (Blake et al. 1996) in relation to the climatological heights of the ~700 and ~500 hPa stable layers; **d** the location of the high-altitude, remote, background-aerosol Ben MacDhui site in relation to the ~700 and ~500 hPa absolutely stable layers over South Africa, together with annual source apportionment of coarse and fine fraction aerosols at Ben MacDhui (Piketh et al. 1996; Tyson and Preston-Whyte 2000)

likewise become concentrated below the layer (Gregory et al. 1996; Tyson and D'Abreton 1998) (Fig. 2.27c).

It is possible to sample well-mixed rural, backgound aerosol concentrations at a high-altitude site on the top of the 3 000 m Ben Macdhui Mountain on the southeastern edge of the Lesotho massif as aerosol-laden air is transported over and from southern Africa to the Indian Ocean and beyond (Piketh et al. 1996). In the coarse aerosol fraction (2.5–10 µm particle diameters), 85% of the particulate matter being transported out to sea is aeolian, surface-derived, mineral dust (Fig. 2.27d). The second largest contribution to total plume loading is industrially derived sulphur at 13%. In contrast, in the fine fraction (particles <2.5 µm in diameter), industrial sulphur constitutes 59%, aeolian dust 36% and particulates from biomass burning only 6% (Piketh et al. 1996). South of around 20° S, biomass burning produces only a small fraction of the aerosol loading of the lower troposphere over southern Africa (Piketh et al. 1999). North of 20° S the amount of burning is much greater (Cahoon et al. 1992; Scholes et al. 1996a,b) and products of biomass burning constitute a higher proportion of the aerosol loading.

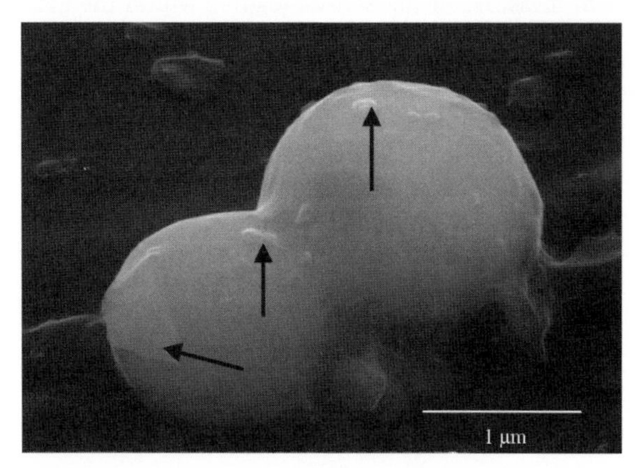

Fig. 2.28. Anthropogenically derived sulphur adhering to an aeolian dust particle transported over Ben MacDhui at an altitude of 3 000 m over South Africa (Piketh et al. 1999)

Sulphur is transported not as elemental or sulphate particles *per se*, but as patchy coatings of precipitated sulphur products on small dust nuclei (Piketh et al. 1999; Piketh 2000) (Fig. 2.28). Although aeolian dust transport is at a maximum in winter, sulphur transport is at a maximum in warmer, moister summer air, when oxidation of SO_x is at its maximum.

The air transport in the plume from South Africa that moves towards Australasia has been modelled empirically by kinemetic trajectory analysis (Sturman et al. 1997) (Fig. 2.29b). Transport of CO_2 in the plume from southern Africa has been simulated using GCM chemical transport models (Rayner and Law 1995) (Fig. 2.29c) and direct satellite observations of dust transport from South Africa to Australia have been made (Herman et al. 1997) (Fig. 2.29d).

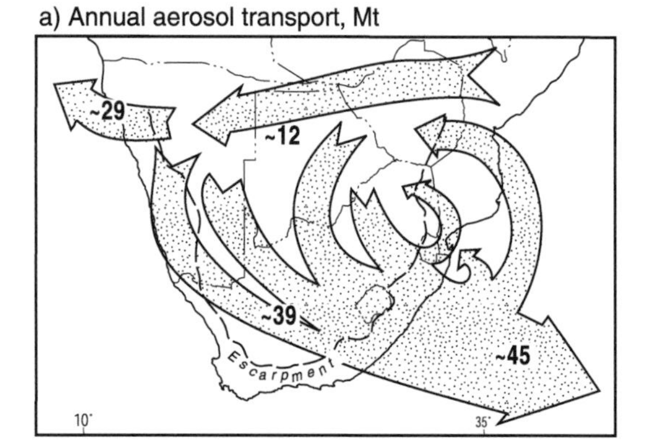

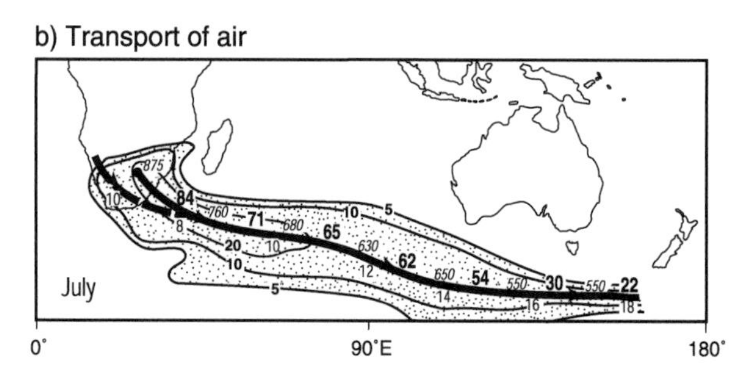

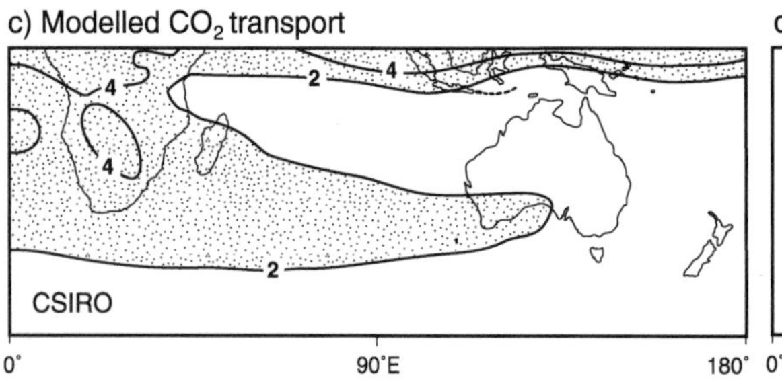

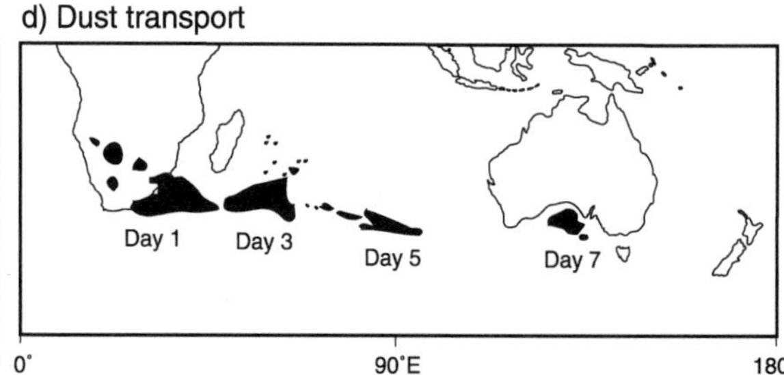

Fig. 2.29. Transport plumes exiting South Africa to the Indian Ocean and beyond: **a** annual mass fluxes of aerosols (Mt a^{-1}); **b** vertically-integrated 850–800 hPa westerly-component, zonal July transport of air across the Indian Ocean from the Highveld of South Africa (contours give percentage transport, the bold line indicates the maximum frequency pathway, bold numbers denote total percentage plume transport across the meridian, italic numbers indicate geopotential heights and Roman numbers times of transport in days) (Sturman et al. 1997); **c** simulated CO_2 transport across the Indian Ocean as modelled by the CSIRO model (peak-to-peak amplitudes are given in ppmv at 500 hPa) (Rayner and Law 1995); **d** a 7-day variation of aerosol transport observed in Nimbus-7/TOMS data (Herman et al. 1997)

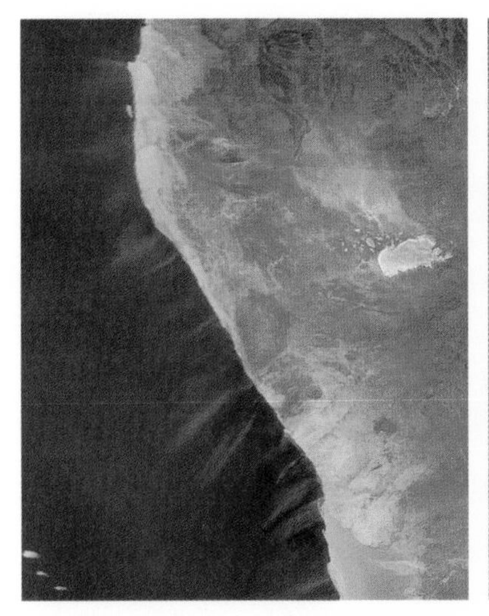

Fig. 2.30. Satellite images to show contrasts in aerosol and trace gas transport patterns off the west and east coasts of southern Africa (NASA 2001). Off the west coast, relatively short, narrow individual dust plumes extend only short distances offshore before subsiding into the ocean surface. Off the east coast, the aerosol and trace gas plume is more integrated, coherent and massive. In the example given, the plume can be seen extending at least 3500 km from the coast to the south-east over the Indian Ocean before disappearing out of the picture. At the widest off the South African-Mozambique coast the plume is ~1500 km wide; south of Madagascar it is 500–800 km wide

The east-coast plume off southern Africa is a major feature of the southern hemisphere circulation field and produces large-scale inter-regional transfers of aerosols and trace gases. Owing to the frequent prevalence of cloud of the east coast, the plume is difficult to capture on satellite images. On certain occasions it may be seen (Fig. 2.30). It is a much larger and more coherent feature than its west coast counterpart. The plume indicates clearly the influence of continentally derived air from Africa on the oceanic surrounding region. In addition to transporting aerosols and carbon dioxide, the plume carries tropospheric ozone far over the Indian Ocean, particularly in spring (Fishman 1991; Fishman et al. 1991). Moody et al. (1991) demonstrate from Global Chemistry Project analyses of rainfall samples collected over the period 1980–1987 on Amsterdam Island, approximately midway between South Africa and Australia at about 38° S, 78° E, that radon, non-sea salt sulphates and nitrates may be transported to the island from southern Africa and Madagascar, a distance exceeding 5 000 km, in as little as 3 days on occasions. With anticylconic systems prevailing over South Africa, the average time of transport to Amsterdam Island is around 6 days (Garstang et al. 1996; Tyson et al. 1996a).

2.5 Recirculation of Trace Gases and Aerosols

A striking feature of the transport of aerosols and trace gases over southern Africa is the degree of recirculation that takes place. This occurs when air, and the contents therein, is not transported directly across or out of the airspace over the subcontinent, but instead is entrained by the prevailing airflow to be recycled before exiting the region. An example of air trapped between the ~700 hPa and ~500 hPa stable discontinuities over southern Africa and recirculating in the enclosed space for more than three weeks is given in Fig. 2.31a. It appears that approximately 44% of all air circulating over southern Africa on fine (no-rain) days has been recirculated on a subcontinental scale at least once (Tyson et al. 1996b). The annual flux of aerosols being recirculated to the west over central areas of northern South Africa and southern Zimbabwe is estimated to be around 11.5 Mt yr^{-1}; that being recirculated to the east over central South Africa at around 32° S to be 17.3 Mt yr^{-1}. Recirculation occurs at a variety of temporal and spatial scales extending from hours to weeks and from tens to thousands of kilometres. Most often it is confined to the haze layer below ~500 hPa.

Whatever the form of transport, direct or recirculated, it usually involves trans-boundary movement of air over southern Africa. Examples of air originating at the Victoria Falls, being transported over southern Angola and Namibia, recurving to the east over South Africa, being transported to the north over the Madagascar region, recurving to the west over Mozambique, moving across Malawi and thence back to the Zimbabwe/Zambia border region have been reported (Garstang et al. 1996). Of air moving from the industrial heartland of South Africa, 43% recirculates over Mozambique, Zimbabwe and Botswana before recurving back over South Africa (Freiman et al. 1999) (Fig. 2.31b). Such recirculating air carries the aerosols and trace gases entrained or emitted into it. It is estimated that more than 30% of all material transported in air over Mozambique and Botswana has recirculated from South Africa (Piketh et al. 1999).

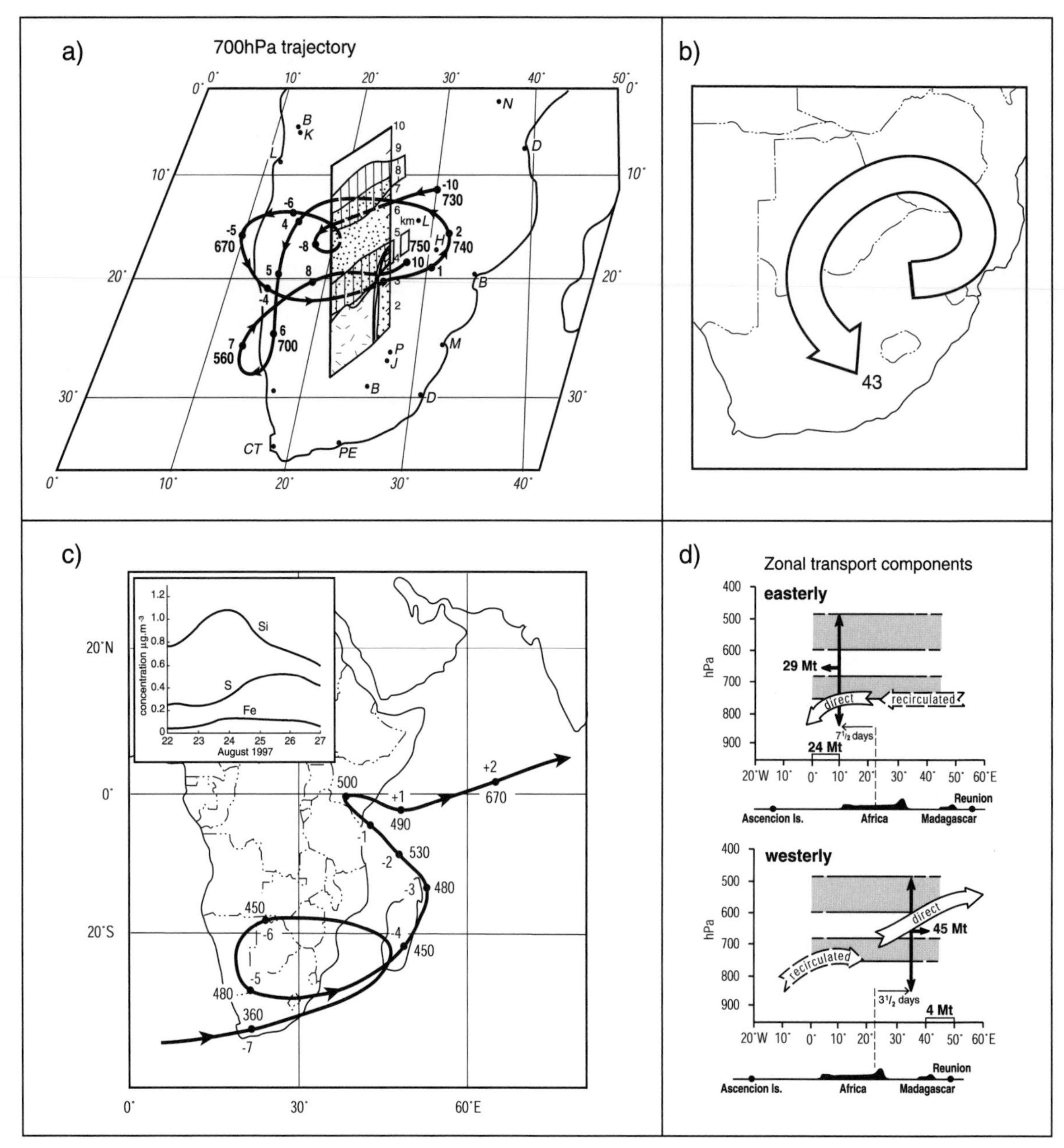

Fig. 2.31. Recirculation of aerosols over southern Africa: **a** 700 hPa kinematic trajectories over a 20-day period to show recirculation trapped between the ~700 and ~500 hPa stable layers over South Africa (days of travel and daily geopotential heights are shown along the trajectory) (Tyson 1997); **b** percentage mean transport between 800 and 700 hPa from the South African Highveld, 1995–1998 (Freiman 1999); **c** transport of silicon, sulphur and iron from southern Africa to Mount Kenya and beyond (the height of the transport plume (hPa) is given at specified times in days (Gatebe et al. 1999); **d** mean annual vertically-integrated 850–500 hPa zonal transport of aerosols (Mt yr^{-1}) over the east and west coasts of South Africa in relation to annual mean positions of the semi-permanent absolutely stable layers

Up to 30% of air reaching Zimbabwe has likewise passed over South Africa before reaching the country from the east. That reaching Zambia and Angola from the same source is up to 15% and that getting to Tanzania and Kenya 5% or less per annum.

That long distance transport of aerosols takes place over the southern African region is not in doubt. Sulphur emitted into air recirculating over Zimbabwe, Botswana, Namibia and South Africa has been shown to be transported over Madagascar before being recirculated inland to be recorded on the upper slopes (at 4 220 m) of Mount Kenya (Gatebe et al. 1999) (Fig. 2.31c). In the example cited, sulphur was trapped within the air below the ~500 hPa stable layer, transport took place at the top of the layer and transit time from South Africa to Kenya was around 5 days. After passing Mount Kenya and crossing the equator, the transport plume moved on towards India.

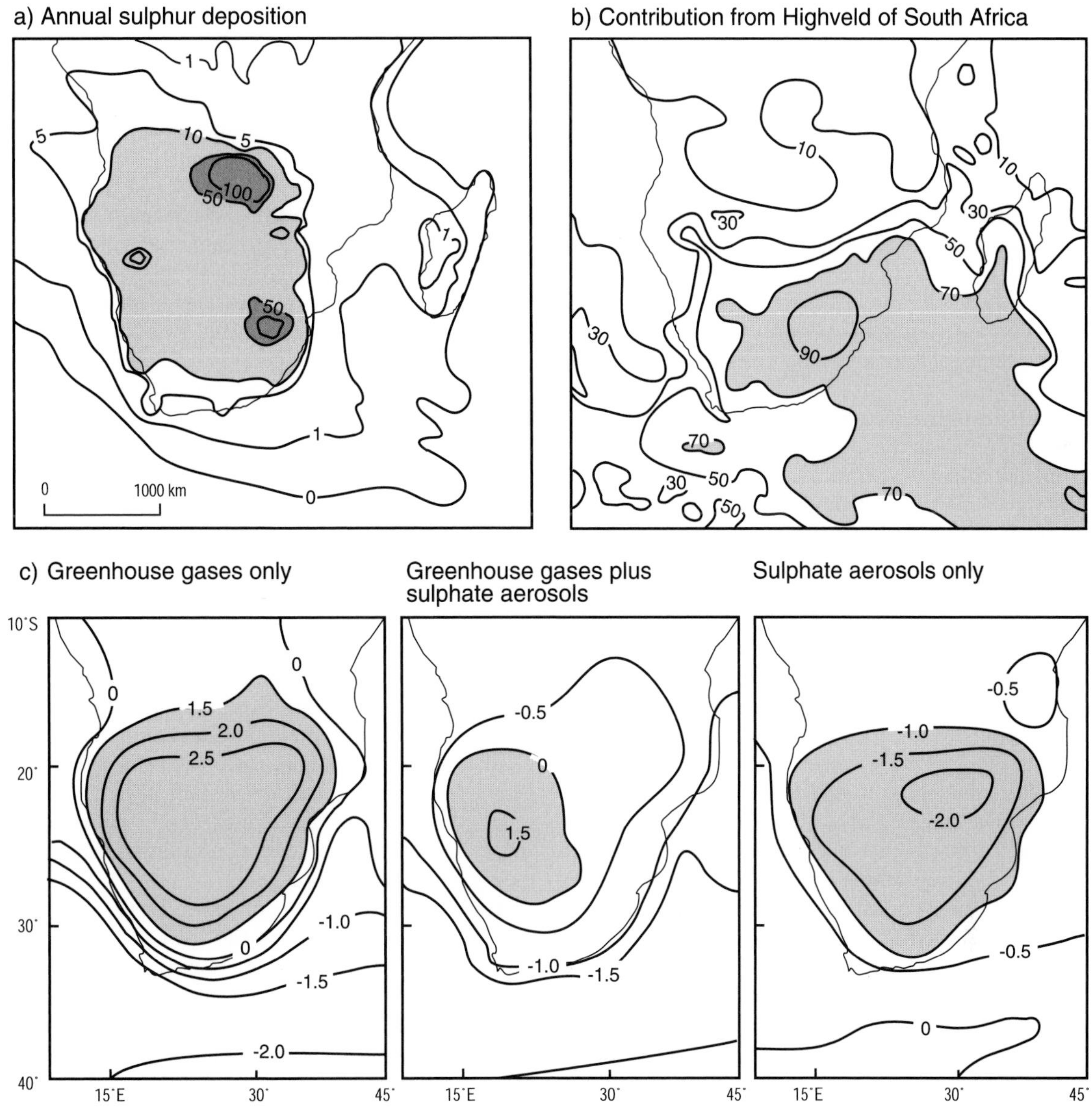

Fig. 2.32. Annual sulphur deposition over southern Africa **a** in mg m^{-2} (after Zunckel et al. 1999); **b** the percentage contribution from the South African Highveld alone (Zunckel et al. 1999), together with **c** HadCM2 coupled ocean-atmosphere climate model estimations of sulphate aerosol cooling of the atmosphere with a doubling of greenhouse gases (Joubert and Kohler 1996)

2.5.1 Sulphur Deposition

The most up-to-date inventories of southern African SO_2 emissions reveal two main areas of emission: the South African Highveld and the Zambian/Congo Copperbelt (Zunckel et al. 1999). Of these, the South African contribution is the greater. The regional Swedish MATCH model (Robertson et al. 1996) has been used to model both wet and dry sulphur deposition over the southern African region (Zunckel et al. 1999). Given that rain falls on so few days per year (around 20% of days for most areas) only dry deposition is illustrated (Fig. 2.32a). Annual dry deposition covers most of the subcontinent, with highest rates of deposition occurring around the major emission areas. The Indian Ocean plume is a feature of the deposition pattern. The percentage contribution by South African emissions to total annual sulphur deposition is greatest over South Africa and adjacent regions (Fig. 2.32b) and constitutes over 70% of the sulphur transported to the Indian Ocean.

2.5.2 Effects of Sulphur Aerosols

Use of the HadCM2 coupled ocean-atmosphere climate model (Joubert and Kohler 1996) allows an estimate to be made of the extent to which global sulphate aerosols are inducing an anthropogenic climatic change over southern Africa. By comparing the modelled climate using greenhouse gas and combined greenhouse gas and aerosol forcing, the aerosol effect may be isolated (Fig. 2.32c). On the basis of the current state of knowledge, and subject to the uncertainties known to be associated with current models (IPCC 1996), it would appear that sulphate aerosols are causing cooling in excess of 1 °C over most of the subcontinent in early winter. During the rest of the year sulphate-induced cooling is much less (Joubert and Kohler 1996).

2.5.3 Regional Aerosol Mass Fluxes

Annual mass fluxes of aerosols being transported over southern Africa in the 4–6 km deep haze layer may be estimated from the volume of air being transported in mean plumes (Tyson et al. 1996a) and from measurements of background ambient aerosol loading (Wells et al. 1987; Taljaard and Zunckel 1992; Rorich and Turner 1994; LeCanut et al. 1996; Maenhaut et al. 1993, 1996; Swap 1996). The annual fluxes are large (Fig. 2.29a). Over the central subcontinent, 12 Mt yr^{-1} is transported over Zimbabwe and Botswana in the direction of the Atlantic Ocean. Over Botswana and Namibia the transport field diverges into a major plume recurving to the south, with a minor plume moving westward. By the time the latter exits the continent to the Atlantic Ocean off Namibia, the flux has increased to 29 Mt yr^{-1}. Recurving anticyclonically towards the east and the Indian Ocean at 30–32° S, the flux in the main transport plume reaches ~39 Mt yr^{-1} over central South Africa. By the time the plume has reached 35° E off the southeast coast the flux is estimated to be ~45 Mt yr^{-1}. These flux estimates are subject to uncertainties and do not take into account wet and dry deposition. They appear to be about half those similarly determined for transport westward out of a much larger Saharan area from northern Africa (D'Almeida 1987; Duce et al. 1991).

The locus of the aerosol plume to the Atlantic Ocean is at about 18° S at a mean height of ~800 hPa, i.e., below the level of the ~700 hPa stable layer (Fig. 2.31d). The mean time of transit to reach 10° E from the central continent is 7.5 days. Of the total amount transported to the west, 24 Mt yr^{-1} is deposited in the sea by trajectories going to surface between the coast and the Greenwich Meridian to the west. In contrast, the integrated 850–500 hPa offshore transport to the Indian Ocean over the east coast of South Africa is about 45 Mt yr^{-1} at 35° E. The locus of the plume as it crosses the coast is at about 31° S with a mean height of about 750 hPa (again below the ~700 hPa stable layer). The transit time from the central interior to 35° E is 3.5 days. Only 4 Mt yr^{-1} is estimated to be deposited by trajectories going to surface between 40° and 50° E. Most deposition of this kind appears to occur over the central Indian Ocean beyond 70° E (Piketh 2000).

2.6 Some Consequences of Aerosol Transport

2.6.1 Nutrient Deposition

Recirculation of aerosols in the atmosphere over southern Africa implies recirculation of particulate nutrients as well. Deposition of such nutrients and their subsequent recycling back into the atmosphere as dust, biomass burning products and industrial pollutants has considerable biogeochemical significance for both terrestrial and marine ecosystems. The effect on the biogeochemistry of the largest inland wetland ecosystem in the region, the Okavango Delta of Botswana, is significant. The Okavango has an annually flooded area of up to 12 000 km^2. The delta is in a semi-arid region (annual rainfall ~500 mm) where annual evaporation exceeds precipitation by a factor of three; in an extreme dry year the annual net radiation is capable of evaporating around ten times the yearly rainfall (Tyson 1986). The water of the delta is supplied by the Okavango River, which rises in subtropical central Angola. The rainy season in the catchment is January–March. Peak discharge into the delta at its head occurs in April; it takes a further five months for the water to traverse the 250 km from head to toe. The base flow in the river sustains at maximum 4 000 km^2 of permanent swamp; the annual flood seasonally inundates up to a further 8 000 km^2. Only 1.5% of the river inflow leaves the delta. The maximum areal extent of flooding occurs in winter (July–August), when aerosol loading and deposition are at their greatest (Piketh et al. 1999).

The annual sediment load carried by the river into the Okavango Delta is about 620 000 t (McCarthy and Metcalfe 1990). The estimated atmospheric deposition from anticyclonic systems alone (which occur on ~80% of days in winter and ~25% in summer) is estimated to be about 250 000 t, i.e., ~40% of the river load (Garstang et al. 1998). The delta is a hyper-oligotropic system (Cronberg et al. 1995). Much of the catchment area is covered with Kalahari sand with a low inherent nutrient status. The general absence of chemical weathering in this environment results in a low dissolved solids load in the water of the river, most of which consists of silica and calcium and magnesium bicarbonates (Hutchins et al. 1976). Nutrient species occur at low concentrations: nitrogen (as nitrate or ammonium ions) is typically 0.08 ppm, phosphorus 0.04 ppm and potassium 2.9 ppm (Cronberg et al. 1995). Atmospheric aerosol deposition is estimated at 0.57 kg ha^{-1} d^{-1} over the delta as a whole (Garstang et al. 1998). In the delta distributary channels, water supplies 90% of the nitrate and phosphate needs of plants. However, in low productivity areas beyond the edges of the channels (which constitute 90% of the area of the delta) aerosols supply up to 52% of phosphates and 30% of nitrates (but only around 10% of potassium needs) (Fig. 2.33). The hitherto not-considered atmos-

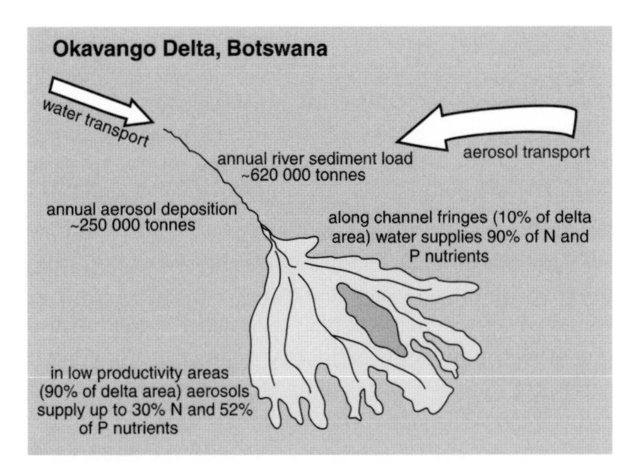

Fig. 2.33. Aerosol transport and atmospheric deposition of N and P over the Okavango Delta wetland ecosystem, Botswana (Garstang et al. 1998)

pheric contribution to nutrient cycling and the nutrient budget of the wetland ecosystem is substantial. The implications for other ecosystems, both terrestrial and marine, may be significant.

2.6.2 Impact on Precipitation

Aerosols affect climate both directly and indirectly. Their effect in ameliorating global warming in the region has already been considered. One way in which to approach possible effect of aerosols on precipitation is to use a simple basin hydrological model to determine their impact on the water cycle. This has been done for Lake Tanganyika.

The hydrology of the lake basin has been modelled and the sensitivity of modelled precipitation to changes in atmospheric transmissivity has been assessed (Tyson et al. 1997). Considering only the effects of changing aerosol loadings on solar transmissivity and the solar radiation balance, and the feedback of these changes on the basin hydrology, has enabled indirect estimates of the effects of transmissivity on rainfall to be made. It appears that the model is more sensitive to changes in atmospheric transmissivity than to changes in surface area of the lake, temperature, albedo or Bowen ratio. The model suggests that a 10% decrease in atmospheric transmissivity over the region of Lake Tanganyika might produce a 15% diminution in rainfall. Such a change in transmissivity falls within the range of inter-annual variability observed within the second half of the twentieth century in southern Africa. Changes in transmissivity of this order are possible in future if GCM predictions for a drier regional future with global warming are realised.

Qualitatively, the results from the simple model are in agreement with those of the aerosol-incorporated GCM simulations for southern Africa. The findings for Lake Tanganyika may be an indication of the kind of precipitation changes increasing the bulk loading of the atmosphere may have in future over a much wider region in Africa.

Another way in which aerosols may affect rainfall is by modifying the microphysical processes regulating condensation and precipitation formation in warm clouds. South African cloud seeding experiments have demonstrated that the convective clouds of the region are very sensitive to the size and chemistry of aerosols in the atmospheric boundary layer (Mather et al. 1996, 1997a,b). Aerosols derived from the crust of the earth and biomass burning, as well as sulphates and elemental material from industrial sources, frequently remain in the haze layer for periods of a week, and on occasions for as long as three, while they recirculate anticyclonically over South Africa before being exported offshore (Tyson et al. 1996b). Particles surviving this long in the lower layers of the atmosphere typically have diameters less than 2 μm. An excess of such small particles will enhance the growth of small droplets in clouds at the expense of large ones, thus inhibiting the development of cumulus clouds in which the coalescence process and rainfall production are optimised. The consequence may be a diminution of regional rainfall (Mather et al. 1997a,b).

2.6.3 Iron Fertilisation of the Central South Indian Ocean

The extent to which atmospheric transport of aeolian dust maintains or enhances ocean productivity and the ocean carbon cycle has remained moot until recently. Iron is known to be an essential ingredient for phytoplankton to flourish even in regions of the world oceans where nutrients are abundantly available at the surface through upwelling (Watson 1999). Areas where phytoplankton blooms are known to occur regularly in the world oceans act as effective sinks of CO_2 (Behrenfeld et al. 1996; Cooper et al. 1996; Boyd et al. 2000; Abraham et al. 2000). Controlled experiments have revealed the extent to which over small areas of the ocean, and short periods of time, iron fertilisation of the surface significantly affects the growth of phytoplankton in waters where iron is otherwise a limiting nutrient (Cooper et al. 1996; Boyd et al. 2000). Over longer time scale the increased photosynthesis in an iron-enriched marine region serves as a biological CO_2 pump (Behrenfeld et al. 1996; Cooper et al. 1996; Boyd et al. 2000; Abraham et al. 2000).

As part of the Joint Global Ocean Flux Study (JGOFS), extensive data on ocean-atmosphere CO_2 fluxes have been collected from around the world. Key areas where the oceans act either as sources or sinks of atmospheric CO_2 have been identified (Takahashi et al. 1997). The

South Indian Ocean between South Africa and Australia has been identified as a major sink region. Both observations (Sturman et al. 1997; Herman et al. 1997) and modelling (Rayner and Law 1995) reveal that the aerosol plume from southern Africa may reach Australasia on occasions. It crosses, and completely covers, the area of carbon sink (Piketh et al. 2000) (Fig. 2.34a). As the plume leaves South Africa it tends to rise owing to the convergence and ascent of air with poleward geostrophic flow and a constant pressure gradient occurring in the mean windfield on the western margin of the South Indian Anticyclone (Tyson and Preston-Whyte 2000). As air is transported towards the centre of the South Indian Anticyclone in the vicinity of 70° E it tends to subside. On occasions, when subsidence is particularly strong, air parcels may subside to the surface. Both case studies and mean climatologies show this to be the case.

From measurements made at the high-altitude, baseline site at Ben Macdhui on the edge of the Lesotho massif (and located beneath the locus of the mean transport plume streaming off South Africa), it has been possible to determine mid-tropospheric mass fluxes of iron-bearing aerosols as they are transported towards Australasia. Deposition of aerosols takes place either by rainout in wet deposition or by dry deposition. The latter can occur either as dust fallout from an elevated plume or by trajectories of air parcels being forced to the surface by atmospheric subsidence. If only the latter form

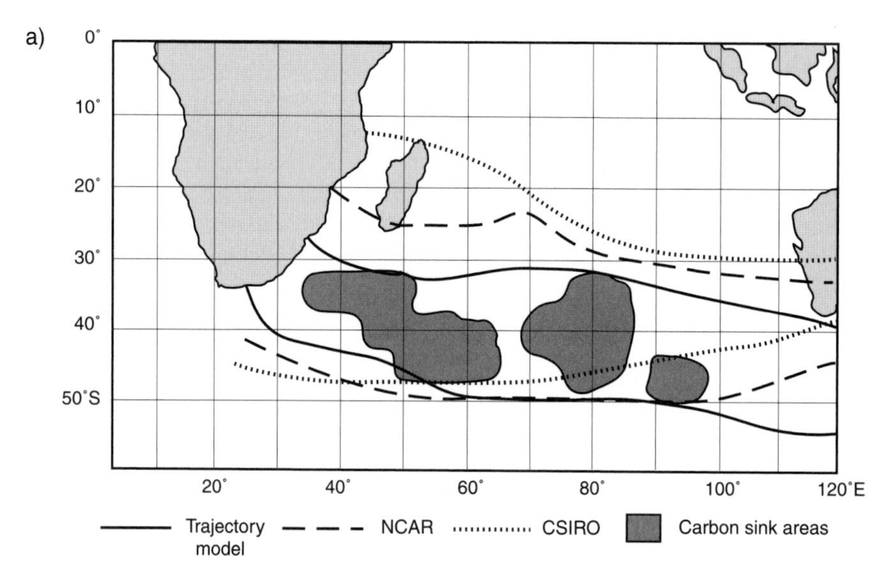

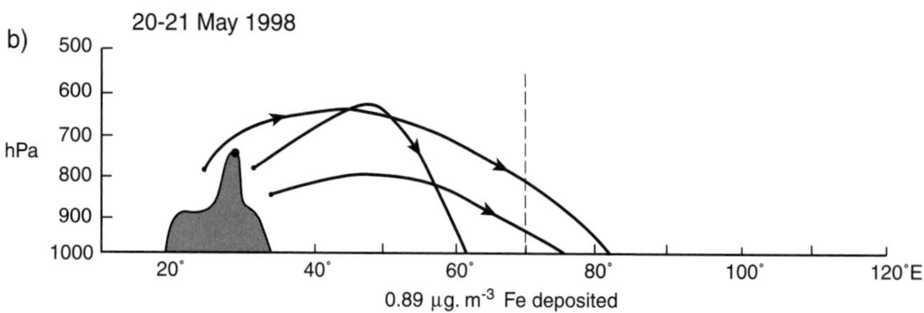

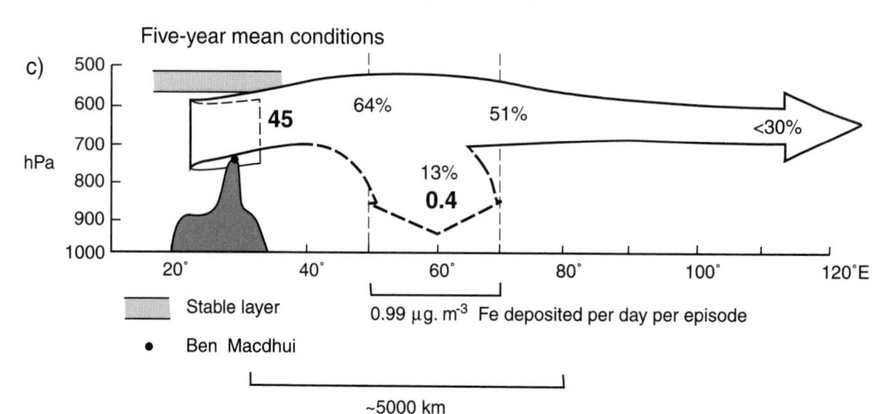

Fig. 2.34. Aerosol transport and iron fertilisation of the South Indian Ocean (Piketh et al. 2000): **a** aerosol and trace gas transport plumes from southern Africa to the South Indian Ocean (shaded areas represent areas of enhanced biological productivity and carbon sinks); **b** iron fertilisation from southern Africa (shaded) on 20–21 May 1998 resulting in 0.89 $\mu g\ m^{-3}$ iron deposited into the ocean between 60° E and 85° E; **c** mean daily deposition of 0.99 $\mu g\ m^{-3}$ into the ocean per iron-transport episode between 50° E and 70° E as estimated from a 5-year trajectory climatology and PIXE measurements of iron loading in the plume passing over Ben MacDhui at 3 000 m in South Africa (approximate annual mass fluxes of total aerosol loading are given as $Mt\ yr^{-1}$ in bold; numbers of trajectories passing given meridians or having subsided to the surface are given as percentages)

of dry deposition is considered, then daily case studies suggest that as much as 0.89 µg m^{-3} of elemental iron may be deposited between 60° and 85° E (Piketh et al. 2000) (Fig. 2.34b). Over long periods of time, as evidenced in a 5-year climatology, on average 13% of trajectories subside to the surface between 50° and 70° E (Fig. 2.34c). Given that iron constitutes 0.7% of the fine aerosol loading at Ben Macdhui, and that high-iron transport episodes occur about every 11 days on average (Piketh 2000), it would appear that the mean daily deposition of iron into the sea in the central South Indian Ocean following a peak-concentration episode over eastern South Africa is around 0.99 µg m^{-3}. Such peak concentrations occur on around 33 days in a year. Given that the average duration of an episode centred on the peak is 3 days, the number of days a year in which iron fertilisation may be significant appears to be around 100.

In an equatorial Pacific Ocean iron enrichment experiment, a region of 64 km^2 was enriched with ferrous sulphate to a final concentration of 2*n*M (0.11 µg m^{-3}) of iron, which was sustained for several days by further enrichment (Behernfeld et al. 1996; Cooper et al. 1996). A significant bloom of phytoplankton took place in the patch. The estimated mean concentration of 0.99 µg m^{-3} deposited over the South Indian Ocean from the southern African aerosol plume is the same order of magnitude. Not all the iron in the plume is likely to be in soluble form. What is will be available for phytoplankton enrichment.

Strong evidence exists to suggest that aeolian transport of aerosols from South Africa, and atmospheric iron fertilisation of marine biota, support enhanced biological productivity and the South Indian Ocean carbon sink in the central ocean between South Africa and Australia. Inter-regional nutrient transfer over distances exceeding 5 000 km have been established and links have been demonstrated between continental, terrestrial ecosystems and their remote marine equivalents.

2.7 The Oceans around Southern Africa and Global Change

The oceanic circulation systems adjacent to southern Africa are diverse and complex (Fig. 2.35). To the south, the Southern Ocean links the oceans of the world and exerts an important influence on global climate. Bordering the east coast, the warm Agulhas Current is an intense western boundary current. Even though its geographic dimensions are limited, it plays a critical role in the transport of warm water in the South Indian Ocean, in regional weather modification and in the inter-ocean exchange of tropical and subtropical water (Lutjeharms 1996). The exchanges it produces have important global climatological implications (Gordon 1986; Broecker 1991). The South Atlantic Ocean to the west of southern Africa is the conduit through which the warm products of the Agulhas Current move equatorward (Gordon and Haxby 1990; Byrne et al. 1995; Garzoli et al. 1997; Goni et al. 1997). Along the west coast of southern Africa is one of the largest coastal, wind-driven, upwelling systems in the world (Shannon 1985; Shannon and Nelson 1996); it exerts a dominant control on coastal-zone ocean productivity. Immediately to the south, the Agulhas retroflection area is one of high ocean-atmosphere fluxes and is characterized by being one of the most variable of all global ocean zones (Cheney et al. 1983).

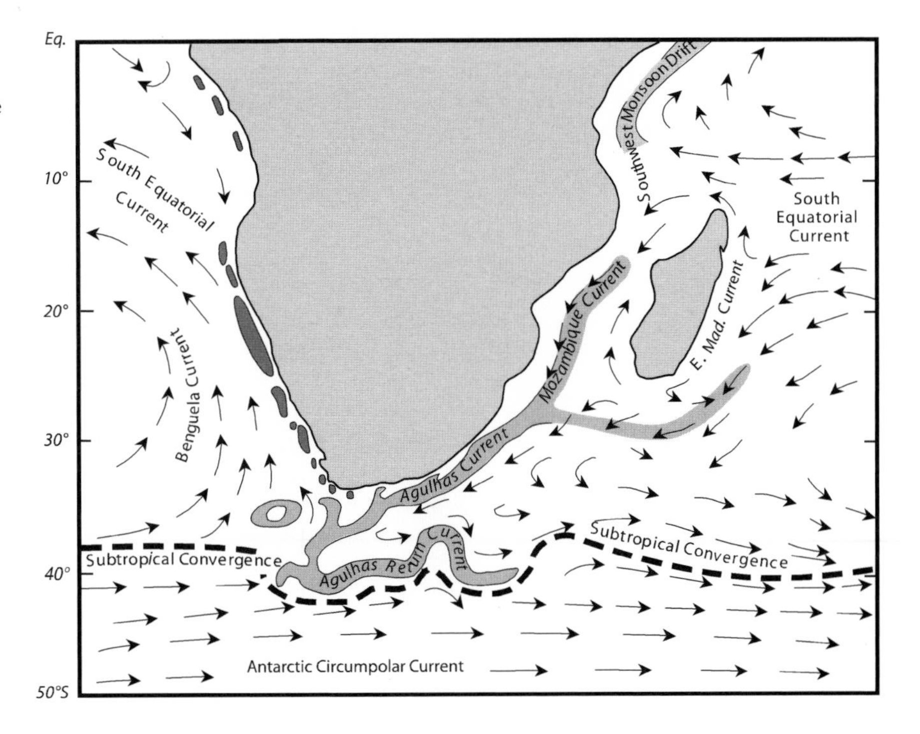

Fig. 2.35. Ocean currents around South Africa (modified after Niiler 1992). The shaded areas off the west coast represent cells of upwelling of cold water in the Benguela upwelling system

At intermediate depths below the surface currents alluded to above is the global thermal conveyor (see Fig. 2.37 to follow) that links changes that may take place in the oceans adjacent to South Africa with changes occuring in the North Atlantic Ocean where the conveyor is commonly taken to have its origin.

2.7.1 The Agulhas Current System

The greater Agulhas Current system (Fig. 2.35) consists of its source waters, the current itself, its outflow, its shedding products and its effect on the circulation on the adjacent continental shelf. Once fully constituted off South Africa, the current extends to a depth of at least 2 000 m (Beal and Bryden 1999) (Fig. 2.36a). In the upper layers, it carries predominantly tropical and subtropical surface water poleward at rates of ~2 m s^{-1}. This water is largely derived by a strong recirculation in the South West Indian Ocean subgyre (Stramma and Lutje-harms 1997), as well as from the eastern side of Madagascar and the Mozambique Channel. A seasonal signal in the current propagates downstream and manifests itself in the Agulhas Current retroflection (Quartly and Srokosz 1993), where the current reverses direction and subsequently lies along the Subtropical Convergence. In the process of retroflection rings of warm water are shed into the South Atlantic Ocean and generate the turbulence and variability so characteristic of the waters to the south of South Africa.

It is possible that with future global change that strengthened wind stress on the southwestern Indian Ocean, together with an increase in surface water temperatures in the subtropics, will result in more and warmer water flowing south in the Agulhas Current. The amount of heat carried by the current and shed into the South Atlantic Ocean may increase and this would have immediate global consequences by influencing the rate of overturning of the thermohaline cell of the Atlantic Ocean as a whole (Weijer et al. 1999). Increased warm surface water accumulation in the Agulhas retroflection region may also influence synoptic weather patterns over southern Africa (Crimp et al. 1998). Locally, ocean-atmosphere fluxes of heat and moisture would increase (Lutjeharms et al. 1986) with positive impacts on the rainfall of adjacent coastal areas (Jury et al. 1993) provided the morphology of the current does not change.

The northern Agulhas Current is unique amongst western boundary currents in having an extremely stable trajectory. The current follows the continental shelf edge closely (Fig. 2.35), with meanders to either side of less than 15 km (Gründlingh 1983). On irregular occasions, but occurring about 4 to 6 times a year, a solitary meander, the Natal Pulse, interrupts this stable flow pattern (Lutjeharms and Roberts 1988). The feature moves down the current at a rate of about 20 km d^{-1} affecting not only the inshore coastal water circulation, but also the downstream behaviour of the Agulhas Current itself. A dramatic consequence of a large pulse is to cause the Agulhas Current to retroflect far upstream at the Agulhas Plateau (Lutjeharms and van Ballegooyen 1988b). In this way the waters in the current do not reach the normal retroflection area and are therefore not available for inter-ocean exchange.

Over the Agulhas Current the moisture flux from the sea surface to the atmosphere is large (Lee-Thorp et al. 1999). The distance of the core of the current from the coastline has a marked effect on the local rainfall in coastal regions (Jury et al. 1993). It controls the locus of maximum cumulus cloud development over the current (Lutjeharms et al. 1986; Lee-Thorp et al. 1998) and the magnitude of the moisture flux to the land with onshore winds (Rouault et al. 2000; Jury et al. 1997). With global change, an increase in wind stress and an increase in current velocity, a significant increase in the number of Natal Pulses per year could noticeably affect east coast climates by increasing the annual frequency with which

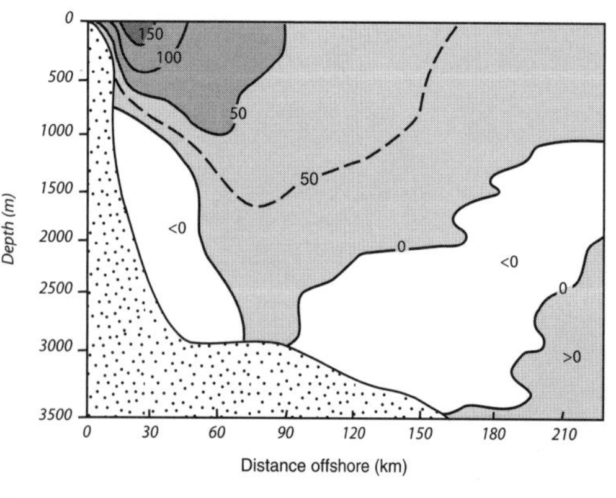

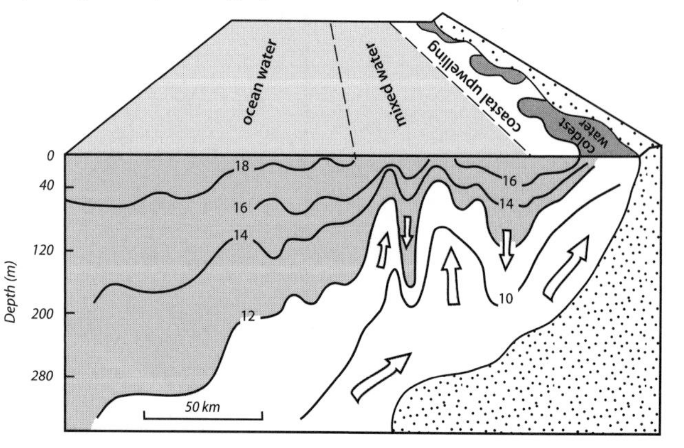

Fig. 2.36. **a** A vertical section through the Agulhas Current off the east coast of South Africa to show the speed of the current in cm s^{-1} (Pollard and Smythe-Wright 1996); **b** a vertical section through the Benguela Current off the east coast of South Africa to show the temperature structure, isotherms in °C, and coastal upwelling (Bang 1971)

the current is moved further offshore to diminish the flux of moisture feeding onshore cumulus convection. Modelling tends to confirm this suggestion and that Natal Pulses may induce noticeable effects downstream from the area of origination (Lutjeharms and de Ruijter 1996).

Where the Agulhas Current retroflects abruptly to the east, about 90% of the current transport is to the South Indian Ocean in the Agulhas Return Current. The exact location of the retroflection appears to be dependent on the mass flux of the current (Lutjeharms and van Ballegooyen 1984), with lower-latitude retroflections occurring the greater the flux. The 10% mass transport to the South Atlantic Ocean takes place by the shedding of Agulhas rings at the retroflection (Lutjeharms and Gordon 1987) and by the removal of filaments from the upper layers of the current (Lutjeharms and Cooper 1996). Rings may have diameters of 300 km (Duncombe Rae 1991), extend to depths of more than 4 000 m and carry heat in excess of 0.045 PW yr^{-1} between the Indian and Atlantic Oceans (Van Ballegooyen et al. 1994). Filaments, on the other hand, are shallow features that appear to lose most of their excess heat to the atmosphere soon after forming. Their annual contribution to inter-basin salt exchange is about 15% that of rings (Lutjeharms and Cooper 1996). The major factor controlling inter-ocean exchange is the dimension of rings and the frequency of their formation.

Each Natal Pulse indirectly causes the shedding of a ring when the pulse reaches the retroflection area (Van Leeuwen et al. 2000). If the frequency of pulses increases with global warming and a coeval increase in ring shedding into the Atlantic Ocean occurs, this may have a marked effect, not only on inter-ocean exchanges south of Africa, but also on the thermohaline circulation of the whole Atlantic Ocean. Many uncertainties relating to this notion remain to be resolved. One such is whether retreat of the retroflection to a more northerly position at the Aghulhas Plateau with global warming may not alter the geometry of the retroflection area to diminish the frequency of ring shedding to the Atlantic.

Within 4 months of being spawned about 30% of Agulhas rings lose their identity completely; the rest lose up to 50% of their dynamical signal (Schouten et al. 2000). This means that extra heat and salt from the South Indian Ocean are not distributed widely over the South Atlantic Ocean, but are initially inserted into a limited area southwest and west of Africa (Fig. 2.37). They are also confined to an area where sea surface temperatures are known to have teleconnections with South African rainfall (Mason 1995).

Water exchanges south of Africa are an important component of the global thermohaline circulation (Gordon 1985, 1986) (Fig. 2.37). The fluxes of heat, salt and anticyclonic vorticity are largely uni-directional from

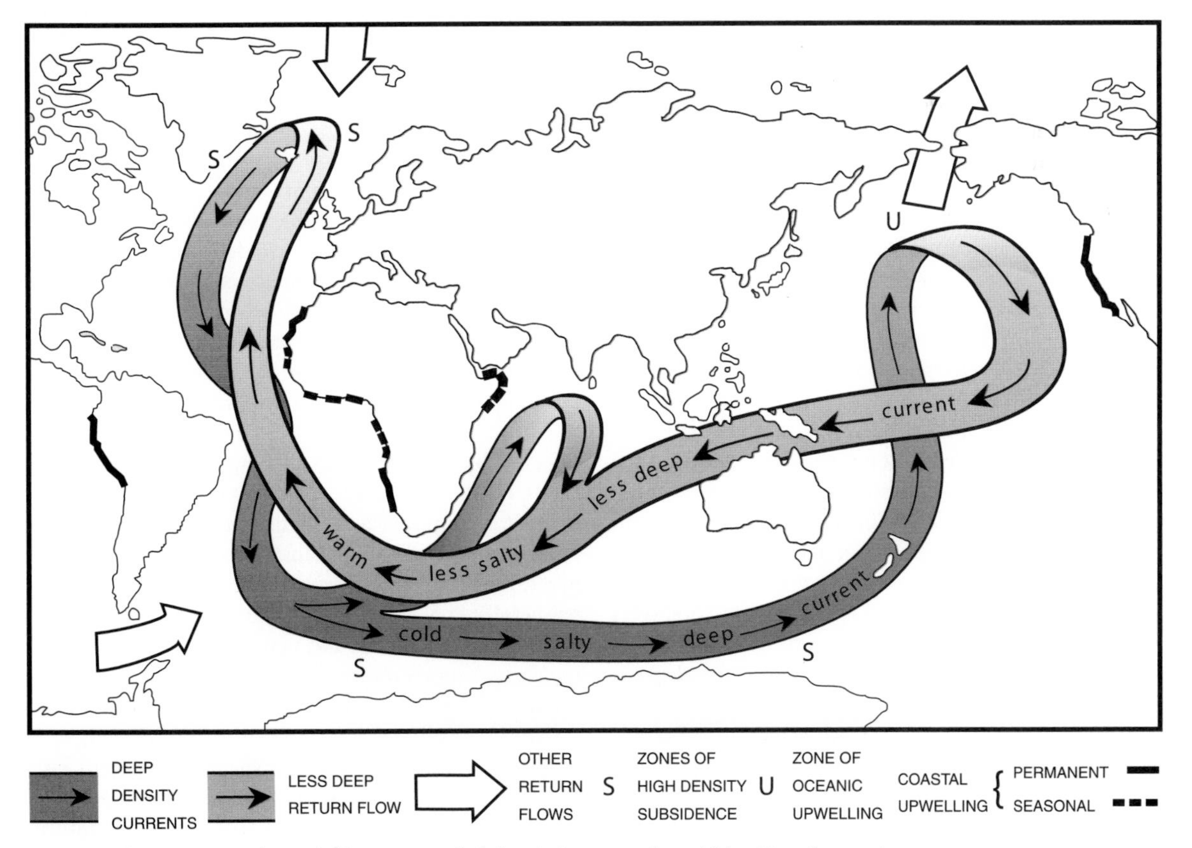

Fig. 2.37. The deep-ocean thermohaline conveyor belt in relation to southern Africa (Broecker 1991)

the South Indian to the South Atlantic Ocean and may have a determining influence on the rate of overturning of the thermohaline cell of the Atlantic Ocean as a whole (Weijer et al. 1999, 2000, 2001). Modelling suggests a persistent increase in the inter-ocean flux will lead to an increase in the rate of overturning with a time lag of 10 years (Weijer et al. 1999). Climatological changes in the rate of inter-ocean exchanges south of Africa could lead to dramatic alterations in one of the major components of the global heat engine. There are strong palaeoceanographic indications that at times the Subtropical Convergence south of Africa moved sufficiently equatorward to have closed off the inter-ocean conduit completely (CLIMAP 1981; Howard 1985; Jousaumme and Taylor 1995). The effect of such closure has yet to be demonstrated unequivocally, but it is likely to be considerable (Weijer et al. 1999). Such conditions would have occurred with glacial conditions. Global warming is likely to produce a poleward rather than equatorward movement of the Subtropical Convergence.

Along the west coast of southern Africa marine biological change is largely a function of changing wind stress on the surface and upwelling. In contrast, in the eastern coastal zone the low-production and high-biodiversity system is more prone to change induced by terrestrial forcing through changing rainfall and runoff. The system exhibits a susceptibility to the direct effects of global warming and sea-level rise (Odada 1996; NOAA 1999). Much of the eastern seaboard of southern Africa is oligotrophic, particularly in subtropical latitudes (Sea Trust 1997). Ocean ecological dynamics are much less significant on the east coast than the west owing to the absence of nutrient cycling by persistent upwelling systems. The main marine biological role of the Agulhas Current is advective, with the global conveyor circulation providing a transport mechanism for Indo-Pacific species to continually colonize the coastal zone (Lutjeharms et al. 2000).

On the east coast, terrestrial influences affect the coastal zone to a considerably greater extent than on the west coast. The discharge of warm, sediment-laden water from numerous rivers, often highly polluted, is a major determinant of the marine ecology. Two of the most important and to some extent mutually dependent coastal ecosystems, mangrove forests and coral reefs are well adapted to the dynamics of the interaction of river and ocean forcing (McClanahan and Obura 1995; MICOA 1998). Whereas the distribution of mangrove swamps is constrained by river fluxes of mud and nutrients, coral reefs (found from northern South Africa to Kenya) are constrained by water temperature and the oligotrophic nature of the warm tropical ocean (McNae 1968; McClanahan and Obura 1995) and to a lesser extent water turbidity. Both ecosystems are highly susceptible to disturbance by global climate, including the anthropogenic effects of direct exploitation and the degradation of river catchment systems and chemical changes to runoff water.

Mangrove systems occur along most of the east coast of southern Africa from north of Durban to Kenya (Sea Trust 1997). They develop in esturine systems where there is marked and regular variation in salinity conditions provided by tidal ebb and flow. The rivers not only modulate the salinity conditions, but provide the essential detrital organic matter and silt to support production and the maintenance of the required mud habitat (Hemminga 1995). Mangroves are important both ecologically and economically, with the latter being strongly dependent on the sustainability of the former (MICOA 1998). The ecological role of mangroves is to create the low turbulence environments where fine sediments (mud and silt) and terrestrial organic detritus can deposit and create the required habitat for a wide range of invertebrates and birdlife and the considerable biodiversity so characteristic of tropical coastlines (McNae 1968). These habitats are important nursery areas for species that colonise coral reefs systems, as well as providing food, construction and fuel resources for humans (MICOA 1998).

Global change threatens mangrove ecosystems through rising sea level (Snedaker 1995; Pethick 1991). This will result in more persistent flooding of the forests by sea water and resulting exposure to increased salinity and sulphates (Snedaker 1995). Any diminution in rainfall will result in a positive feedback by reducing the flux of mud and silt necessary to maintain accretion in the forest habitat stands, as well as to maintain the necessary nutrient fluxes. Decreased runoff will also lead to silting of the mangrove environments with more sterile coastal sand sediments, otherwise regularly flushed out to sea. Increased stress, lower biodiversity and ultimate demise of species may follow. This effect is already noticeable near the mouth of the Zambezi River owing to the building of the Kariba and Cahora Bassa dams (MICOA 1998). The current degraded state of mangroves forests on the East coast of Africa is predominantly due to excessive exploitation, with a contribution from pollution and changing land use practices in the catchments. In Maputo Bay the mangrove forest cover has been reduced by 15% over the past 15 years (MICOA 1998).

Coral reefs are among the first ecosystems to be affected by global climate change (Wilkinson 1996; CORDIO 1999). The most important habitat requirements for coral reef development and maintenance are water clarity to sustain photosynthetic activity and temperatures in the range of 25–30 °C (Sea Trust 1997). Mean temperatures of the world's oceans are increasing. Corals on most equatorial, and many tropical, reefs exist at their upper temperature tolerance limits during the season of highest water temperatures. At high temperatures the coral-zooxanthellae symbiosis is highly sensitive to temperature-related increases in the photosynthetic rate, resulting in breakdown of the symbiosis through explosion of zooxanthellae. If anomalously high temperatures persist, coral bleaching occurs as white coral skeletons become visible through the transparent coral tissue and reefs begin to suffer on a large scale. This is what happened along the east coast of Africa and the western Indian Ocean during the 1997–1998 El Niño, when estimates of coral mortality commonly exceeded 90% (CORDIO 1999). After the event rapid colonisation of algae and bio-eroding organisms occurred in the reefs. A key factor in the survival of the southern African corals will be the rate at which sea surface temperatures will change. Indications are that adaptation takes longer than a few decades and requires hundreds to thousands of years (Pittock 1999).

Off east Africa, corals have developed in an area of relatively high freshwater input to the coastal zone and water turbidity and nutrient levels are high in reefs areas (McClanahan and Obura 1995). The structure of river-mouth reefs reflects this long-term interaction. Increasing ocean turbidity following greater sediment transport to the coastal zone as a consequence enhanced erosion in inland watersheds poses a real threat to coral reefs, for instance off Kenya (van Katwijk et al. 1993). Any processes of global change affecting such processes, e.g., an increase in tropical cyclone frequency and intensity in the Mozambique Channel (as happened during the late summer of 2000), will accelerate reef degradation. It is possible that the coral reefs of southern Mozambique will have been severely damaged by the massive increase in sediment load borne by the floodwaters associated with tropical cyclone Eline.

2.7.2 The Benguela Upwelling System

The upwelling system off the west coast of South Africa and Namibia has not only the general characteristics of such systems worldwide (Shannon 1985; Shannon and Nelson 1996), but also some unusual distinguishing patterns (Fig. 2.35). The upwelling system extends from the Angola-Benguela frontal zone off Angola to the north to Cape Agulhas on the south coast of South Africa. It consists of a number of distinct upwelling cells (Lutjeharms and Meeuwis 1987), the most intense one occurring off Lüderitz. The central cells are active all year round, while the southern cells are strongly seasonal. Intense upwelling occurs in the austral summer when the South Atlantic Anticyclone is displaced equatorward. The northern extremity of the system is also highly sensitive to the wind regime (Kostianoy and Lutjeharms 1999). The upwelling regime consists of a contiguous upwelling strip, largely over the shelf region, and an adjacent band consisting of wisps, filaments, eddies and vortex dipoles (Bang 1971; Lutjeharms and Stockton 1987) (Fig. 2.36b). Some of these filaments may extend 1000 km offshore (Lutjeharms et al. 1991); others may interact with near-shore passing Agulhas rings (Duncombe Rae et al. 1992). At present, such interaction is uncommon, since the trajectories of rings are too far offshore (Goni et al. 1997). With global warming, however, increased wind stress may carry cold filaments further offshore to increase the interaction.

The contribution of each upwelling cell to the overall flux of upwelled water is highly variable in both space and time (Monteiro 1997). Of the total flux of 0.6 SV, the northern Benguela sector accounts for 78%, the southern sector the remainder (Monteiro et al. 2000). The largest contribution from a single cell, that off Lüderitz in the northern sector, contributes 42%. In contrast the Namaqualand cell in the southern sector contributes only 5% to the total upwelling of the system. This spatial heterogeneity suggests that the Benguela upwelling system will be sensitive to any future changes in the wind-stress field that act to enhance or diminish the sharp differences in upwelling fluxes between the Namibian and South African parts of the Benguela upwelling system.

Long time series of sea surface temperature are not available for assessing the temporal variability of upwelling fluxes. The Lüderitz and Cape Columbine cells are the two most active from the northern and southern sectors of the system (Monteiro 1997). The asymmetric response to wind forcing is revealed by the limited data for 1980–1986 (Fig. 2.38). It is also apparent that the interannual variability within Lüderitz cell is larger than the

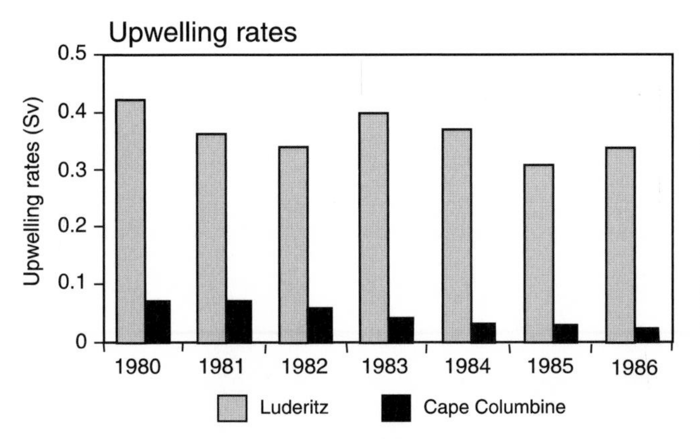

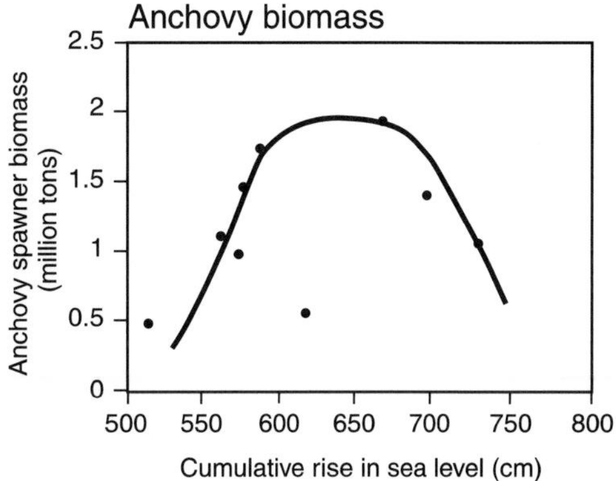

Fig. 2.38. Upwelling rates at Lüderitz and Cape Columbine (Monteiro 1997) and variation of anchovy spawner biomass with cumulative rise in sea level (Waldron et al. 1998a)

absolute magnitude of the upwelling flux in the Cape Columbine cell. A small decline in the upwelling flux off Cape Columbine appears to be unmatched off Lüderitz, and may be insignificant.

The sudden collapse of the Angola-Benguela front allows a flow of warm water poleward along the coast to cause Benguela Niños (Shannon et al. 1986) (Fig. 2.39). They effect considerable modifications to the biological productivity of the upwelling system and result in large-scale death of organisms in the coastal zone. The short-term changes to the carbon cycle are preserved in carbonaceous deposits in bottom sediments and replicate the variability associated with Benguela Niños over more than one glacial cycle (Kirst et al. 1999). Since the extent and intensity of the coastal upwelling is dependent on the local wind stress, climatological changes in this parameter will have immediate consequences on the upwelling system and on ocean sea surface temperatures of the region.

Persistent increases in wind stress over the South Atlantic Ocean will result in cold upwelled water being advected further offshore. However, intensification of the South Anticyclone is likely to be associated with increased atmospheric subsidence, less cloud, increased insolation and higher air temperatures. The resultant warming of the surface water may negate the increased upwelling and offshore transport of cooler water. Considerable uncertainty attaches to which of the two effects will predominate. Either way, the resultant changes in the sea surface temperature fields are likely to affect rainfall over parts of the adjacent subcontinent (Walker 1990; Mason 1995). Inception of a Benguela Niño is dependent on winds in the equatorial South Atlantic Ocean. Any future changes to these winds will have noticeable affects on the temperature structure off the west coast of southern Africa.

Despite increased upwelling with persistently greater wind stress, there may be a limit to how cold the water may become (Lutjeharms and Valentine 1987). The depth from which upwelled water is derived is to some extent a function of the depth of the continental shelf edge (Fig. 2.36b). This imposes a limit on the final temperature that may be reached. The west coast upwelling system is characterized by nutrient-limited high rates of phytoplankton new production, which provides the primary source of energy for the food web and the economically important fishing industry (Hutchings 1992; Hutchings et al. 1995; Crawford 1998). In the absence of deposition from the atmosphere, nutrient supply is driven by upwelling, which in turn is forced by the equatorward wind stress (Shannon and Nelson 1996).

Carbon fluxes in the system are driven by nutrient inputs, mainly of nitrogen and phophorus, which are dependent on upwelling rates (Monteiro 1997; Monteiro et al. 2000). Any changes in these affect nutrient and carbon cycling and the ecologically important planktonic foodwebs and fish production (Monteiro et al. 1991). Carbon flux modelling of the main biogeochemical reservoirs shows that new production, which provides the energy supply to the food web, is 11.7 Mt yr^{-1} and 1.9 Mt yr^{-1} in the northern and southern sectors of the Benguela system respectively (Monteiro 1997). The carbon export flux available for zooplankton production is estimated to be 8.6 Mt yr^{-1} and 1.7 Mt yr^{-1} in the northern and southern sectors of the system (Verheye et al. 1992).

The relationship of carbon production and demand by the foodweb suggests that in the southern Benguela the grazer food web, which provides the main food base for pelagic fish, may be limited by the magnitude of the carbon flux generated by the upwelling in that sector. In contrast, the northern sector has a significant surplus of carbon (~25% of total flux), which is not consumed

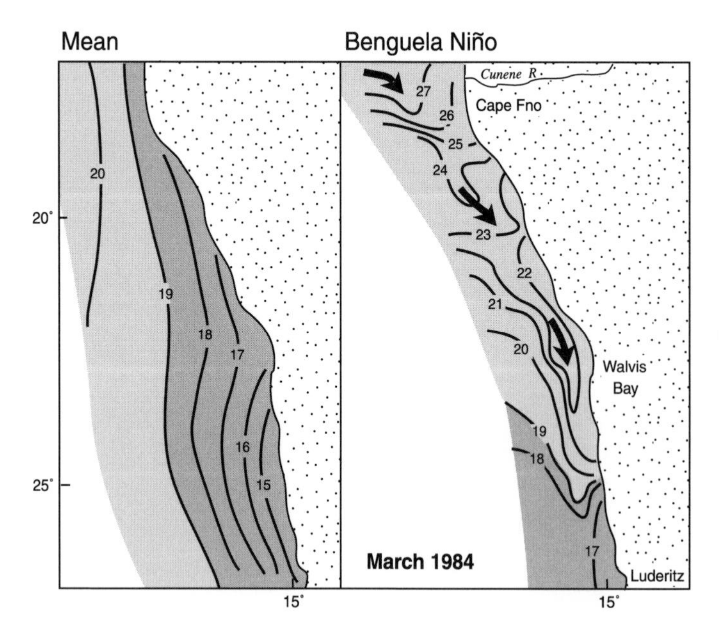

Fig. 2.39.
The mean sea surface temperature field (°C) off the west coast of Namibia and the occurrence of a Benguela Niño event (Shannon et al. 1986)

and is deposited on the inner shelf where it contributes to the organic-rich anoxic sediments on the ocean floor (Bremner 1983; Monteiro 1997). Modelling indicates that the net burial fluxes in the northern and southern sectors are 0.74 and 0.09 Mt C yr^{-1} respectively, figures which are in good agreement with observation (Monteiro 1997). The close dependency between carbon export production and food web energy requirements in the southern Benguela sector is supported by several independent food web studies. Zooplankton variability at St Helena Bay and Cape Columbine from 1951–1996 showed an increase in biomass of crustacean zooplankton individuals from 3.74×10^3 to 7.03×10^5 m^{-2} (Verheye and Richardson 1998). The increase, together with a commensurate increase in size structure, is in agreement with an observed overall increase in the local upwelling index for this part of the system (Shannon et al. 1992) that subsumes the small decrease shown in Fig. 2.38 for Cape Columbine. Despite zooplankton production exceeding pelagic fish requirements, the differences are small and may not be significant (Richardson et al. 1998). Fish recruitment and biomass show a high degree of variability attributed to environmental forcing (Shannon et al. 1992; Hutchings and Hampton 1998). Anchovy and sardine spawner biomass tend to vary inversely in a ~40-year cycle off the west coast of southern Africa (Verheye 2000) that is attributed to environmental impacts in the early stages of life. Biological changes in food supply and physical losses across the open-ocean boundary of the upwelling system are thought to be responsible (Lutjeharms et al. 2000).

It would appear that the southern Benguela ecosystem (and the fisheries dependent on it) is more susceptible to future changes in equatorward surface wind stress than the northern sector (Monteiro 1997). However, the situation is complicated by the fact that successful fish recruitment is dependent on the balance between food supply and losses across the open ocean boundary, both of which are a function of the wind stress (Boyd et al. 1992; Monteiro 1997). The impact of this balance on fisheries is illustrated by the optimal environment window notion, which relates annual fish recruitment to upwelling determined from accumulated sea level variability (Waldron et al. 1998a) (Fig. 2.38, right). The best recruitment appears to occur in intermediate upwelling conditions, when the relationship between food supply and loss across the open ocean boundary is optimal. The interaction introduces a non-linearity into the predicted relationship of equatorward wind strength, the food web and resulting recruitment of fish into a fishery. The economic yield of the fishery will respond negatively in the southern Benguela if climate change pushes the system to either extreme. The relationship is likely to be more difficult to identify in the northern sector because of the excess of new production over the food web and resultant fish population. To the north, it is likely that only changes greater than 25% in the upwelling flux will produce major productivity responses. The main environmental effect producing decreased upwelling in this sector is the southward intrusion of the Angola front with Benguela Niños. Under such conditions, mainly in late summer, extensive extreme hypoxia associated with oxygen depletion causes mass mortalities of fish, as occurred on occasions between 1994 and 1996 (Bailey 1999). The main region prone to extreme hypoxia occurs off Angola.

Fluctuations in total organic carbon in ocean sediment cores preserve a record of past changes in the ocean carbon cycle and productivity. They also allow a reconstruction to be made of past conditions in the Benguela upwelling system. Cores taken on the continental shelf off Walvis Bay in a line transverse to the coast (Kirst et al. 1999), allow comparisons to be made between coastal zone and offshore ocean environments over the past 150 000 years and not far from the present-day major upwelling cell near Lüderitz. For sediments in the Benguela upwelling area, it has been shown that fluctuations in total organic carbon are similar to the variability of marine biomarker contents (Muller et al. 1997; Hinrichs et al. 2000). Consequently, total organic carbon reflects input from marine export production. Enhanced upwelling injects cold, nutient-rich water into the surface mixed layer and enables high export productivity of organic carbon. In the Benguela system, changes in productivity appear to be primarily responsible for late Quaternary variations in total organic carbon, not preservation/remineralization processes (Summerhayes et al. 1995). Periods of high carbon burial, when coinciding with low sea surface temperatures, are indicative of strong upwelling (Lyle et al. 1992). Over the past 150 000 years, total organic carbon of the sediment cores off Walvis Bay decreases markedly from the coastal zone of upwelling on the continental shelf to the deep ocean (Kirst et al. 1999) (Fig. 2.40). Concentrations in the coastal zone were substantially higher in the period 35 000–50 000 years ago. Sea surface temperatures were commensurately lower at that time (Kirst et al. 1999) and upwelling, and hence wind stress on the surface with invigorated southwesterly winds, would have been much stronger than at present. During the last interglacial maximum at around 135 000 B.P., upwelling, wind stress and coastal zone ocean productivity were at a minimum, while sea surface temperatures reached their maximum in 150 000 years.

Present-day onset and persistence of harmful algal blooms in the Benguela system have been shown to be closely linked to occurrence of low water turbulence, enhanced heating and stratification in late summer and autumn (Pitcher and Boyd 1996). Such conditions allow opportunistic dinoflagellates to out-compete diatoms, which are dominant in normal turbulent upwelling conditions (Smayda 1998). Any climate change that leads to a dimi-

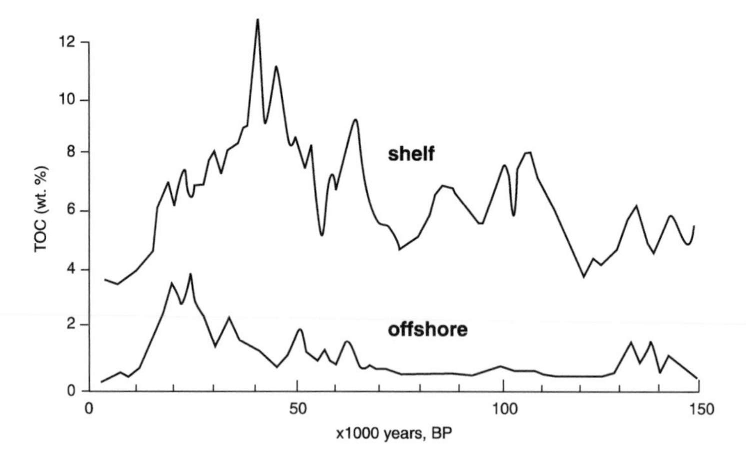

Fig. 2.40.
Variation in total organic carbon in ocean sediments off Walvis Bay, Namibia over the past 150 000 years (Kirst et al. 1999)

nution in upwelling will lead to an increase in the incidence and persistence of harmful algal blooms. In turn, the economic viability of shellfish aquaculture and the rock lobster industry will be adversely impacted by hypoxic conditions (Pitcher et al. 1996; Pitcher and Cockroft 1998).

As in all upwelling systems, nitrates constitute a primary nutrient supporting the food web. Consequently nitrogen and carbon fluxes are closely linked. Carbon fluxes have been estimated in a section across the Continental Shelf for the southern Benguela system, south of the Orange River (Waldron et al. 1998) (Fig. 2.41). Doubling values gives a conservative first estimate of fluxes in the whole Benguela system, which includes the large and very active upwelling centre off Lüderitz in southern Namibia. The estimates of carbon flux are based on remotely sensed sea surface temperature to give areas of upwelled water, a tight correlation between temperature and nitrate, and Redfield ratios of nitrate to carbon. Nitrate is imported from the open ocean by upwelling (3.90 carbon equivalents), is enriched by shelf decompositon and oxidation processes (1.70 carbon equivalents) to give a total of 5.60×10^{13} g C yr^{-1} of nitrate-based potential new production. Of this, the majority is sequestered into shelf sediments (3.70 units), whilst 0.20 units are exported to the open ocean. The remainder (1.50 units) is utilised over the shelf in biological activity, such as the food web processes leading to fish production. The budget suggests that the Benguela upwelling system may be a net sink for carbon, which is deposited into the shelf sediments. This finding is not without controversy and must be tested further.

A big question for the future is the extent to which deposition of nutrients from the atmosphere will need to be incorporated into nutrient budgets for the Benguela system, as has been shown to be required in the case of the Okavango wetland system. The present estimate is that the latitudinally integrated flux of aerosols across the west coast in a plume ~1 500 km wide with a mean axis of maximum transport at ~18° S is ~29 Mt yr^{-1} (Tyson et al. 1996a). Trajectory analysis reveals that most air parcels carrying the aerosols are taken to surface by

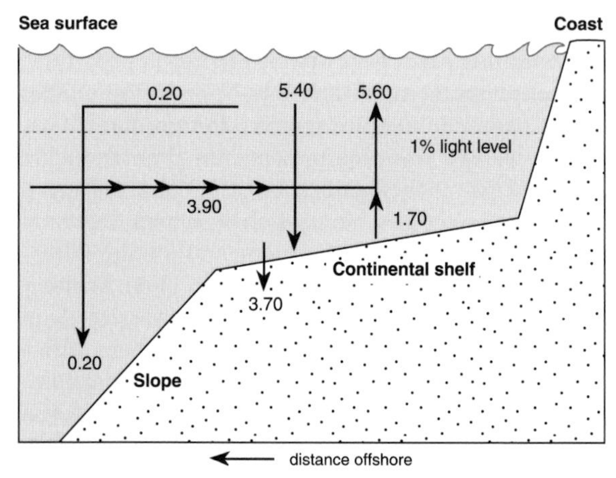

Fig. 2.41. Carbon fluxes in a section across the continental shelf for the southern Benguela system, south of the Orange River, South Africa (Waldron et al. 1998b)

the strong subsidence after leaving the west coast. About 25 Mt yr^{-1} appears to be deposited in the sea in this way between the coast and the Greenwich Meridian. The fraction deposited in the narrow (~200 km-wide) coastal upwelling zone is not known. Relative to upwelling ocean nutrients, the proportion may be small and it may be that only in the oligotrophic surface waters beyond the upwelling front is atmospheric deposition of significance. Further research is needed to resolve these issues.

2.7.3 The Southern Ocean and Southern African Climate

Owing to its very size, nature and intense interaction with the overlying atmosphere, the Southern Ocean exerts a significant influence on global climate. The generic border between the Southern Ocean and oceans to the north, the Subtropical Convergence, significantly influences the climate of southernmost Africa. South of Africa, the convergence is associated with considerable mesoscale turbulence to beyond 60° E (Cheney et al. 1983; Lutjeharms and Valentine 1988) (Fig. 2.42). The

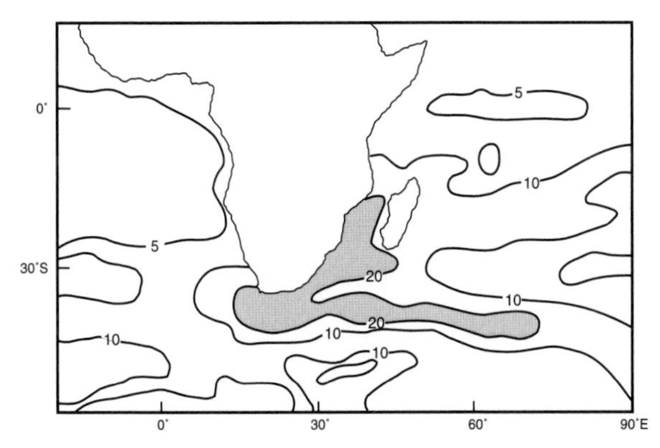

Fig. 2.42. Sea level variability in the oceans adjacent to southern Africa from October 1992 to October 1994 as determined from the TOPEX/POSEIDON altimeter (after Wunsch 1994)

variability is enhanced owing to the proximity of the Agulhas Return Current transporting warm water eastward (Lutjeharms and Ansorge 2001). Strong meridional shear causes increased meandering and shedding of eddies (Lutjeharms 1987). The heat flux from these eddies exceeds 800 W m^{-2} (Rouault and Lutjeharms 2001) and is sufficient to modify weather systems in their eastward passage. Tracks of severe weather systems frequently coincide with the location of the Subtropical Convergence (Jury and Majodina 1997). Consequently, any future global change inducing a latitudinal displacement of the convergence zone will have marked effects on the climate of South Africa and countries to the north.

On a larger scale, Antarctic sea ice coverage and its modification of proximate ocean waters and atmosphere have been shown to influence the climate of southernmost Africa (Hudson 1998). Likewise, the effect of the Antarctic circumpolar wave (White and Peterson 1996) cannot be discounted. Future modification, by global change, of these effects will have regional ramifications. More generally, changes in the general circulation of the atmosphere over the southern oceans will have substantial feedback effects on southern Africa, an issue so broad that it can only be touched on briefly.

Changes in the global temperature fields modulate southern African rainfall through global and local teleconnections in two major ways. The first is through the inter-annual fluctuation of the ocean-atmosphere system and ENSO; the second is via variations in sea-surface temperatures in the South Atlantic and Indian Oceans (Harrison 1983; Lindesay 1998; Mason and Tyson 2000). El Niño and La Niña are usually associated with droughts and floods in the region south of the equatorial zone (Walker 1990; Jury et al. 1993; Mason 1995; Landman and Mason 1998) and the impacts may be severe. The droughts of the early 1990s and the Mozambique floods of early 2000 are examples.

As has been pointed out earlier, one of the effects of global change is likely to be a change in extreme conditions, such as droughts and floods. Such changes are likely to be much more significant than changes in mean conditions and will continue to have major regional ramifications that will heighten the vulnerability of at-risk communties and the environment. Impacts on national economies may be devastating and are usually accompanied by many indirect consequential regional and local effects (Benson and Clay 2000). The impacts of changes in the mean conditions and extremes will affect not only agriculture and health, but also regional hydrology, water supply and management.

2.8 Change and the Hydrological Cycle

2.8.1 Vulnerability under Present Climatic Conditions

Many rural people in southern Africa populate valleys of ephemeral rivers with high runoff variability in countries that are unable to provide assured water supplies to these areas. The unreliability of the climate and hydrological regimes is a severe constraint on self-sufficiency in food production and socio-economic development often cannot be achieved (Desanker and Magadza 2000). In assessing the impact of global change on the water cycle it is necessary to consider secondary, and often highly significant, hydrological consequences of changes in the magnitude and frequency of extreme events, and in seasonality of runoff in relation to intra-seasonal water demand and supply of water. In addition, costs of water storage, changes in reservoir storage drawdown, changes in groundwater recharge, land use changes and shifts in agricultural and livestock farming areas associated with inter-annual climate variability, and the effects of upstream actions on downstream users are further important secondary effects. Higher order impacts of changing water resources need to be considered as well (Desanker and Magadza 2000). These include changes to water quality and associated impacts on purification costs and human health, impacts of climate and land use change on acquatic ecosystems and on environmental integrity, and finally, potential conflicts over shared water resources as water scarcity increases.

Southern Africa's rainfall climate ranges from an arid southwest Namibia and South Africa (<100 mm per annum), through the semi-arid belt Kalahari (<400 mm) and parts of Kenya, to humid areas experiencing in excess of 1 000 mm of rain in parts of Zambia, Tanzania, Uganda and Mozambique. Rainfall is, however, variable and highly so in places. Its inter-annual coefficient of variation is an important factor in water storage planning and ranges from a low 15% over parts of Zambia to 20–30% over much of Zimbabwe and Botswana to 50% and more over large parts of Namibia and western South Africa (Hulme 1996; Schulze 1997a). Furthermore, rainfall is often highly seasonal and concentrated into short

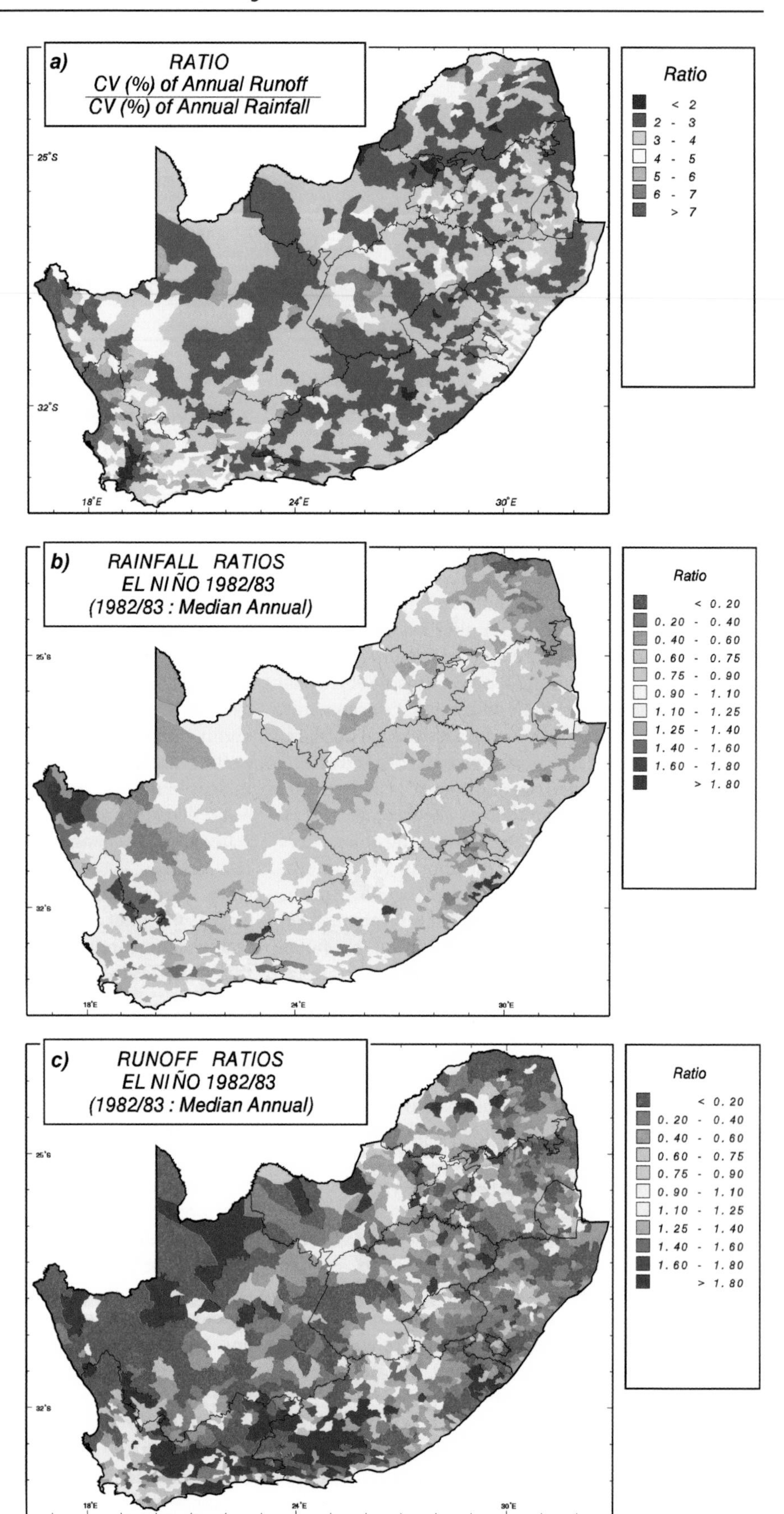

Fig. 2.43.
Ratios of coefficients of variability: **a** of mean annual rainfall and runoff, **b** the 1982/83 El Niño rainfall to median annual rainfall, **c** the 1982/83 El Niño runoff to median annual runoff (Schulze 2000)

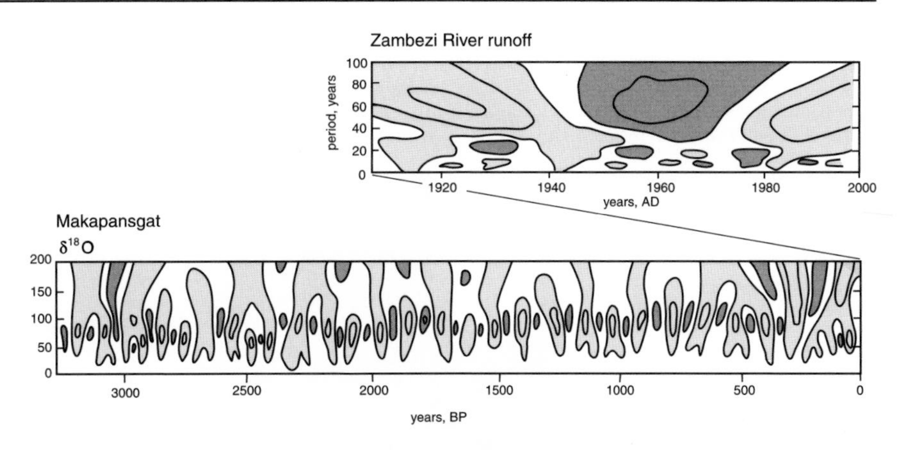

Fig. 2.44.
Wavelet analysis to show a ~100-year fluctuation in Zambezi River runoff and oxygen isotope ratios derived from a Cold Air Cave stalagmite, Makapansgat, South Africa

periods within the year (Schulze 1997a). The region's high annual potential evaporation exceeds 1 200 mm over large areas, with open water evaporation rising to over 2 500 mm over significant tracts of Mozambique, Namibia and South Africa. In many places annual potential evaporation exceeds rainfall by a factor of 3; in some drier areas it exceeds rainfall by a factor of more than 10 and the resulting conversion of rainfall to runoff is low.

The water regime is even more variable than that of rainfall on time scales ranging from the individual event to inter-annual. Inter-annual runoff variability varies from less than 50% in the wetter eastern to more than 300% in the more arid western parts of South Africa (Schulze 1997a). Simulations using the daily time-step deterministic ACRU hydrological model (Schulze 1995) show that over South Africa the runoff variability is considerably greater than that of rainfall. A unit change in rainfall results in tripling of runoff over much of southernmost Africa, but with significant spatial variation. In certain areas runoff may be quintupled or more (Schulze 2000; Schulze and Perks 2000) (Fig. 2.43a). During the 1982/83 El Niño drought in southern Africa low rainfall resulted in even lower runoff. Whereas much of the summer rainfall region of southeastern southernmost Africa received 50–80% of its long-term median rainfall during the drought, the runoff in the same locality yielded only 20–50% of its corresponding median; a considerable area of the region received less than 20% of its normal runoff (Schulze 1997b) (Fig. 2.43b and c).

The variability in the hydrological regimes of southern Africa occurs not only at inter-annual time scales. Significant rainfall variability resides in bands across a spectrum of fluctuations on decadal to century scale and longer. Variability at about 18 years has been reported for South African rivers (Abbott and Dyer 1976; Alexander 1995). Variability associated with an oscillation of ~100 years has been reported in the flow of the Zambezi River at Victoria Falls and shown by wavelet analysis to be in phase with a similar oscillation in proxy temperature and rainfall derived from the Makapansgat valley $\delta^{18}O$ record in the north-eastern interior of South Africa (Fig. 2.44). The ~100-year variability appears to have been present and stable for at least 3 000 years. Over the past millennium rainfall and hydrological responses have been 180° out of phase between South Africa and Kenya. Modelling of southern African hydrological regimes in decades and centuries to come will need to account for such variability and its spatial manifestation.

Extreme hydrological events in southern Africa (such as the devastating regional drought of 1997/98 or the destructive Mozambique floods of 2000) tend to be severe. The ACRU model has been used to assess the risk of droughts and floods. The annual runoff in the driest year in 10 is about one quarter of median runoff in the wetter eastern half of South Africa, whereas western parts generate zero runoff in the driest year in 10 (Fig. 2.45a). For flood events, the 1 in 50 year simulated flood volume is 3–10 times that of the annual one-day flood volume in the wetter parts and as high as 20–100 times as high in the more arid west (Fig. 2.45b).

2.8.2 Hydrological Uncertainties and Climate Change

Anticipated spatial changes in many of the components of the hydrological system have been modelled for southern Africa (Reynard 1996; Meigh et al. 1998; Schulze and Perks 2000). Based on the HadCM2 model (Murphy and Mitchell 1995), projected mean annual temperatures increases of 1–2 °C over much of southern Africa translate into mean annual potential evaporation increases of 5–20% south of 10° S (Hulme 1996). Tropical regions of Kenya, Tanzania, Uganda and northeastern Zambia are likely to experience increased rainfall and runoff with doubling of greenhouse gases. In contrast, subtropical southern Africa is likely to experience diminished rainfall and negative hydrological consequences (Meigh et al. 1998). However, it is changes in extremes rather than in mean conditions that will have greatest impact (IPCC 1996). One simulation suggests an increase of about 100% in the frequency of rainfall events with >12.5 mm per rain day over the winter rainfall region of South Africa and an increase of up to 40% in parts of

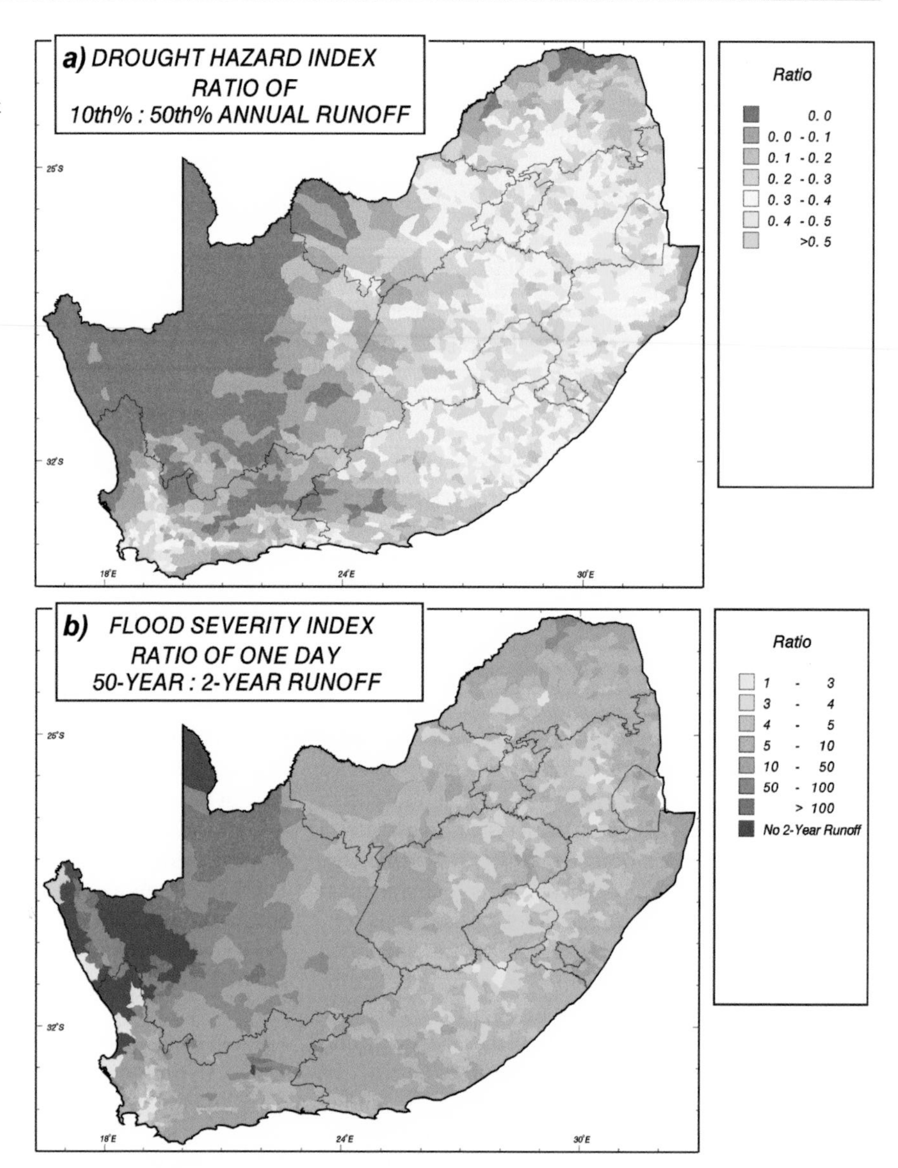

Fig. 2.45.
a Drought hazard index (ratio of 10:50% percentile annual runoff); **b** flood severity index (ratio of one-day 50-year: 2-year runoff)

the summer rainfall region (Joubert 1995). Such a change would cascade through the hydrological system exacerbating stormflows, changing sediment yields and shortening return periods of extreme events. At the same time, the simulation suggests that a reduction in the frequency of low-magnitude rain days will result in changing patterns of demand for irrigation. Some models suggest future increases in tropical cyclone intensity and frequency along southern Africa's east coast (König et al. 1993). Were this to occur, the ramifications for regional design flood estimations would be considerable. Model predictions of increasing extreme condition in future are given some credibility by suggestions that climate extremes in southernmost Africa are in fact increasing (Mason 1996; McClelland 1996; Mason et al. 1999; Poolman 1999).

Catchment-scale hydrological models that are driven by GCMs are associated with many uncertainties. Nevertheless, they are powerful and extremely useful, particularly in assessing the sensitivities of systems to future change. Uncertainties are associated with both the internal dynamics of the models themselves and with the nature of the driving forces (Schulze and Perks 2000). Incorporation of sulphate aerosol forcing in addition to that of greenhouse gases produces different spatial responses in mean annual runoff over southernmost Africa (Fig. 2.46). Further uncertainties concern the rate of CO_2 feedback that acts on stomatal resistance and consequent transpiration rates, and so soil moisture status and runoff potential.

Despite the uncertainties, it is clear that changes in the inter-annual variability of streamflow are likely to be of greater consequence than those in mean runoff. Variability is likely to increase throughout the region in widely different climatic regimes by 2050 (Reynard 1996) (Table 2.3). This will need to be taken into considera-

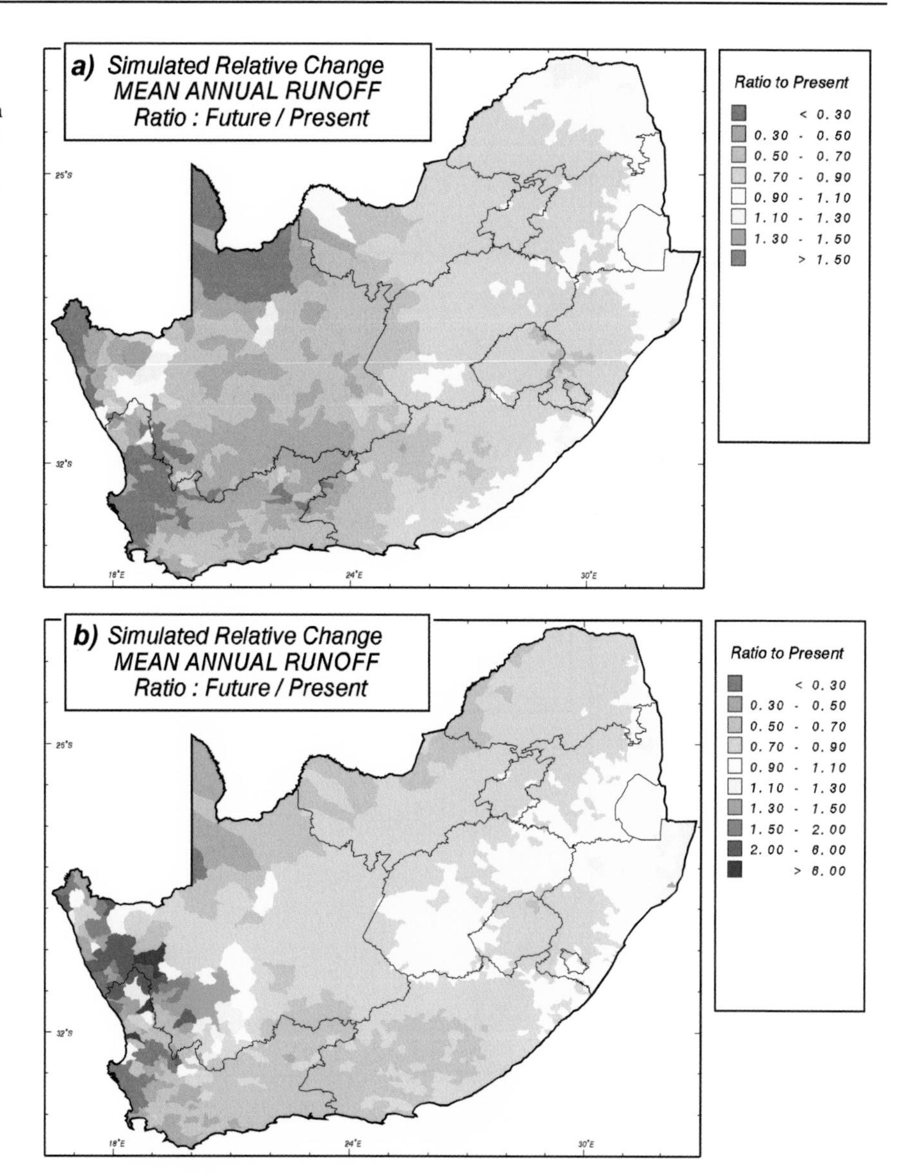

Fig. 2.46.
Modelled ratios of future-to-present mean annual runoff using the ACRU model driven by HadCM2 transient run scenarios **a** with greenhouse gas forcing only, **b** with combined greenhouse gas and sulphate aerosol forcing (Schulze and Perks 2000)

Table 2.3. Annual runoff statistics for 10 selected GCM grid cells for baseline (1961–1990) conditions and a 2050 climate change (UKTR95) scenario (after Reynard 1996)

Selected GCM grid cell	Mean and median () annual baseline runoff (mm)	Inter-quartile (25th–75th %) range of annual runoff (mm)	Coefficient of variation (%)		Change (%) in annual runoff for 2050 scenario
			1961–1990 baseline	2050 scenario	
1 Western Tanzania	377 (395)	275 – 475	38	57	-20 to -10
2 Coastal Tanzania	585 (615)	415 – 710	31	42	10 to 40
3 Western Zambia	197 (155)	120 – 270	51	69	-10 to 10
4 Central Zambia	709 (715)	610 – 820	25	37	-10 to 10
5 N. Mozambique	503 (480)	410 – 600	30	42	-20 to 10
6 Kalahari	6 (6)	0 – 9	36	50	-40 to -10
7 Southern Zimbabwe	47 (35)	30 – 60	70	96	-40 to -20
8 Eastern South Africa	89 (70)	60 – 115	66	83	-40 to 20
9 Central South Africa	8 (8)	0 – 15	34	45	-20 to 10
10 Southwestern S. Africa	45 (15)	10 – 25	275	270	-40 to -20

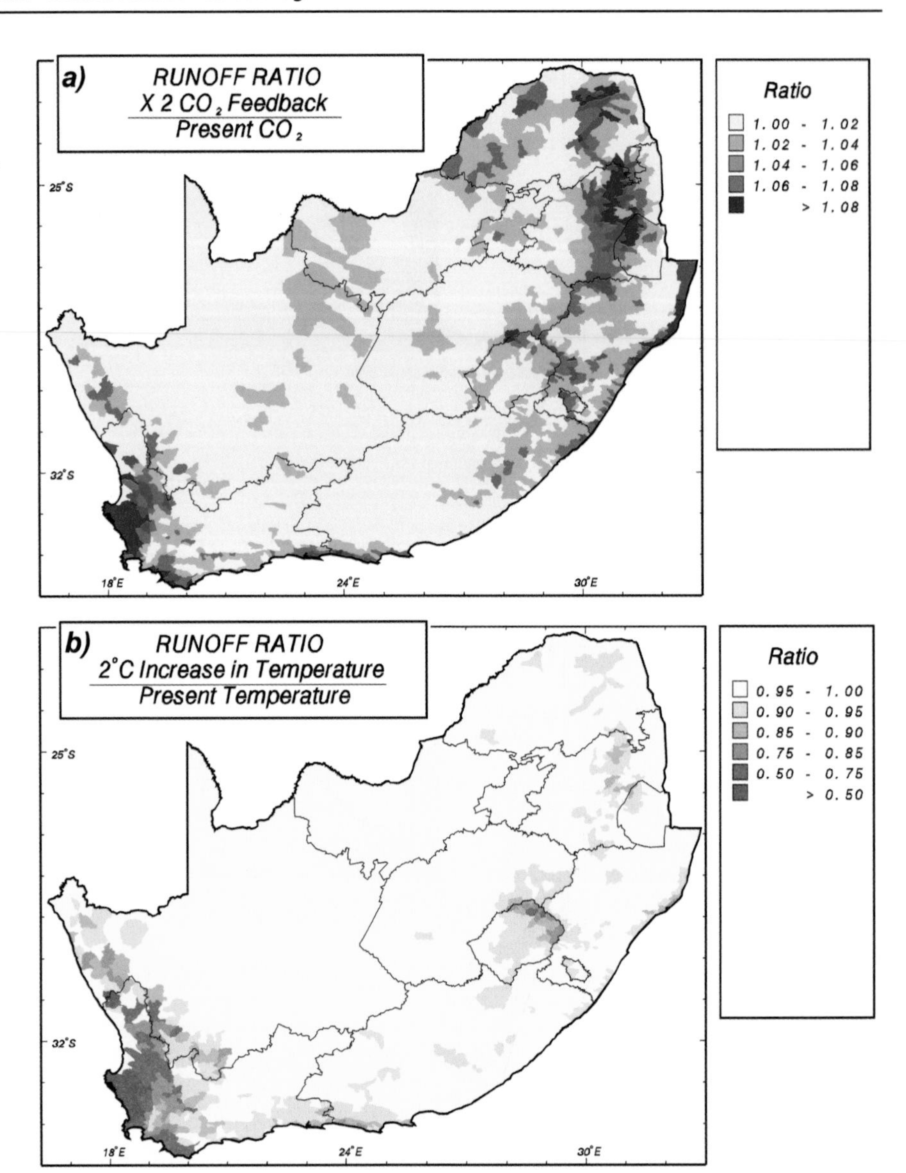

Fig. 2.47. ACRU-modelled ratios of **a** runoff with doubled CO_2 feedback to present CO_2 and **b** runoff with a 2 °C increase in temperature to present-day temperature (Schulze and Perks 2000)

tion when planning and operating reservoirs for assured water supplies in future.

Sensitivity analyses using the ACRU model show that crop yields are sensitive to CO_2 feedback. The same is not the case in the hydrologically system (Fig. 2.47a). In most areas of southernmost Africa long-term runoff varies by less than 4% in response to the CO_2 feedback. Similarly, the sytem is relatively insensitive to temperature changes that affect evaporation. An increase of 2 °C reduces mean annual runoff over most of summer-rainfall South Africa by only 5% (Fig. 2.47b). In the winter rainfall region the response is more dramatic, where a 2 °C increase in temperature may produce a reduction runoff in excess of 50% (under present climatic conditions evaporative losses are relatively low from the moist soils in winter; with the anticipated increase in temperature replenishment of drying of soils between rainfall events will significantly reduce runoff).

Hydrological responses to rainfall changes do not generally mirror the latter (Schulze 2000; Schulze and Perks 2000). In many instances a local change in rainfall translates into a higher order hydrological response, such as a change in sediment yield or the rate of percolation from irrigated fields or groundwater recharge. The resulting spatial pattern of hydrological varaibility is different from its rainfall variability counterpart as a result of the complexity of the hydrological system and its local dependence on the underlying substrate. Furthermore, regional changes occurring in response to global forcing will not take place simultaneously over the whole of southern Africa, but will be modulated by factors operating differentially throughout the region. Thus a 10% modelled change in mean annual runoff induced by greenhouse warming may occur by as early as 2015 in the west of southernmost Africa, but by only 2060 in eastern areas (Schulze and Perks 2000) (Fig. 2.48).

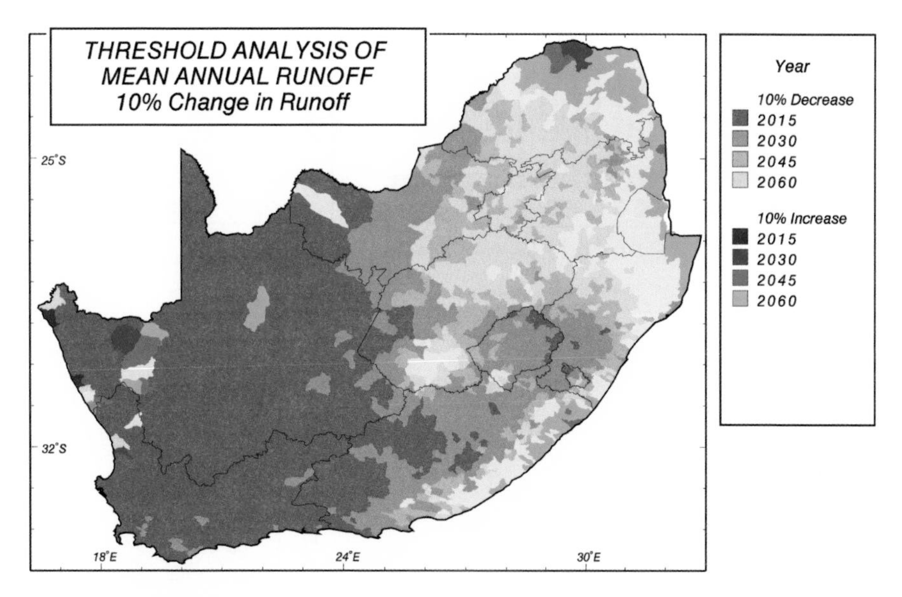

Fig. 2.48.
The ACRU model forced by transient-run HadCM2 model doubling of greenhouse gases to show years by which either a 10% increase or decrease in rainfall thresholds may occur (Schulze and Perks 2000)

2.8.3 Added Complexity and the Human Dimension of Change

Humans add an additional dimension of complexity to the hydrological system owing to changes in land use over time, to a variety of population pressures and to changes in water demand in relation to availability. The impact of changing land use on water availability is illustrated by a case study from eastern South Africa, the 4 079 km^2 Mgeni catchment, which supplies nearly 6 million people with water. The Acocks (1988) veld types have been taken to represent the undisturbed, baseline vegetation cover and the impact of the change to the present-day land use on streamflow has been modelled (Kienzle et al. 1997) (Fig. 2.49). Reductions in streamflow exceeding 60% have occurred, mainly in areas of intensive sugarcane cultivation and exotic afforestion, whereas increases exceeding 100% have been observed in overpopulated areas where livestock overstocking and land degradation have occurred, as well as in urban areas. Future runoff will be affected, not only by the present land use, but also by the changes certain to take place in time to come as population pressures on the land increase (and as climate continues to change). The HadCM2 model suggests that with doubling of greenhouse gases catchment runoff is likely to diminish to 70% of present in the upper Mgeni basin, while remaining relatively unaffected in the lower catchment (Schulze and Perks 2000). In the lower reaches of the basin increased runoff induced by deteriorating land use will offset the diminution induced by the lower rainfalls predicted.

Human activities likewise affect water quality substantially. Four South African catchments illustrate the point (Fig. 2.50). The electrical conductivity of water is a measure of its salinity. In industrial and mining areas, such as the Spook Spruit near Middelburg where coal-mining activities pollute water substantially, the conductivity of the water has increased steadily from 1980 onwards (Ross 1999). The same is true of water in an impoundment on a small tributary of the Sand River in the gold mining area of Klerksdorp (Herold et al. 1996). The salinity of the Vaal River at Vereeniging has shown a steady increase since 1935 (du Plessis and van Veelen 1991). In contrast, the conductivity of unpolluted water in the forested rural Jonkershoek catchment is low (an order of magnitude lower than that of the Vaal River in recent years) and has been constant over the past few decades (Ross 1999). Where human population pressures increase on the land with the development of informal urbanisation and squatter settlements with poor sanitation, water quality deteriorates and the health hazard increases accordingly. Simulations of the distribution of the pathogen *Escherichia coli* have been made for the Mgeni River basin (Fig. 2.50) using livestock densities and numbers of humans living under poor sanitary conditions in close proximity to streams as input to the ACRU model (Kienzle et al. 1997). The distribution of the *E. coli* health hazard is seen to be excessively high in places in the Mgeni catchment and is likely to increase in future unless land use can be controlled effectively. The issue of water quality is one of considerable concern. By UN estimates two-thirds of humanity will face shortages of clean water by the year 2025; in Africa the proportion is likely to be higher.

Present agricultural water demand in southern Africa (mean ~71%) outweighs all other forms of usage (domestic ~20%; industrial ~8%) (Table 2.4) (FAO 1995). This highlights the high dependence of the welfare of the countries of southern Africa on the vagaries of rainfall and runoff variability. High water demand occurs largely in large irrigation projects or around conurbations. As urbanisation increases so will demand to the point of exhaustion of conventional supplies.

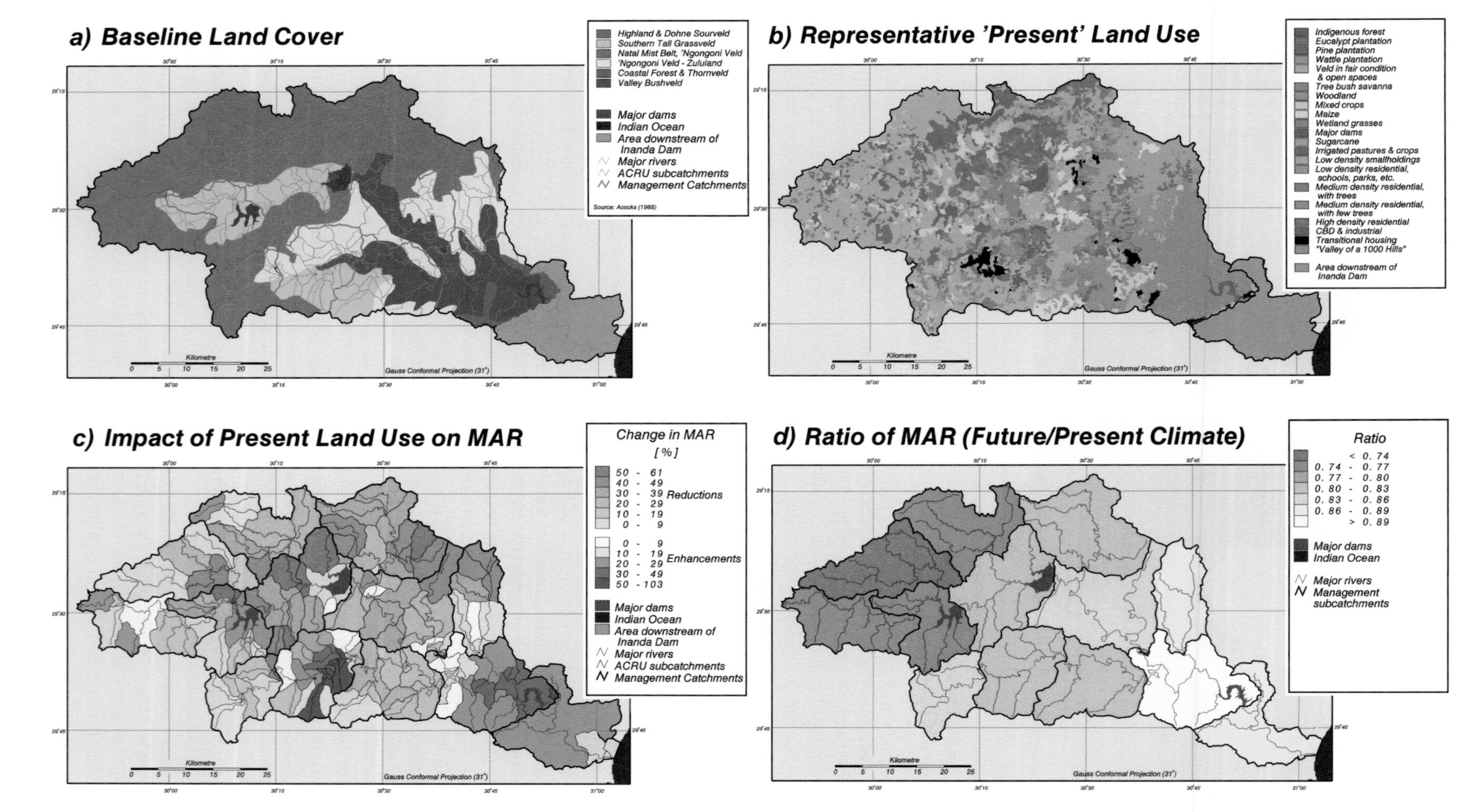

Fig. 2.49. ACRU model simulations of the effect of land use changes on Mgeni River catchment runoff: **a** baseline land cover as indicated from Acocks' veld types, **b** present land use, **c** the impact of present land use on runoff, and **d** ratio of present to future mean annual runoff modelled with forcing derived from transient-run HadCM2 model doubling of greenhouse gases (Kienzle et al. 1997; Schulze and Perks 2000). The location of the Mgeni catchment is given in Fig. 2.53

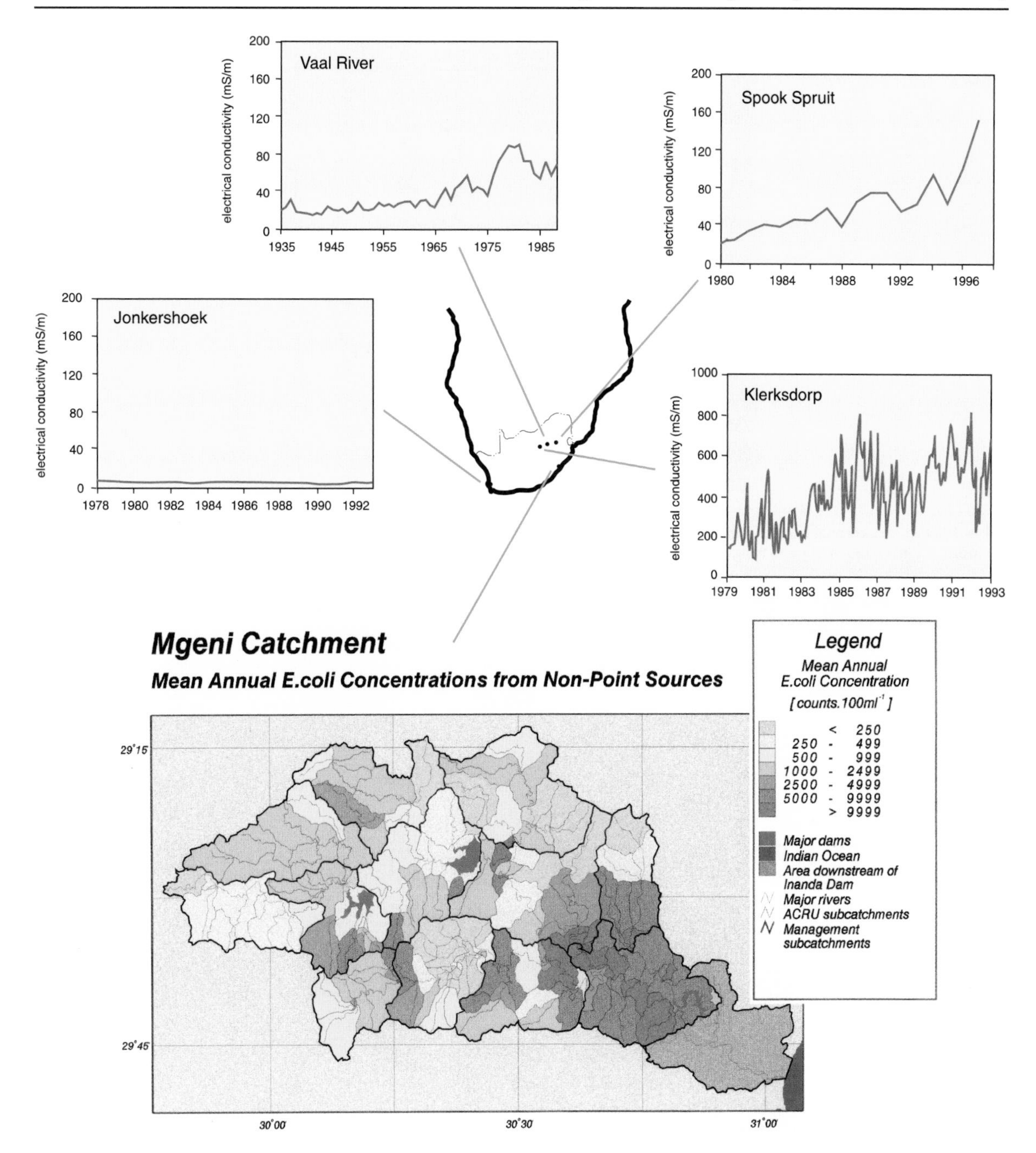

Fig. 2.50. Variations in water quality at various localities in South Africa during the twentieth century (Ross 1999; Kienzle et al. 1997)

Present and future (2050) water demands in southern Africa have been compared to water availability using the HadCM2 model with greenhouse gas forcing and incorporating a water availability index recognising both seasonal and inter-annual variations in supply and groundwater availability (Meigh et al. 1998). Demand takes account of differential water sector usage as well as varying uses by urban and rural populations. Built into the 2050 scenario are projected urban: rural population changes, future demands according to projected levels of development and impacts of climate change on water supplies. At present it is estimated that 19% of southern Africa experiences water shortages (Table 2.5). The model suggests that by 2050 the area having shortages will have increased by 29%. The area of the region likely to experience severe shortages (i.e. a water availability index ≥0.75) is likely to increase from 7 to 14%. Largest expansions of water-short areas are likely to be in Mozambique (expansion by ~167 000 km^2), Tanzania (by ~121 000 km^2) and South Africa (by ~106 000 km^2).

The multiple causes and consequences of environmental change are so closely intertwined and mutually

Table 2.4. Withdrawals of water by major sector (after FAO 1995)

Country	Withdrawals by sector			Annual withdrawal (10^6 m^3)
	Domestic	Industrial	Agriculture	
Botswana	32	20	48	113
Kenya	20	4	76	2050
Lesotho	22	22	56	50
Malawi	10	3	87	936
Mozambique	9	2	89	605
Namibia	29	3	68	249
South Africa	17	11	72	13309
Swaziland	2	2	96	656
Tanzania	9	2	89	1156
Uganda	32	8	60	200
Zambia	16	7	77	1706
Zimbabwe	14	7	79	1220
ESA regional means	20.3	8.4	71.3	

Table 2.5. Country comparisons of water availability between present and 2050 scenarios (after Meigh et al. 1989). The water availability index is defined as WAI4 and ranges from –1 (shortage of water) through 0 (demand = supply) to +1 (water excess)

Country	Area (km^2)	Total No. of cells	Present situation				Year 2050 scenario (2050 climate; high demand)					
			Cells with shortage (WAI4 ≤ 0)		Cells with severe shortage (WAI4 ≤ –0.75)		Cells with shortage (WAI ≤ 0)		Cells with severe shortage (WAI ≤ –0.75)		Increase in cells with shortage	Increase in cells with severe shortage
			No.	%	No.	%	No.	%	No.	%		
Botswana	569797	205	6	3	0	0	10	5	1	0	4	1
Kenya	582646	191	58	30	17	9	63	33	29	15	5	12
Lesotho	30355	11	10	91	3	27	11	100	9	82	1	6
Malawi	118484	38	1	3	0	0	11	29	6	16	10	6
Mozambique	799380	293	11	4	4	1	77	26	23	8	66	19
Namibia	824292	303	7	2	0	0	24	8	0	0	17	0
South Africa	1221037	483	267	55	103	21	309	64	180	37	42	77
Swaziland	17364	5	5	100	4	80	5	100	5	100	0	1
Tanzania	945203	316	28	9	2	1	76	24	26	8	48	24
Uganda	241139	79	38	48	14	18	43	54	36	46	5	22
Zambia	752614	251	2	1	0	0	11	4	2	1	9	2
Zimbabwe	390272	134	11	8	10	7	26	19	13	10	15	3
Region	6492583	2309	444	19	157	7	666	29	330	14	222	173

reinforcing that it is necessary to consider them together as far as is possible. In all water-related aspects of regional change, human activities have been shown to be a major driving force. The same is true in many other respects including ecosystems, land use cover and land use change.

2.9 Terrestrial Ecosystems and Change

Ecosystems respond to the entire complex of changes taking place in their environment. The future pathway of this interactive system is relatively predictable in some aspects, but remains highly unpredictable in others. Some of the uncertainty about the future is inherent in the dynamics of non-linear systems: they have the characteristic of behaving in ways that are inherently unpredictable. In southern Africa, a key uncertainty is the magnitude and sign of future changes in rainfall and its seasonality. A further major uncertainty relates to how human societies, with their changing population growth, technological capacity, political and economic organisation, will respond to changes in the natural environment in southern Africa.

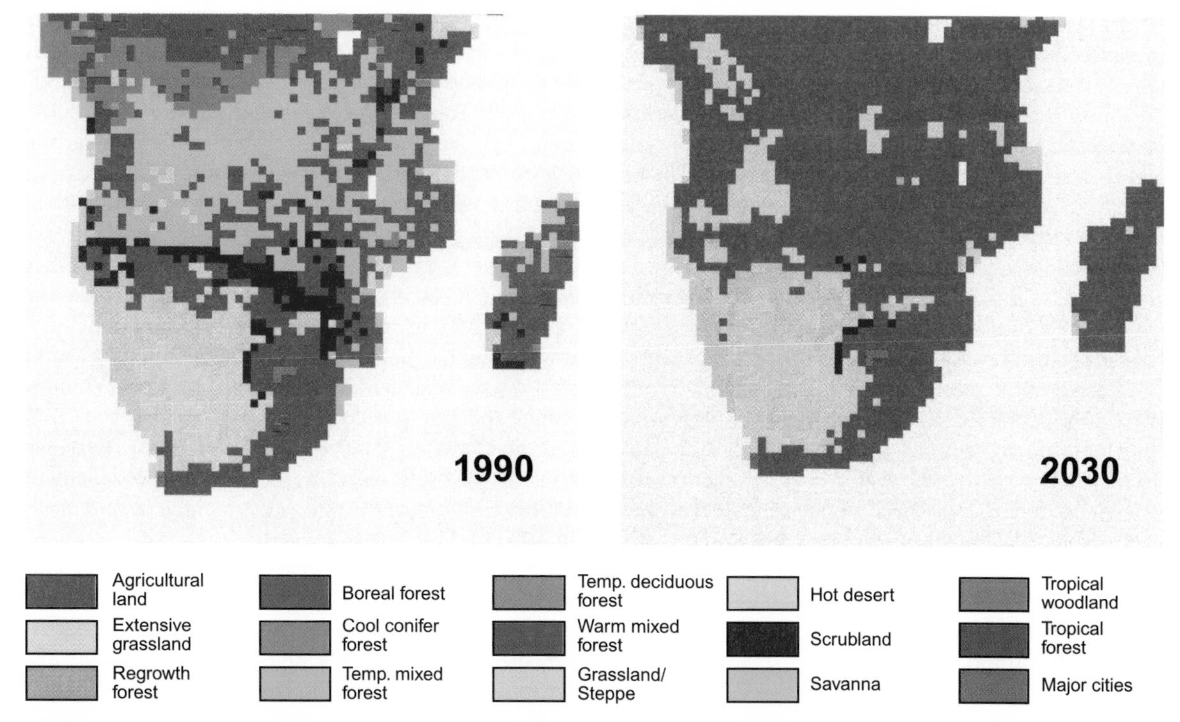

Fig. 2.51. Changes in land use over southern Africa between 1990 and 2030 as modelled using the IMAGES 2.1 model and Baseline a scenario (based on IPCC IS92a scenario with medium population and economic growth) (Alcamo et al. 1998)

2.9.1 Land Use Changes

Over a long period of time humans have altered African land cover as both deliberate and unintended outcomes of the use of the land. Pre-agrarian and pre-colonial human activity, particularly in relation to the use of fire and domesticated animals, began a process of change in the African landscape. By the beginning of the twentieth century land and environment were almost certainly not in equilibrium. So-called natural systems were far from being so; the distinction between natural and human-induced is often untenable in the context of Africa.

In future it will be increasingly difficult to identify natural systems. Modelling suggests that profound land use changes will occur in southern Africa this century (Alcamo et al. 1996) (Fig. 2.51). One scenario is based on population continuing to increase at a rate in excess of economic growth, with a large proportion of the poorer rural population resorting to areal extension of agriculture rather than technological intensification. As a result, much of the semi-natural vegetation in the arable parts of south central Africa will be transformed to low-input agriculture or pastoralism. Increasing demand for veld products, such as medical potions and bushmeat, will target useful species and drive them to local and global extinction (Scholes et al. 1999). Increasing economic polarisation between a relatively better-off urban population and a poverty-stricken rural population is likely. Demand for biofuels for domestic energy will continue to result in thinning of woodlands within accessible distance of settlements.

Examples of such processes are visible already in Kenya, Zimbabwe and Malawi. How rapidly they will spread as population growth continues is not known. Conservative estimates suggest population will double by the latter twenty-first century, despite the effect of AIDS. There is some evidence of declining birth rates associated with urbanisation and increased levels of income and education, which suggest that the southern African population may peak during this century. Extreme land use change outcomes are not inevitable providing new agricultural technology can be applied to support even a double population off the current cropped and grazed land area.

In the hottest parts of southern Africa plant and animal production are periodically constrained by high temperatures. Future increases in temperature may threaten the continued viability of isolated plant communities, particularly high-elevation communities in the biodiverse Cape Floral Kingdom and the Afromontane phytochorion (Midgley and O'Callaghan 1993). Changing temperature conditions may also cause vegetation distribution change through reducing the incidence of frost in the interior region of South Africa, but very few data are available to quantify this. Bioclimatic modelling of the distribution envelopes of key plant species under present and possible future climates indicate that they are potentially quite sensitive to changes in rainfall and temperature of the magnitude which may occur (Rutherford et al. 2000).

The key climatic uncertainty for terrestrial ecosystems in southern Africa relates to future trends in net moisture (Scholes 2001, pers. comn.). Virtually all plants and micro-organisms in southern Africa, and the animals that depend on them, are water-limited for a portion of their life cycle, and thus sensitive to changes in water balance. The net outcome of a concomitant increase in temperature (which would increase evaporative demand), an increase in CO_2 (which reduces plant transpiration) and a putative decrease in rainfall is difficult to determine with any confidence. The modelling done to date suggests that the effects may cancel each other (Scholes 2001, pers. comn.).

Changes in the seasonal distribution of rainfall in relation to temperature have the potentially to profoundly alter both fire regimes and plant phenological cues, especially in the southern Cape. However the chances that seasonality will change much in future appear remote with current GCM scenarios (Joubert and Kohler 1996).

2.9.2 Changes in Ecosystem Composition and Distribution

Biome boundaries in southern Africa, as in other parts of the world, are strongly related to climatic indices, and would be expected to move in the long term in response to climate changes. An example is the savanna/grassland boundary, which is highly correlated to mean dry-season temperature, although the proximal mechanism is unknown (Rutherford and Westfall 1986). Modelling suggests that in South Africa the Nama-Karoo and Savanna Biomes will advance with continued global change at the expense of the grassland (Ellery et al. 1991) (Fig. 2.52). Trees, with a greater dependency on carbon investment in wood than grasses, may benefit more from the effects of CO_2 fertilization (Midgley et al. 2000; Bond and Midgley 2000). However, the competitive interaction between trees and grasses is complex and, other than the effects of climate, involves factors such as the frequency, season and intensity of fire, the intensity of grazing and harvesting policies. It is remains unclear to what degree trees will be advantaged by global change, despite the fact that current grassland areas may well become climatically more suitable for trees. There has been a widespread increase in tree cover and density in southern Africa over the past century in areas not heavily harvested for fuelwood.

Regional climate change in southern Africa threatens the near-extinction of the Succulent Karoo Biome, the world's richest arid hotspot, and will cause extensive species loss from the diverse Fynbos Biome (Rutherford et al. 2000). Aridification in the western half of southern Africa threatens the floristic depauperization of semi-arid ecosystems such as the Nama-Karoo and Savanna Biomes. With changing climate, future distributions of specific species may alter considerably, as is illustrated by modelled changes in the distribution envelopes of *Aloe marlothii* (Fig. 2.53). The major unknown

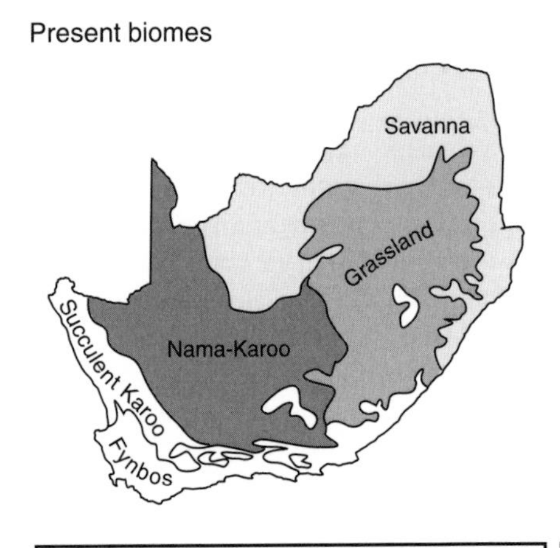

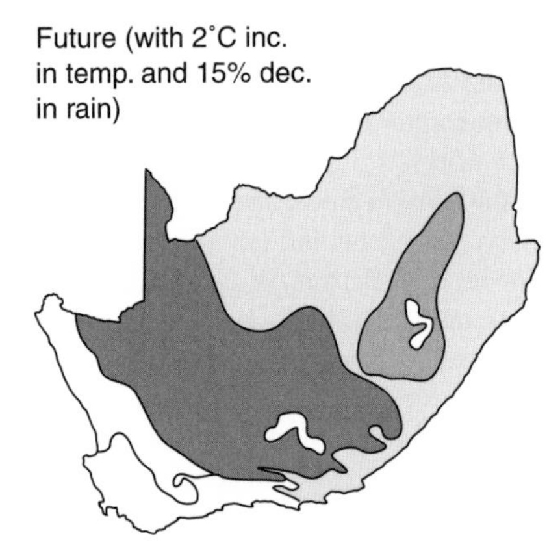

Fig. 2.52. Possible changes in major biomes with a 2 °C increase in mean annual temperature and a 15% decrease in rainfall over southernmost Africa (Ellery et al. 1991)

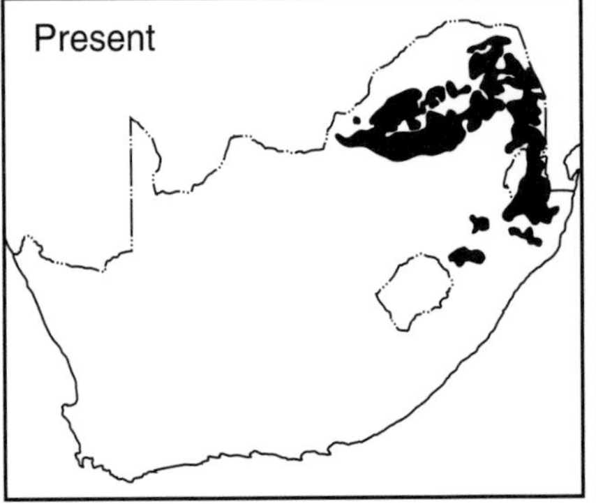

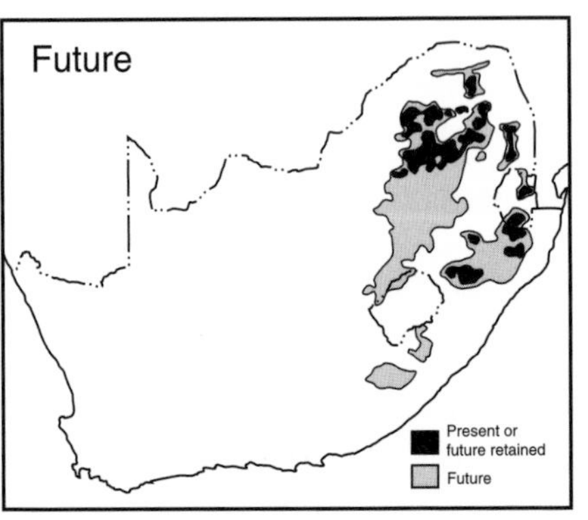

Fig. 2.53. Changing distribution of *Aloe marlothii* with global warming modelled from scenarios developed from the HadCM2 model using greenhouse gases and sulphate aerosols (Rutherford et al. 2000). The future distribution shows the future retained distribution as well as the future potential extended distribution

is whether the plant and animal populations will have sufficient mobility to move as assemblages through landscapes highly fragmented by human use. Palaeobotanical evidence suggests that organisms migrate at different rates in response to directional forcing, so compositional changes within the biomes may be anticipated. Likewise as changes occur, equilibria between organism niches and climate will break down and system instabilities and outcomes, such as caused by insect predation, invasion by alien species and fire, can also be expected.

2.9.3 Impacts on Ecosystem Function

2.9.3.1 *Net Primary Production*

Net primary production underpins agricultural, forestry and pastoral yields, as well as a range of other natural resources. About 80% of Africa is covered by arid grassland, shrubland or savannas. Within these formations, net primary production shows a strong positive relationship with the duration of water availability within the soil profile for plant use (Scholes 1993). Below an annual rainfall of about 700 mm this relationship is nearly linear. For much of the region, a variety of models suggest rainfall is likely to decrease during the growing season in future with global warming (Joubert and Kohler 1996). Increases in temperature will increase evaporative demand, which in conjunction with lower rainfall, will reduce soil moisture. However, increasing CO_2 will have a strong positive effect on plant water use efficiency, especially in C_4 grasses (Wand et al. 1999a). In areas where growth is water-limited, this may either extend the growing season, or increase above-ground production. There is much less intrinsic difference in CO_2 responsiveness between C_3 and C_4 grasses than initially suggested (Wand et al. 1999b). Increased production in C_4 grasslands is not unlikely, given the observed stimulation of southernmost African C_4 grasses in elevated CO_2 experiments (Wand et al. 1999a). Preliminary modelling provides an indication of how net primary production may change in future. It would appear that the water use efficiency increase will counterbalance almost exactly the effect of anticipated reduced rainfall and increased temperature, leaving grass production unchanged (Scholes 2001, pers. comn.). Experimental evidence in savanna mesocosms under realistic rainfall regimes suggests a moderate increase in net primary production of grasses, under increased temperature and constant rainfall, as CO_2 increases to a concentration of about 550 ppm (Scholes 2001, pers. comn.). Above this value no fertilization effect is observed, probably owing to nutrient limitation, which is widespread in southern Africa. Species of the nutrient-limited Fynbos Biome show very low CO_2 responsiveness, even when fertilised (Midgley et al. 1995, 1999).

The Integrated Biosphere Simulator (IBIS) model (Foley et al. 1996) forced with the HadCM2 GCM in transient mode, has been used to estimate the present and future distribution of net primary production over southern Africa (Shannon 2000) (Fig. 2.54). In the areas of southern Africa where rainfall is currently sufficient for essentially continuous plant growth, or where rainfall is predicted to increase, net primary production is likely to increase moderately due to a combination of climatic and CO_2 effects. Land cover changes in the tropical forest areas, where rainfall may increase or remain the same, may nevertheless lead to overall decreases in production and carbon storage.

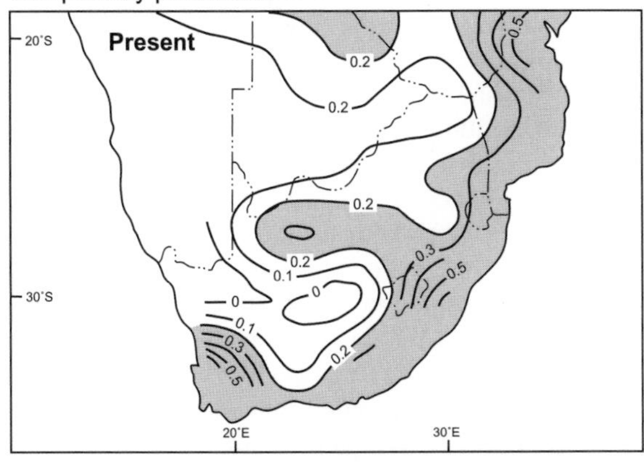

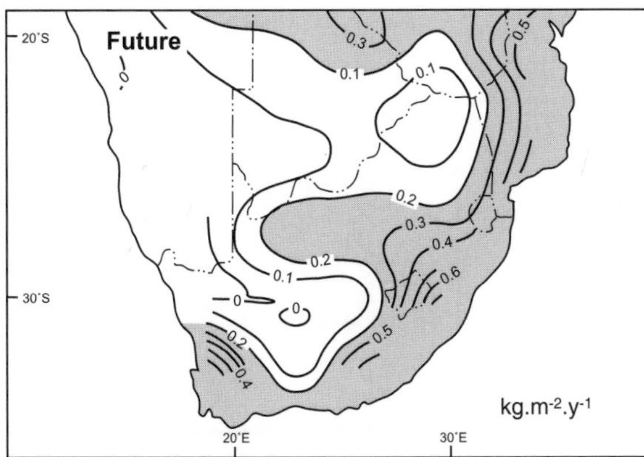

Fig. 2.54. Modelling present and future distributions of net primary production over southern Africa using the Integrated Biosphere Simulator (IBIS) model forced by the HadCM2 GCM in transient mode (Shannon 2000)

2.9.3.2 *Secondary Productivity*

The yield of animal products does not necessarily reflect changes in the growth of plants (Scholes 1993). Over most of Africa, animal production is limited by the quality (nitrogen content) rather than quantity of grass. Increased grass growth due to elevated CO_2 almost always decreases the nitrogen content (Mooney et al. 1999). This does not seem to have a strongly negative effect on *in vitro* digestability, perhaps because in grasses much of the additional carbon is stored as starch, which is readily digestable. Browser carrying capacity is controlled by tree leaf biomass and (inversely) by the content of secondary leaf compounds, such as tannins; both may increase in future, but with opposite effects. Consequently, it is difficult to predict a general increase or decrease in animal carrying capacity with future regional change. However, attempts have been made to model possible future scenarios for ungulates in general (Hulme 1996), but little credibility can be attached to the detailed results except that ungulate populations and species composition are sensitive to climate.

2.9.4 The Effects of Pests and Diseases

With current knowledge and modelling skills it is difficult to incorporate the effects of pests and diseases into integrated scenario development. Instead their effects tend to be considered in isolation. An illustration of the way in which global change may impact on patterns of disease occurrence in animals and humans is given by considering possible future changes in mosquito, tick and tsetse fly distributions. This has been done using statistical modelling and future climate scenarios developed from a UK Meteorological Office GCM (Hulme 1996). Mosquitoes transmit a range of viral, protozoal and filarial diseases to humans, among which malaria is the most significant. About 90% of world malaria cases occur in Africa where the disease is hyper-endemic (Hulme 1996). Major malarial vectors in southern Africa belong to the *Anopheles gambiae* complex of species. The modelling suggests that mosquitoes will retain their stronghold in equatorial and lowland regions and increase in many parts of the region with continued global warming (Fig. 2.55). Tsetse flies (*Glossina* spp.) transmit human and animal trypanosomiasis, causing sleeping sickness in humans and nagana in domestic animals, both of which may be fatal. In contrast to the possible changes in mosquito distribution, in future tsetse flies may tend to lose much of their present distribution, particularly in the central parts of the region. Ticks (*Rhipicephalus* spp.) are widespread and locally abundant vectors of livestock diseases. In future it may be that they become less abundant over southeastern parts of the region, while increasing somewhat over northwestern parts.

2.10 Human Acceleration of Regional Change

The interaction between humans and the environment is complex, both directly and indirectly. To unravel the interactional processes and gain a better understanding of environmental change in the region it is necessary to examine further some of the human factors that play a role in regional change in southern Africa.

African economies depend on natural resources. Changes in availability and access to land and water, for example, usually impact across several sectors (e.g. agriculture, mining and industry) and scales (e.g., regional, local area or household level). Land is a critical resource and through agriculture contributes about 40% of regional GDP and employs more than 60% of the labour force (World Bank 1998). Land degradation (including erosion, desertification and declining soil fertility) impacts negatively on the production potential of the land and also has an important feedback into the earth-atmosphere system.

Transformation of the landscape by humans is governed by many factors including government actions, institutional policy, farming actions, decisions made in households or by individuals, and so on. These human-environment interactions produce various landscape configurations rooted in historical, social, political and economic settings. Within the agricultural sector, for example, the livelihood activities of the rural poor (Chambers and Conway 1992) and capital-intensive commercial farming shape, drive and transform landscapes in various ways.

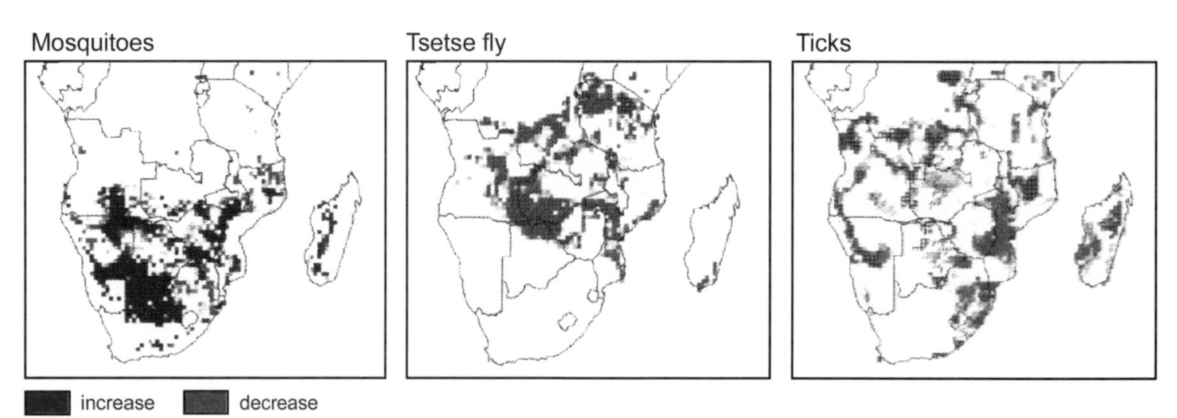

Fig. 2.55. Changes in the distributions of mosquitoes, tsetse fly and ticks over southern Africa that may result from global change to the year 2050 (after Hulme 1996)

2.10.1 Poverty and Rural Subsistence

In rural systems, poverty is both a major cause and primary consequence of environmental change. Lack of access to physical resources (land and water in particular) and the economic means to improve their situation determines how poor rural people in the region interact with their environment. Environmental and global change issues therefore cannot be divorced from issues of basic development and livelihood security (UNDP 1999; UNEP 1999).

In an integrated approach to assessing regional aspects of global change, it is necessary to consider such issues as rangeland degradation, natural resource use, resilience to change and links between household wealth and resource use. A substantial corpus of literature relates to these issues in Africa (see for example: Anderson and Grove 1987; Leach and Mearns 1988, 1998; Behnke et al. 1993; Seely and Jacobson 1994; Hoffman 1995; Scoones 1995; Dahlberg 1996; Clarke et al. 1996; Campbell 1996; Lipton et al. 1996; Odada et al. 1996; IPCC 1996).

One way of examining the complex interactions between agents and actors of enironmental change across land cover and land use types is by focussing on case studies. Rangelands, for example, cover large areas of southern Africa and provide the livelihood of many millions of people (e.g. de Bruyn and Scogings 1998; Mokitimi et al. 1998; May et al. 1998). The two cases considered here, Matsheng in Botswana and Paulshoek in Namaqualand, South Africa emphasize the interactions between the human and biophsyical drivers that shape environmental change in rangelands.

2.10.1.1 *Change around Matsheng, Botswana*

A good example of pressure exerted on rangelands in semi-arid areas is that of Matsheng in the northern Kgalagadi (Kalahari) of Botswana (Arntzen et al. 1998). A mix of biophysical and socio-economic drivers explains much of the environmental change that is occurring in the area. Matsheng is part of the driest region of Botswana and receives an annual rainfall of about 300 mm with high seasonal variability (Bhalotra 1985). Water scarcity is exacerbated by evapotranspiration that exceeds mean precipitation by more than a factor of three (Bhalotra 1985). Natural vegetation consists of open savanna with xerophytic woody plants, adapted to aridity and frequent droughts. Undisturbed parts of the area support dense grass cover making the area an excellent rangeland environment for wildlife. The area has many pans, which are a source of water for both wildlife and livestock during the summer. All major settlements (Hukuntsi, Lehututu, Tshane and Lokgwabe) are located close to the margins of these pans, where wells have been dug for potable and livestock water. Within 30 km of villages mixed farming is practiced. Beyond 30 km, in wildlife management areas, wildlife utilisation is the main form of land use and income generation (Arntzen et al. 1998).

Wildlife and other natural resources, such as veld products, have traditionally been abundant in the area. In the nineteenth century and earlier, the people of Matsheng obtained their livelihood from local natural resources through hunting and gathering, pasture-based livestock and in some cases arable agriculture. In the past, the state of the rangelands was always of critical importance to welfare and livelihood security. The present situation is very different. While rangelands are still used for the gathering of berries, edible plants and wood, their contribution to the welfare of families has become limited. Livestock production, mainly cattle, sheep and goats, is increasingly becoming an important socio-economic activity and source of wealth. The livelihood base is narrowing as a consequence, with increasing numbers of livestock being owned by relatively few wealthy groups (Arntzen et al. 1998; Chanda et al. 1999).

Sedentary living, increasing numbers of population settlements and the opening up of the rangelands through the provision of borehole water, together with overstocking, particularly in times of climate stress, are factors accelerating regional change in arid areas such as the Matsheng of Botswana. Progressive degradation of the rangeland environment is occurring at present around Matsheng and several other villages in the area (Ringrose and Matheson 1991; Perkins and Thomas 1993; Arntzen et al. 1998; Chanda et al. 1999). Deteriorating grass cover and bush encroachment of species such as *Terminalia sericea*, *Acacia mellifera* and *Grewia flava*) are indicative of the inexorable changes taking place as population increases. Grass cover is virtually absent in some areas around pans and settlements owing to overgrazing by cattle (Fig. 2.56), a trend occurring throughout Botswana (Blair-Rains and Yalala 1972; Scarpe 1986; Arntzen et al. 1994; Ringrose et al. 1996). The problem is compounded by an attitude among local people that it is climate, not overgrazing, that is responsible for degradation (Arntzen et al. 1994; Ringrose et al. 1996; Chanda 1996; Chanda et al. 1999).

At present rangeland degradation is occurring mainly because of the cattle ranching activities of relatively few people. The Matsheng area, like other remote rural areas in the country, is one of poverty (BIDPA 1997). This poverty is offset by the remittance of government pensions and welfare programmes and other off-farm incomes derived typically in urban or industrial areas. Were the national economy of Botswana to weaken, welfare to be reduced and people to be forced in increasing numbers back on the land for survival, the problem of environmental degradation would be compounded. How the Botswana rangeland communities will be able to with-

Fig. 2.56.
Degradation of the rangeland environment near Tshane Pan in Marsheng, Botswana (Photo: R. Chanda 1998)

stand shocks associated with socio-economic changes wrought by HIV/AIDS, droughts and impacts of conflict in the wider region remains moot.

2.10.1.2 *Change at Paulshoek, South Africa*

Paulshoek is part of the Leliefontein communal reserve in Namaqualand, South Africa. It is located in a semi-arid area receiving 150–200 mm rain a year. Nonetheless it is an area rich in plant diversity on sandy plains surrounded by rocky granite hills. There are currently eight Namaqualand communal areas that form more than 25% of the local region and support over 40% of the area's population (Todd and Hoffman 1999). The people, livestock, grazing practices and the conservation of Namaqualand are inextricably linked (Todd and Hoffman 1999). In the Paulshoek area, much like the aforementioned case in Botswana, remittances from government, welfare cheques, pensions, welfare support and employment (either permanent or self-employment) contribute substantially more to people's livelihoods than does livestock and other forms of agricultural production. It is estimated that only 2% of cash income is derived from livestock; approximately 52% of total village income is derived from wages and remittances from permanent jobs (mostly outside of Paulshoek); pensions, disability allowances and social security payments for child support constitute 29% of total income and self-employment and casual labour 17% (Global Change and Subsistence Rangelands in Southern Africa 1999).

Despite the reduced reliance on farming for a major income source, livestock grazing remains one of the major drivers of environmental change in the area. Past land tenure policies, lack of access to markets and financial resources among communal farmers, and a declining source of income from mines in the wider Namaqualand area have all encouraged villagers to seek alternative means of survival. One of these has been increased usage of ecological capital, including grazing of the commons by the locals. When these drivers of environmental change are imposed on an area that receives a variable supply of rainfall, undergoes extremes of temperature and other seasonal changes, then a clearer picture of degradation in certain areas becomes possible.

Over the past 30 years the commons have been grazed at about twice the stocking rate of the neighbouring commercial farms. Changes in plant community composition and species richness across fencelines demarcating heavily-grazed communal and lightly-grazed commercial farms are distinctive (Fig. 2.57a). The commons have been altered in certain areas from a mixture of perennial succulents and palatable woody shrubs to a mixture of annuals and poisonous perennial shrubs such as kraalbos, *Galenia africana* (Todd and Hoffman 1998, 1999). In times of drought those areas that are overgrazed become increasingly vulnerable to soil erosion and further degradation. In addition, increased collection of biofuels from the land further adds to the depletion of the vegetation.

Sharp changes in land cover and land use gradients that are configured by different land management policies rather than changes in biophysical environments or climate are are not confined to local occurrences, but may be just as evident on regional scales. Such changes may reflect a suite of land use management decisions ranging from regional and national policy to household decisions (Thomas and Sporton 1997). Example of pronounced regional land cover and land use gradients often occur across international boundaries. Such is the case across the western Lesotho-South Africa border (Fig. 2.57b).

Fig. 2.57. Land cover discontinuities resulting from different land use practices: **a** across a boundary fence in the Paulshoek area of Namaqualand, South Africa (*left:* commercial; *right:* communal rangeland stocked at about twice the recommended rate for the past 30 years) (Photo: S. Todd 1996); **b** across the South Africa-Lesotho national boundary

2.10.2 Unravelling Human and Biophysical Causes of Change

Much work has been done in relation to the role of land cover and land use change and other human drivers of environmental change in southern Africa (Mentis et al. 1989; Perkins and Thomas 1993; Shackleton 1993; Scoones 1996; de Bruyn and Scogings 1998; Ward et al. 1998). Notwithstanding contestation, there is evidence that changes in land cover and land use are important factors contributing materially to environmental change. In addition, other human factors such as single or multiple livelihood strategies, variations in environmental and agricultural knowledge systems, complexities of land tenure and land ownership and the role of policy intervention are producing distinctive responses in rural landscapes (Hoffman et al. 1999). Likewise, increasing population densities, influx of people to peri-urban areas, lack of infrastructure, poor education, poor run-off control, inappropriate agriculture and animal husbandry practices all leave their marks on the landscape and contribute in one way or another to soil, veld and other environmental changes. Overall land degradation, for example, exhibits complex spatial and temporal variability, as the southernmost African experience shows (Hoffman et al. 1999) (Fig. 2.58).

The interactions between the biophysical and human systems are complex, making it difficult to isolate the predominant driver of change. On the one hand, land cover and land use changes have immediate feedback effects on climate if they occur on large enough scales,

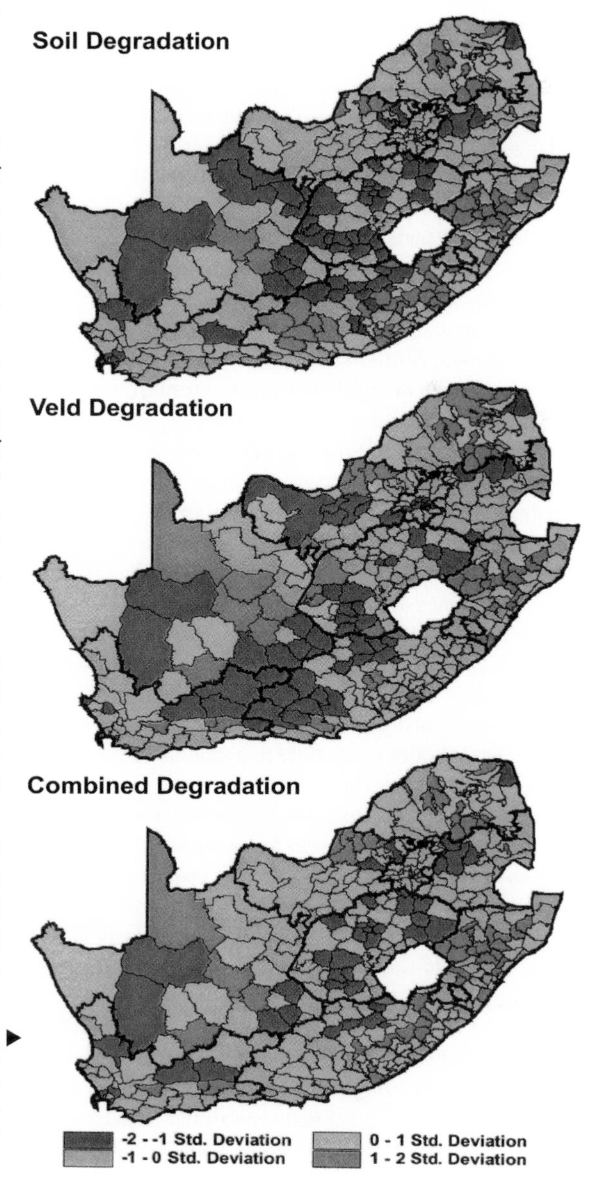

Fig. 2.58. ▶
Soil and veld degradation over South Africa between 1988 and 1998 and combined degradation as perceived by agricultural extension and nature conservation officers (Hoffman et al. 1999). Degradation is expressed as standard deviations about the mean for South Africa as a whole

mainly through the effects of altered surface albedo. On the other, biophysical factors, including periods of severe drought, expose, trigger or exacerbate latent social and economic drivers of regional change. Climatic conditions, for example, not only influence agricultural production, but have a more general consequence for land degradation. This has been especially so since the 1970s (Hoffman et al. 1999). Periods of drought in particular impact negatively on livelihood activties. Lack of water triggers a variety of adaptation stratgies that in turn exert positive feedbacks to accelerate environmantal degradation (Valentine 1993; Scoones 1996; Kinsey et al. 1998; Vogel et al. 2000).

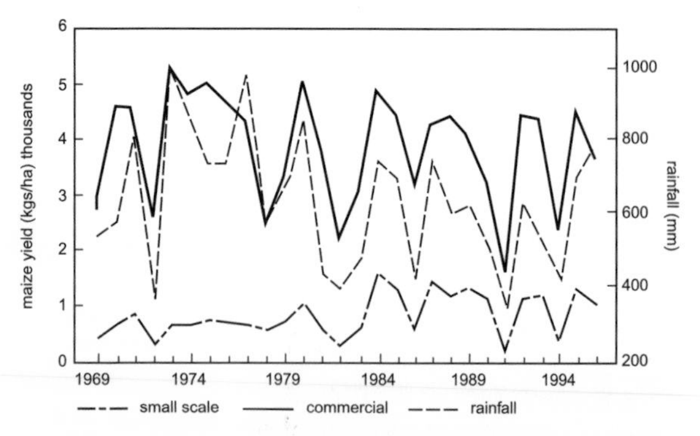

Fig. 2.59. The effect of rainfall variability on both commercial and small-scale maize production in Zimbabwe during the second half of the twentieth century (Unganai 2000, pers. comn.)

2.10.3 Adaptation to Environmental Change

Despite the complexity of the human-biophysical system, the changes that are taking place in the rangelands of Botswana and South Africa appear to be distinctive and reflect wider changes in the region. Changes in land cover and species composition are leading to increasing degradation. Given the likelihood of increasing frequencies and magnitudes of extreme events in the region with global warming (IPCC 1996; Joubert et al. 1996), there is a growing need for adaptation to such changes and the need to develop strategies of mitigation to ameliorate the multiplicity of consequential occurrences (Freeman 1984; Moorsam et al. 1995; Vogel 1995, 2000; Von Kotze and Holloway 1996; Zeidler et al. 1999). Mitigation and adaptation to periods of shock in African communities traditionally encompassed a range of strategies, including reliance on kinship ties (Freeman 1984; Downing et al. 1989; Vogel 1995; Davies 1996, 2000). Increased diversification of survival strategies will be required in future, including the holding of small and large livestock, multi-purpose resource use, and possible increased dependence on cash incomes and remittances.

Livelihood diversification among the rural poor is often underestimated in the region. The diversification is increasing. Craft activities, collection of medicinal plants, multi-purpose use of rangelands and commons and other activities contribute to household incomes to a much greater extent than in the past (IIED 1995; Campbell 1996; Cousins 1998; Shackelton 1996). An enhancement of such activities is an important adaptation option for communities in the face of future climate change (Hulme 1996), but can lead to environmental degradation if carried to excess, particularly where wood products provide the basis for craft industries (Shackleton and Shackleton 1997).

The link between changing climate and human systems is not always simple. Frequently poor land management grossly exaggerates climatic effects; in other instances the impact of climate variability and droughts on crop and rangeland production may be direct, as the agricultural productivity of small farmers in Zimbabwe illustrates (Fig. 2.59). In that country, climate affects both small farmers and the commercial farming community alike, but the former more so given their limited access to capital. Small farmers appear to have become more susceptible to climate variability from the end of the 1970s onward. Commercial farming enterprises throughout the region are much affected by climate variability at present and equally susceptible to future changes.

2.11 Commercial Farming, Forestry and Environmental Change

Within the developed sector of national economies in the region, human activities play an equally important role in shaping patterns of future regional change. In the commercial farming sector, the over-application of fertilizers may act to alter the nutrient balance of inland waterways and groundwater reservoirs. The conversion of grassland and savanna to pasture and cropland may alter the surface albedo, which in turn may alter the local convective regime and initiate a change in climate. Expansion of organized agriculture into marginal areas has negative consequences once the climate begins to vary. Burning of agricultural waste adds to aerosols and trace gases in the atmosphere. All these activities may act, internally within the region, to accelerate change. At the same time, global change is acting externally to impact on the region. Nowhere is this seen more clearly than in commercial agriculture.

In greater southern Africa, the agricultural sector is a significant contributor to the gross national product of economies. It is most significant in Tanzania (46%), Malawi (38%) and Mozambique (34%) and least in Botswana (4%) and South Africa (4%). Throughout the subcontinent, the annual variability in agricultural productivity is closely related to changing weather and climate. In Zimbabwe, where agriculture contributes 18% towards GNP, both capital- and technology-intensive commercial and low-budget small-farm agricultural

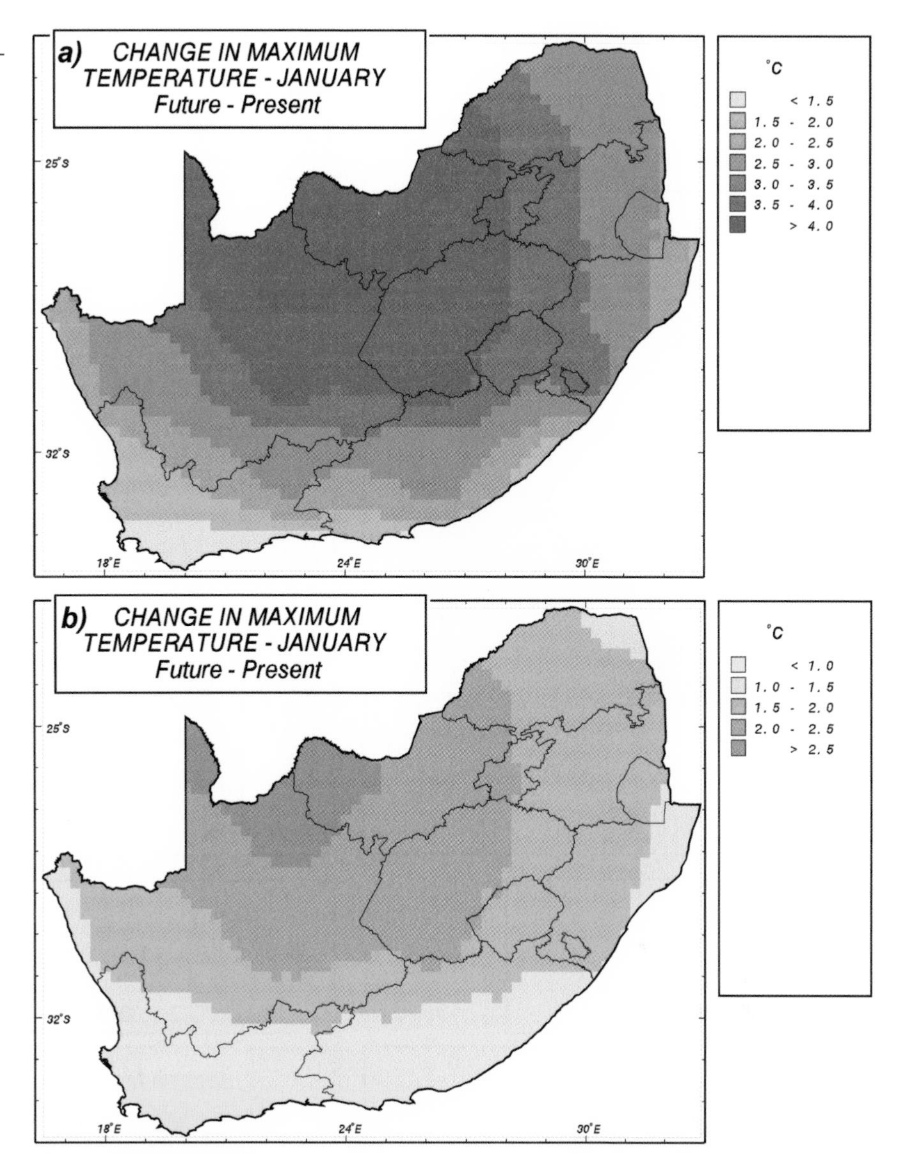

Fig. 2.60.
The effect of including sulphate-aerosol-induced cooling on future January maximum temperatures derived from the transient-run HadCM2 model using: **a** with greenhouse gases only; **b** with combined greenhouse gases and sulphate aerosols (Perks et al. 2000)

productivity vary significantly with climate (Unganai 2000, pers. comn.) (Fig. 2.59). In South Africa, increasing population places demands on the agricultural sector to increase production by approximately 3% per annum to maintain self-sufficiency (Schulze et al. 1995). In countries to the north, similar or greater pressures exist at a time when the regional effects of global warming provide both opportunities for increased production and constraints on expansion of agriculture. Some of these positive and negative factors are illustrated from impact studies carried out in southernmost Africa.

All impact models are sensitive to the nature of the driving GCMs and may compound the uncertainties associated with them. The effect of incorporating aerosols in the driving GCMs is significant, as was shown earlier (Fig. 2.32) and is further illustrated by comparing HadCM2 model increases in January maximum temperature using greenhouse gas forcing alone and greenhouse gas and aerosol forcing together (Fig. 2.60). Maximum temperatures, and degree-days, enhance photosynthetic development during a crop's growing season and are used as an index of the energy available for plant growth (Schulze 1997a). Degree-days are likely to increase with global warming, in many places in southernmost Africa by more than 500 for the October to March growing season (Perks and Schulze 2000) (Fig. 2.61a). At the same minimum temperatures will rise, bringing about a reduction of nocturnal frosts by more than 30 days in a winter season over most of southernmost Africa (Schulze 1997a; Perks and Schulze 2000) (Fig. 2.61b). Offsetting these generally positive factors influencing agricultural productivity will be negative ones, such as a reduction in winter chilling. This will adversely affect the deciduous fruit industry (Fig. 2.61c), since winter chilling is a requirement for deciduous trees to complete their seasonal dormancy and for flowering and

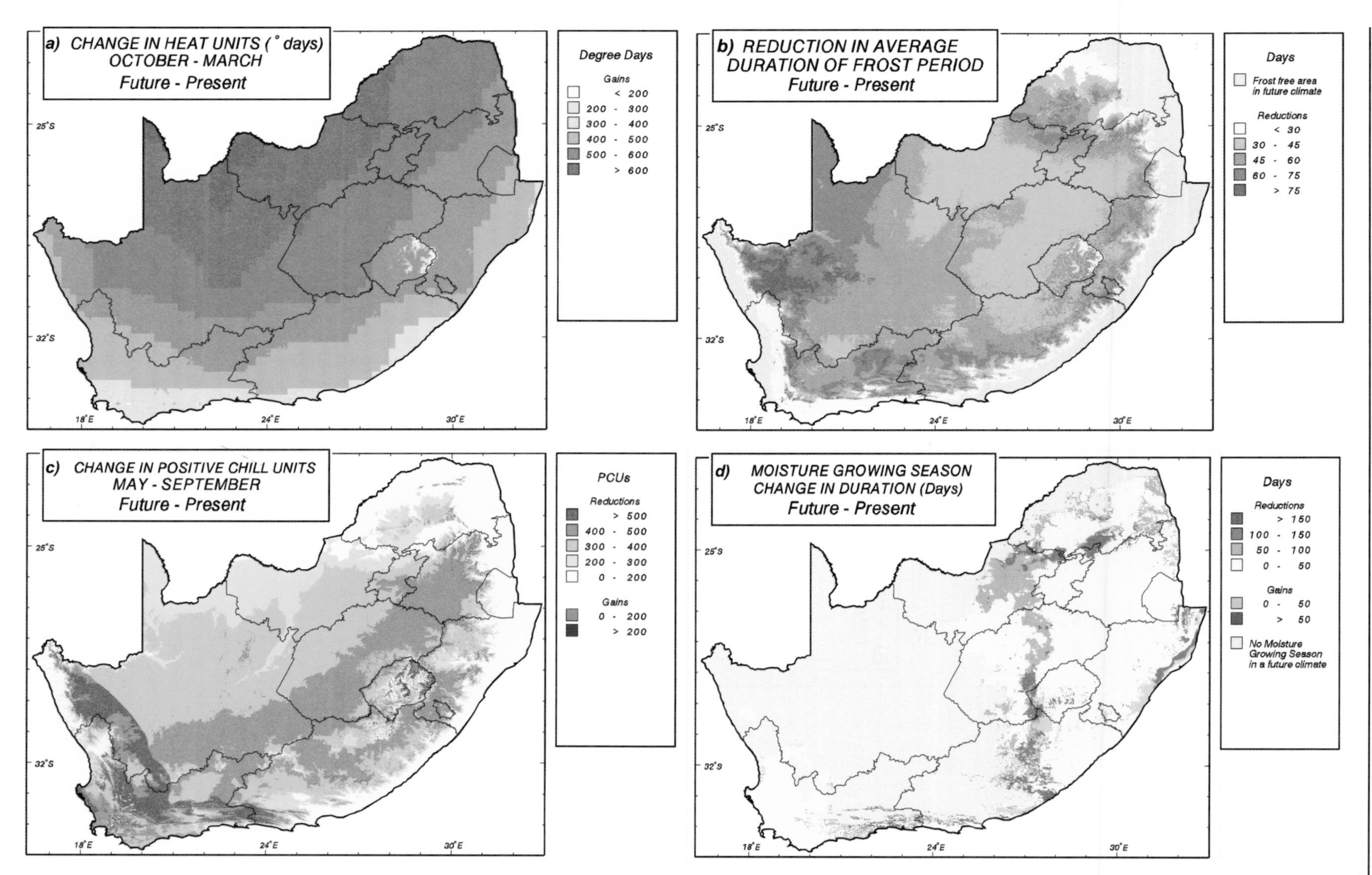

Fig. 2.61. ACRU model simulations, forced by transient-run HadCM2 model doubling of greenhouse gases, to show possible future changes in: **a** heat units (degree-days); **b** duration of average frost period (days); **c** positive chill units from May to September; **d** duration of the moisture growing season (days) (Perks and Schulze 2000)

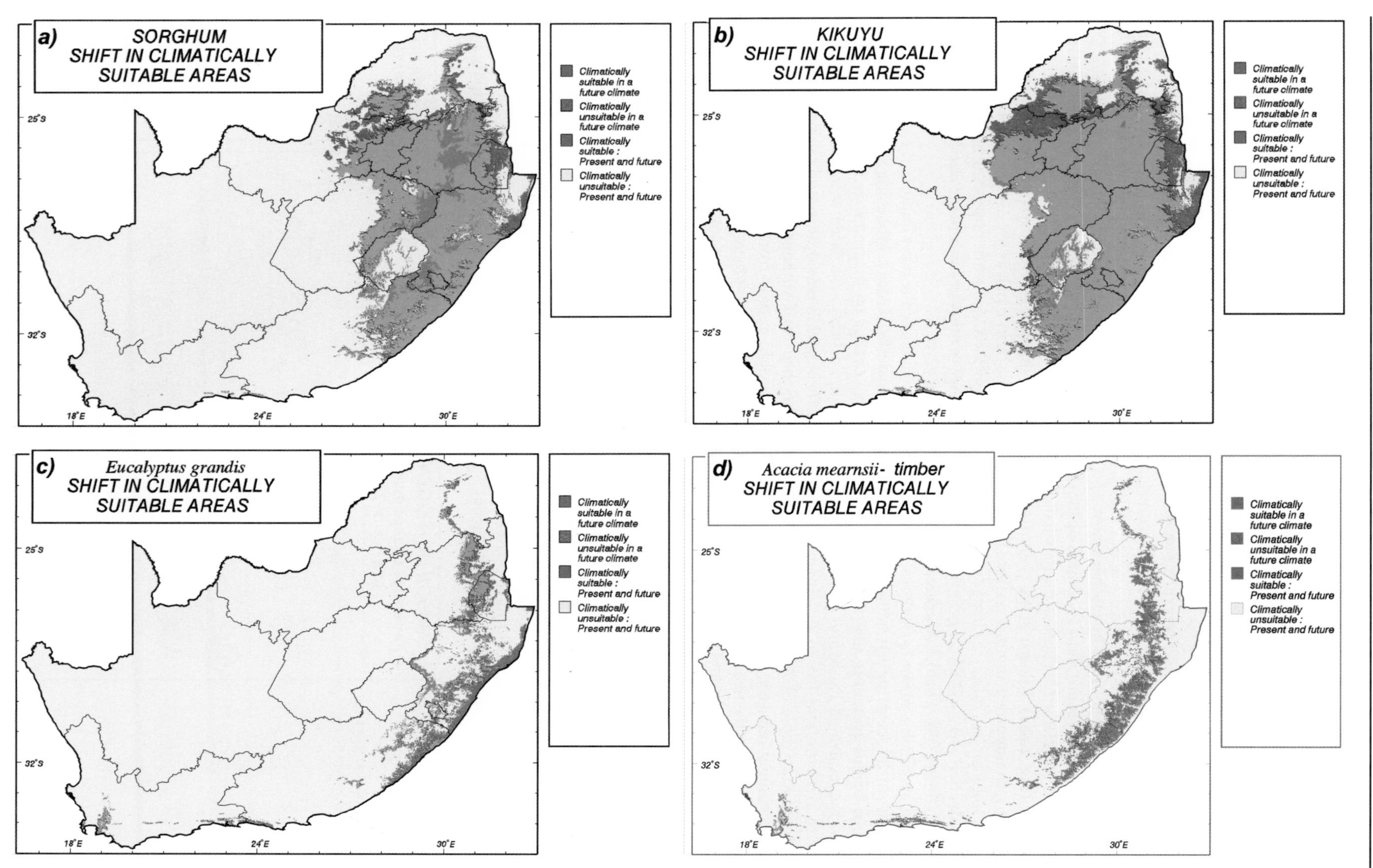

Fig. 2.62. ACRU model simulations, forced by transient-run HadCM2 model doubling of greenhouse gases, to show possible future shifts in climatically suitable areas for the cultivation of: **a** sorghum; **b** kikuyu pasture grass; **c** *Eucalyptus grandis* timber plantations; **d** *Acacia mearnsii* timber growing (Perks and Schulze 2000)

fruit setting. A reduction in the moisture-growing season and in available soil moisture will have a more general adverse effect, especially in many areas already marginal for commercial agriculture (Fig. 2.61d).

Agricultural crops grow optimally within ranges of temperature and rainfall specific to given crops. A consequence of future climate change is that agricultural production belts for different crops may shift. Some belts are robust and are unlikely to change much in future; others are highly sensitive to change. Using Smith's (1994) climatic criteria for crops in southernmost Africa, Perks and Schulze (2000) show sorghum's optimal growth areas to be relatively robust to anticipated climate change (Fig. 2.62a). Kikuyu grass (*Pennisetum clandestinum*), an important summer pasture for cattle grazing, is somewhat less robust to change and in many present production areas will no longer grow optimally in a warmer climate (Fig. 2.62b). Commercial afforestation of exotic species is a major agricultural activity in southernmost Africa, covering 1.35 million ha in South Africa alone. Of this area, over 40% is devoted to the growing of the hardwood *Eucalyptus grandis*. Its optimal growth areas are sensitive to anticipated climate change (Fig. 2.62c); those of *Acacia mearnsii* (wattle), used in the tanning industry, are even more sensitive (Fig. 2.62d). Modelling the responses of a wide range of forestry species (Fairbanks and Scholes 1999) shows them all to be sensitive to the predicted magnitude of climate change, but in different ways. The time scales of timber rotations, and the relative technological sophistication of the forestry industry make it likely that species or cultivars can be substituted to maintain productive forestry in the region in the face of change. Modelling further suggests that overall productivity of individual stands will not be strongly impacted, since the effects of elevated CO_2 approximately cancel future possible decreases in precipitation (Fairbanks and Scholes 1999).

Changes in productivity are not confinded to changing geographical areas of optimal crop production and forest plantations, but also apply to the productivity of grazing conditions. The animal carrying capacity of southern African rangelands is more strongly influenced by forage quality than quantity (East 1984). Forage quality is a function of plant C:N ratios. Sweetveld with a low C:N ratio is indicative of relatively good plant digestibility; sourveld with high C:N ratios has a much poorer forage quality. Modelling the rates of plant assimilation of C and N in relation to air temperature, CO_2 levels and soil fertility allows the a future scenario of rangeland palatability to be developed (Kunz et al. 1995). With a doubling of greenhouse gases, forage quality may deteriorate by as much as 8% over large areas of central southernmost Africa, while improving by about the same percentage in others mainly to the northwest and southeast (Fig. 2.63).

Maize is one of the most important crops, and staple foods, in southern Africa. Modelling the regional impact of global change on its productivity illustrates further the complications associated with doing this kind of work. If changes in temperature and rainfall are considered on their own, without CO_2 fertilisation feedback, then it would appear from ACRU model simulations (Schulze 1995) that with global climate change, the area under maize cultivation in southernmost Africa is likely to shrink, mainly along the westerns margin of the present-day distribution of production. Translated into profits/losses per hectare of cultivation, however the picture is more complicated (Perks and Schulze 2000) (Fig. 2.64a). Losses are likely to be incurred throughout the region, with the greatest likely to occur in the wetter eastern areas. However, when the positive feedback effect of carbon dioxide fertilisation is taken into account, then the picture is very different. In many northeastern areas of southernmost Africa, maize productivity may

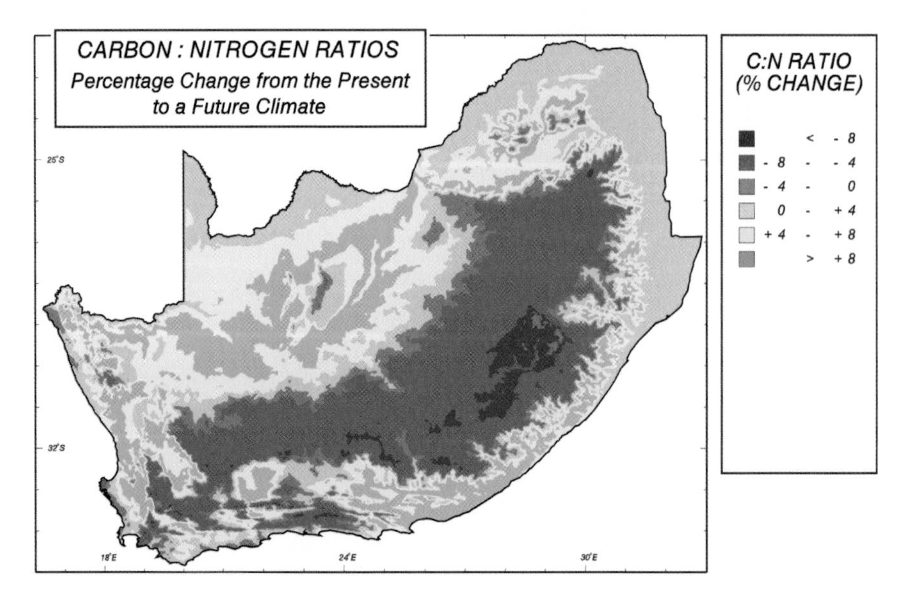

Fig. 2.63. Changes in C:N ratios and forage quality of southernmost African rangelands with a doubling of greenhouse gases (Kunz et al. 1995)

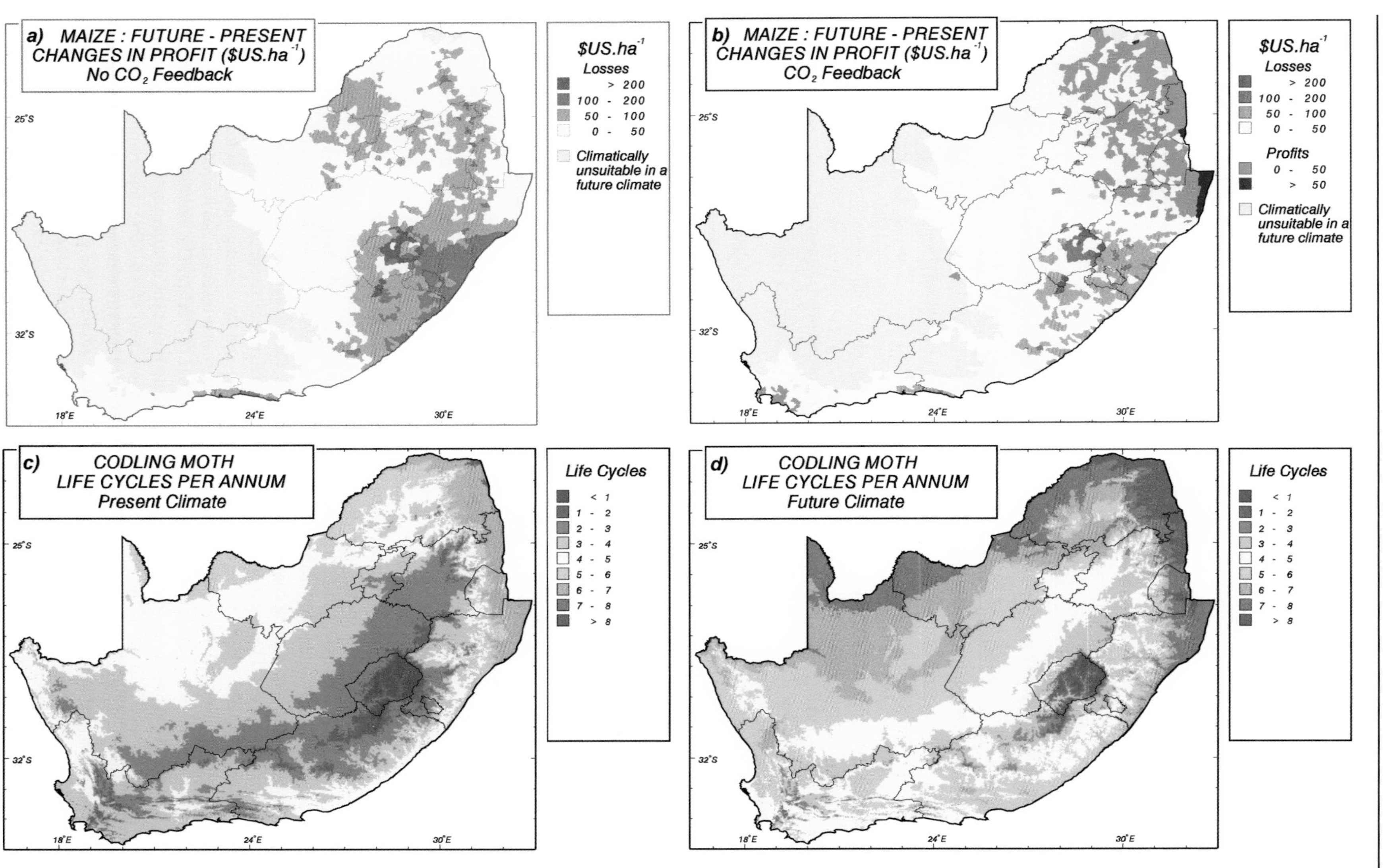

Fig. 2.64. ACRU model simulations, forced by transient-run HadCM2 model doubling of greenhouse gases, to show: **a** future changes in maize yield profits (US$ per hectare) with no CO_2 feedback; **b** future changes in maize yield profits (US$ per hectare) with CO_2 feedback; **c** the present number of life cycles of codling moth per annum; **d** possible future numbers of life cycles of codling moth per annum (Schulze 1997a; Perks and Schulze 2000)

increase with a resultant increase in profits (Fig. 2.64b). Such a scenario takes no account of possible future changes in weeds and pests that will accompany regional global warming. The effect of enhanced weed growth and pest activity may be considerable. One example will be considered. Codling moth affects crop production. With a doubling of greenhouse gases, the present distribution of the moth and its annual number of life cycles will change (Schulze 1997a) (Fig. 2.64c). Not only will the area experiencing the pest increase, but more importantly, the number of life cycles within a year will increase almost everywhere (Schulze 1997a) (Fig. 2.64d). How this will affect food productivity has yet to be determined. In the meantime, it is clear the effect will not be negligible.

2.11.1 Urban Growth and Environmental Change

Urbanization is proceeding rapidly throughout the region as a result of intrinsic growth within urban areas, as well as through the influx of people to cities from rural areas. Industrial activity within towns and cities, and on their fringes, is rapidly adding to aerosol and trace gas emissions (Held et al. 1996). Hardly a city or town in the region is now without a major peri-urban fringe that includes squatter settlements without electrical power or sanitation. The consequent air and water pollution is a major source for concern, both locally and in a wider regional context.

One way in which this urbanisation is adding to regional environmental change, apart from the injection of aerosols and trace gases into the atmosphere, is through the generation of urban heat islands. For a variety of reasons, sensible and latent heat fluxes over cities are altered by the urban fabric. One result is the generation of heat islands over and plumes downwind from cities. These may be several degrees warmer than the surrounding air, may be deep and extend for long distances downwind. An example for Johannesburg is given in Fig. 2.65. This was the situation in the early 1970s (Tyson et al. 1972). Since then the city and its neighbouring urban places have grown to the point where they have become contiguous. An extensive conurbation now exists and the combined urban and peri-urban areas of the Pretoria-Witwatersrand-Vereeniging complex cover an area exceeding 5 000 km^2. The present-day heat plume from this conurbation is likely to be considerably larger than the one illustrated. Whereas urban heat islands are not a major source of human modification of the lower troposphere at present in southern Africa, their future growth in specific parts of the region will contribute increasingly to warming of the environment.

When considering in general the human acceleration of global change within the region, it is necessary to recognise that at all times the effect of climate variability is modulated by human activities. This modulation varies

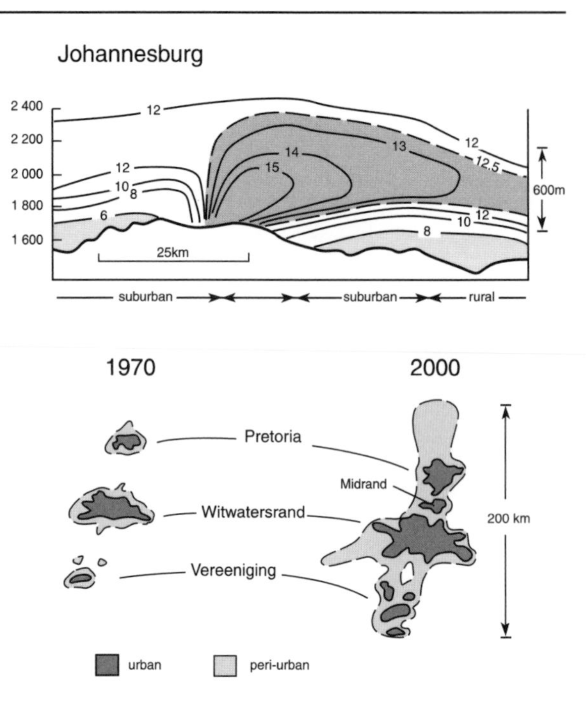

Fig. 2.65. Heat island development and local modification of the thermal structure of the lower atmosphere over the greater Johannesburg region (Tyson et al. 1972)

with the degree of development of the different sectors of national economies. In more developed sectors, climate impacts may be ameliorated; in the less developed, they may never be. In some instances human activities in southern Africa are adding substantially to the regional expression of global change; in others, regional global change is impacting human activities, such as in the cases of water and agriculture. The effect of global change on commercial agriculture offers an example of the complexity of the issue of human activity and environmental change.

2.12 Conclusions

Environmental change has had an important effect on the evolution of humankind in southern Africa. It also has influenced, through changing climate, the modes of living and the livelihoods of people for millions of years. At all times human activities have had a feedback effect on the environment, particularly through the use of fire for hunting and for clearing bush and forest. However, until recently that feedback has been negligible in forcing a change of climate. The situation began to change with the development of urbanisation and industrialisation in southern Africa. More importantly, it began to change when activities elsewhere in the world became sufficient to influence the global climate and when the changes in the global climate system began to impact the region during the late nineteenth and early twentieth centuries. Southern Africa is at present responding to those changes and at the same time influencing them by what is currently happening locally. The region is one in which horizontal fluxes in the atmosphere, in rivers and in the coastal zone have important regional consequences. Human activities in rural areas and the spread of urban centres, with all the attendant effects of urbanisation, are likewise potent agents for accelerating regional environmental change. Untangling the skein of cause and effect in the response to global change is complicated and will take time to resolve. In the meantime, this preliminary attempt at synthesising the available literature to determine the effects of global change in southern Africa has come up with some intriguing and significant findings and many other useful results. These are summarised as follows:

Surprise findings:

1. Recirculation of aerosols and trace gases, trapped beneath stable layers in the atmosphere, occurs on a subcontinental scale to an extent hitherto unforeseen. Over 40% of all aerosol-laden air recirculates over southern Africa at least once.
2. The response to environmental change may be dramatic. In the case of Lake Victoria, a whole universe of cichlid fishes evolved in the last ~12 000 years in response to drying and reflooding of the lake.
3. Climatic teleconnections patterns, similar to those resulting from present-day ENSO variability, have occurred for at least a thousand years on centennial time scales and clearly transcend the ENSO phenomenon. These teleconnections are out of phase between the equatorial/tropical region and the subtropics. The same applies to variability centred around 100 years and extending back over 3 500 years with little phase or amplitude modulation.
4. Over the last two glacial cycles, for reasons not fully understood, inshore warming of the coastal zone waters in the Benguela upwelling system off Namibia has preceded post-glacial warming by ~20 000 years.
5. Atmospheric transport and deposition of N and P nutrients appear to play a significant role in sustaining the wetland ecosystem of the Okavango Delta and, by extension, may play equally important roles in terrestrial and marine ecosystems.
6. Aeolian transport of aerosols from South Africa, and atmospheric iron fertilisation of marine biota, supports enhanced biological productivity and the South Indian Ocean carbon sink in the central ocean between South Africa and Australia.
7. Sulphate aerosols are transported as oxide adhering to small dust nuclei; those originating from the industrial heartland of South Africa may be transported over long distances across the Indian Ocean or, less frequently, by inter-hemispheric transport into the northern hemisphere over Kenya.

Other important findings include:

1. Small changes in the anticyclonicity of airflow and subsidence over the region produce large effects that modulate environmental change.
2. The plume of aerosols and trace gases moving out of South Africa to the Indian Ocean at ~30° S is large, persistent and a major feature that has hemispheric implications; the plume to the Atlantic Ocean is small and infrequent by comparison.
3. Production of aerosols and trace gases by biomass burning is only significant in the tropics; south of 20° S biomass burning products constitute an insignificant fraction of the total aerosol loading in the haze layer that extends to the ~500 hPa level in the atmosphere and blankets the entire region. Industrially derived sulphate aerosols contribute significantly to the haze-layer loading, especially in summer.
4. In the savannas of southern Africa, fluxes of trace gases from soil appear to be significantly greater than in savannas elsewhere in the world and approximately equal trace gases resulting from vegetation fires.
5. Biogenic fluxes of NO exceed pyrogenic fluxes for a period of a few weeks following the onset of the first rains.
6. It is possible that global change could induce changes in the Agulhas and Benguela Currents off southern Africa that may be of significance, not only to the region, but possibly even globally.
7. Over the past 150 000 years the carbon budget of the Benguela upwelling system has varied markedly; at present first estimates suggest the system may be a net carbon sink.
8. Inter-annual variability of hydrological regimes in southern Africa is much greater than inter-annual variability in rainfall.
9. Changes in land use have profound effects on runoff and water quality.
10. Net primary productivity appears unlikely to change much with increasing global change, but this may change as trends in future precipitation become clearer.
11. Significant loss of biodiversity is likely in the savannas, woodlands and forests and in the Cape Floral Kingdom with future changes in temperature and rainfall regimes; the effect of human depredation is certain to increase the loss unless strict conservation is practised.
12. Much of the savannas, woodlands and forests are likely to be converted to cropland and pasture during the twenty-first century. The main driver of this change is likely to be population growth coupled with unrelieved poverty, rather than climatic change.
13. Poor land management, expansion of agricultural land into marginal areas, overstocking, accelerating urbanisation and increasing industrialisation will all contribute negatively to the environment in the region, unless timely ameliorative action is taken.

14. Commercial agriculture and forestry will have to adapt, often at considerable cost, to the changing environmental conditions. However, not all changes will be negative and new opportunities will be presented for the enterprising to exploit.
15. With increasing urbanisation, and the development of large conurbations, urban heat islands will increasingly contribute significantly to the warming of the lower troposphere and thus increase regional contributions to global warming.

The southern African region is important in the global system in respect of aerosol and trace gas transport and ocean circulation, but appears to be less so in respect of the terrestrial carbon cycle. As things stand at present, industrial development notwithstanding, the region does not appear to be likely to become a major contributor to global change. On the other hand, given the degree to which the subtropical atmospheric circulation over much of the region tends to trap and recirculate material before it is exported beyond, any changes in the functioning of the earth system could have profound effects on the region. One of the most important potential future impacts is on water cycle in which the hydrology amplifies regional and local change to a considerable extent.

The linkage between the environment and humans has long been strong in southern Africa and rural societies remain much more vulnerable to future environmental change than do their urban counterparts. However, as elsewhere, both rural and urban human activities have become a major driver of environmental change in the region and this trend will increase in the future. Within the region, in the northern humid tropics natural ecosystems may be better able to buffer change than will be their southern subtropical counterparts, where changes are likely to be amplified.

Looked at holistically, future change in the region will reflect the summation of natural and anthropogenic variability in the system, will be a function of the adaptations to change that can be effected globally and within the region, and will depend on the extent to which population increase and land use change can be managed. A sustainable future depends on all these things and more, as well as an ability to predict them. It is difficult to foresee low-probability events that will produce high-impact consequences in the region. One such, is the possibility of regional anthropogenic activity exacerbating a natural severe drought to such an extent that a threshold is crossed and feedback effects ensure the prolongation of the event, with major adverse consequences, long beyond its natural course.

Understanding the nature of regional change, and its prediction, remain tasks of formidable difficulty. A small beginning has been made for southern Africa; much remains to be done. This can only be accomplished by holistic, systems-based approaches integrating the work of natural and social scientists. In this endeavour it will be necessary to develop more compatible methodologies and to rely more on quantitative regional modelling. To understand global change in future, it will be necessary to understand local and regional processes and their outcomes and to integrate them to give a global picture. A start has been made in this respect and it is clear that many of the southern African regional findings have clear global implications and consequences.

Acknowledgements

Thanks are extended to Wendy Job for preparing the figures. The synthesis is a contribution to IGBP/WCRP/IHDP START regional global change science.

References

Abbott MA, Dyer TGJ (1976) The temporal variation of rainfall and runoff over the summer rainfall region of South Africa. South African Journal of Science 72:276–278

Abraham ER, Law CS, Boyd PW, Lavender SJ, Maldonado MT, Bowie AR (2000) Importance of stiring in the development of an iron-fertilized phytoplankton bloom. Nature 407:727–730

Acocks JPH (1988) Veld types of South Africa. Botanical Survey of South Africa, Memoirs 57. Pretoria, Botanical Research Institute

Alexander WJR (1995) Floods, droughts and climate change. South African Journal of Science 91:403–408

Alcamo J, Kreileman GJJ, Bollen JC, Born GJ van den, Gerlagh R, Krol MS, Toet AMC, Vries HJ de (1996) Baseline scenarios of global environmental change. Global Environmental Change 6:261–303

Alcamo J, Leemans R, Kreileman E (1998) global change scenarios of the 21st century. Results from the IMAGE 2.1 Model. London, Pergamon and Elsevier Science

Allan RA, Lindesay JA, Parker D (1996) El Niño, Southern Oscillation and climatic variability. Australia, CSIRO

Alley RB, Meese DA, Shuman CA, Gow AJ, Taylor KC, Grootes PM, White JWC, Ram M, Waddington ED, Mayewski PA, Zielinski GA (1993) Abrupt increase in Greenland snow accumulation at the end of the Younger Dryas event. Nature 362:527–529

Alley RB, Mayewski PA, Sowers T, Stuiver M, Taylor KC, Clark PU (1997) Holocene climatic instability: A prominent, widespread event 8200 yr ago. Geology 25:483–486

Anderson BE, Grant WB, Gregory GL, Browell EV, Collins JE, Sachse GW, Bagwell DR, Hudgins CH, Blake DR, Blake NJ (1996) Aerosols from biomass burning over the tropical South Atlantic region: distributions and impacts. Journal of Geophysical Research 101:24117–24137

Anderson DM, Grove R (eds) (1987) Conservation in Africa: People, policies and practice. Cambridge, Cambridge Unversity Press

Anderson IC, Poth MA (1998) Controls on fluxes of trace gases from Brazilian cerrado soils. Journal of Environmental Quality 27:1117–1124

Andreae MO (1997) Emissions of trace gases and aerosols from Southern African savanna fires. In: Wilgen BW van, Andreae MO, Goldammer JG, Lindesay JA (eds) Fire in Southern African savannas. Witwatersrand University Press, Johannesburg, 161–183

Andreae MO, Anderson BE, Blake DR, Bradshaw DJD, Collins JE, Gregory GL, Sachse GW, Shipham MC (1994) Influence of plumes from biomass burning on atmospheric chemistry over the equatorial Atlantic CITE-3. Journal of Geophysical Research 99:12793–12808

Andreae MO, Andreae TW, Annegarn H, Beer F, Cachier H, Elbert W, Harris GW, Mauenhaut W, Salma I, Swap R, Wienhold FG, Zenker T (1998) Airborne studies of aerosol emissions from savanna fires in southern Africa: 2. Aerosol chemical composition. Journal of Geophysical Research 103:32119–32128

Arntzen J, Chanda R, Musisi-Nkambwe S, Ringrose S, Sefe F, Vanderpost C (1994) Desertification in the mid-Boteti area, Botswana. Case study for the Intergovernmental Negotiating Committee on the Convention to Combat Drought and Desertification. Gaborone, Ministry of Agriculture

Arntzen J, Chanda R, Fidzani H, Lapologang M, Moffat S, Skarpe C, Tacheba G, Totolo O (1998) Global change and subsistence rangelands in Southern Africa: resource variability, access and use in relation to rural livelyhoods and welfare. Preliminary Literature Review, Botswana (*http:/www.cazs.bangor.ac.uk/rangeland*)

Bailey GW (1999) Severe hypoxia and its effect on marine resources in the southern Benguela upwelling system. International workshop on monitoring of anaerobic processes in the Benguela Current ecosystem off Namibia. Ministry of Fisheries and Marine Research, Swakopmund, Namibia

Baker BH (1967) Geology of the Mount Kenya area. Geological Survey, Report 79, Nairobi

Bang ND (1971) The southern Benguela Current region in February 1966. Part II: Bythythermography and air-sea interactions. Deep Sea Research 18:209–224

Beal LM, Bryden HL (1999) The velocity and vorticity structure of the Agulhas Current at 32° S. Journal of Geophysical Research 104:5151–5176

Behnke RH, Scoones I, Kerven C (eds) (1993) Range Ecology at Disequilibrium: New Models of Natural Variability and Pastoral Adaptation in African Savannas. London, ODI, IIED and Commonwealth Secretariat

Behrenfeld MJ, Bale AJ, Kolber ZS, Aiken J, Falkowski PG (1996) Confirmation of iron limitation of phytoplankton photosynthesis in the equatorial Pacific Ocean. *Nature* 383:508–511

Benson C, Clay E (2000) The economic dimensions of drought in sub-Saharan Africa. In: Whilhite DA (ed) Drought, Vol. 1, A Global Assessment. London, Routledge, 287–312

Beuning KRM, Kelts K, Stager JC (1998) Abrupt climatic changes associated with the Younger Dryas interval in Africa. In: Lehman JT (ed) Environmental change and response in East African lakes. Dordrecht, Kluwer Academic Publishers, 147–156

Bhalotra YPR (1985) Rainfall maps of Botswana. Gaborone, Department of Meteorological Services

BIDPA (1997) Study on poverty and poverty alleviation in Botswana. Vol. 1. Gaborone, Botswana Instiutute for Development Policy Analysis, Ministry of Finance and Development Planning

Blair-Rains A, Yalala AM (1972) The Central and Southern State lands, Botswana. Range Resource Study 11. Tovworth, DOS

Blake NJ, Blake DR, Sive BC, Chen T, Rowland FS, Collins JE, Sasche GW, Anderson BE (1996) Biomass burning emissions and vertical distribution of atmospheric methyl halides and other reduced carbon gases in the South Atlantic region. Journal of Geophysical Research 101:24151–24164

Bond WJ, Midgley GF (2000) A proposed CO_2-controlled mechanism of woody plant invasion in grasslands and savannas. Global Change Biology 6:865–870

Boyd A, Taunton-Clark J, Oberholster GPJ (1992) Spatial features of the near-surface and midwater circulation patterns off western and southern South Africa and their role in the life histories of various commercially fished species. South African Journal of Marine Science 12:189–206

Boyd PW, Watson AJ, Law CS, Abraham EA, Trull T, Murdoch R, Bakker DCE, Bowie AR, Buesseler KO, Chang H, Charette M, Croot P, Downing K, Frew R, Gall M, Hadfield M, Hall J, Harvey M, Jameson G, LaRoche J, Liddicoat M, Ling R, Maldonado MT, McKay RM, Nodder S, Pickmore S, Prodmore R, Rintoul S, Safi K, Sutton P, Strzepek R, Tanneberg K, Turner S, Waite A, Zeldis J (2000) A mesoscale phytoplankton bloom in the polar Southern Ocean stimulated by iron fertilization. Nature 407:695–702

Bremner JM (1983) Biogenic sediments on the South West African (Namibian) continetal margin. In: Thiede J, Suess E (eds) Coastal upwelling: Its sedimentary record, Part B: Sedimentary records of ancient coastal upwelling. New York, Plenum Press, 73–103

Briffa KR, Jones PD, Bartholin TS, Eckstein D, Schweingruber FH, Karlén W, Zetterberg P, Eronen M (1992) Fennoscandian summers from A.D. 500: temperature changes on short and long time-scales. Climate Dynamics 7:111–119

Broecker WS (1991) The great ocean conveyor. Oceanogr 4:79–89

Browell EV (1993) TRACE-A airborne DIAL ozone and aerosol data. Hampton, VA, NASA/Langley Research Center

Burgess RL, Jacobson L (1984) Archaeological sediments from a shell midden near Wortel Dam, Walvis Bay, Southern Africa. Palaeoecology of Africa and the Surrounding Islands 16:429–435

Byrne DA, Gordon AL, Haxby WF (1995) Agulhas eddies: a synoptic view using Geosat ERM data. Journal of Physical Oceanography 25:902–917

Cachier H, Ducret J, Brémond M-P, Yoboué V, Lacaux J-P, Gaudichet A, Baudet J (1991) Biomass burning in a savanna region of the Ivory Coast. In: Levine JS (ed) Global biomass burning: Atmospheric, climatic and biospheric implications. Cambridge, Massachusetts, MIT Press, 174–180

Cachier H, Liousse C, Pertuisot M-H, Gaudichet A, Echalar F, Lacaux J-P (1996) African fire particulate emission and atmosphric influence. In: Levine JS (ed) Biomass burning and global change. Cambridge, Massachusetts, MIT Press, 428–440

Cahoon RC, Stocks BJ, Levine JS, Cofer WR, O'Niel KP (1992) Seasonal distribution of African savanna fires. Nature 357:812–815

Campbell B (ed) (1996) The Miombo in transition: Woodlands and welfare in Africa. Bogor, CIFOR

Carter TR, Hulme M, Crossley JF, Malyshev S, New MG, Schlesinger ME, Tuomenvirta H (2000) Climate change in the 21st century – interim characterizations based on the new IPCC emission scenarios. Helsinki, Finnish Environment Institute

Chambers R, Conway G (1992) Sustainable rural livelihoods: Practices and concepts for the 21st century. IDS Discussion Paper 296, Institute of Development Studies, University of Sussex

Chanda R (1996) Human perceptions of environmental degradation in a part of the Kahalari ecosystem. GeoJournal 39:65–71

Chanda R, Setshogo M, Skarpe C, Tacheba G, Totolo O, Arntzen J (1999) The state of the rangelands in the Matsheng area, Kgalagadi North, Botswana. Draft Research Report, Global Change and Subsistence Rangelands in Southern Africa Project. Global Change and Subsistence Rangelands in Southern Africa: resource variability, access and use in relation to rural livelyhoods and welfare (*http:/www.cazs.bangor.ac.uk/rangeland*)

Cheney RA, Marsh JG, Beckley BD (1983) Global mesoscale variability from collinear tracks of SEASAT altimeter data. Journal of Geophysical Research 88:4343–4354

Clarke J, Cavendish W, Coote C (1996) Rural households and miombo woodlands: use, value and management. In: Campbell B (ed) The Miombo in Transition: Woodlands and Welfare in Africa. Bogor, CIFOR:101–136

CLIMAP (1981) Seasonal reconstructions of the earth's surface at the Last Glacial Maximum. Geological Society of America Map and Chart MC-36

Cockcroft MJ, Wilkinson MJ, Tyson PD (1987) The application of a present-day climatic model to the late Quaternary in southern Africa. Climatic Change 10:161–181

Cohen AL, Parkington JE, Brundritt GB, Merwe NJ van der (1992) A Holocene marine climate record in mollusc shells from the southwest African coast. Journal of Quaternary Research 38:379–385

Conrad R, Seiler W (1982) Arid soils as a source of atmospheric carbon monoxide. Geophysical Research Letters 9:1353–1356

Conrad R, Seiler W (1985a) Influence of temperature, moisture and organic carbon on the flux of H_2O and carbon between soil and atmosphere: field studies in subtropical regions. Journal of Geophysical Research 90:5699–5709

Conrad R, Seiler W (1985b) Characteristics of a biological carbon monoxide formation from soil organic matter, humic acids and phenolic compounds. Environmental Science and Technology 19:1165–1176

Cooper DJ, Watson AJ, Nightingale PD (1996) Large decrease in ocean-surface CO_2 fugacity in response to in situ iron fertilization. *Nature* 383:511–513
CORDIO (1999) Coral reef degradation in the Indian Ocean. Status Report and Project Presentations. Stockholm, Stockholm University, SAREC Marine Science Programme
Cosijn C, Tyson PD (1996) Stable discontinuities in the atmosphere over South Africa. South African Journal of Science 92:381–386
Cousins B (1998) Invisible capital: The contribution of communal rangelands to rural livelihoods in South Africa. In: Bruyn TD de, Scogings PF (eds) Communal rangelands in Southern Africa: A synthesis of knowledge. Alice, Department of Livestock and Pasture Science, University of Fort Hare, 16–29
Crawford RJM (1998) Long term variability in populations of forage fish in the Benguela system – information from predators. International Symposium on Environmental Variability in the Benguela Current. Swakopmund, Ministry of Fisheries and Marine Research
Crimp SJ, Lutjeharms JRE, Mason SJ (1998) Sensitivity of a tropical-temperate trough to sea-surface temperature anomalies in the Agulhas retroflection region. Water SA 24:93–101
Cronberg G, Gieske A, Martins E, Prince Nengu J, Stenstrom I-M (1995) Hydrobiological studies of the Okavango Delta and the Kwando/Linyati/Chobe River, Botswana. I Surface water quality analysis. Botswana Notes and Records 27:151–226
Crutzen PJ (1995) Overview of tropospheric chemistry: Developments during the past quarter century and a look ahead. Faraday Discuss 100:1–21
Dahlberg AC (1996) Interpretations of environmental change and diversity: a study from North-East district, Botswana. Unpublished PhD thesis, Department of Physical Geography, Stockholm University, Sweden
D'Almeida GA (1987) Desert aerosol characteristics and effects on climate. In: Leinen M, Sarnthein M (eds) Palaeoclimatology and palaeometeorology: Modern and past patterns of global atmospheric transport. Dordecht, Kluwer Academic Publishers, 909 pp
Davies S (1996) Adaptable livelihoods: Strategic adaptation to food insecurity in the Malian Sahel. Basingstoke, Macmillian
Davies S (2000) Effective drought mitigation. In: Whilhite DA (ed) Drought, Volume II, A global assessment. London, Routledge
De Bruyn TD, Scogings P (1998) Communal rangelands in Southern Africa: A synthesis of knowledge. Proceedings of a symposium on policy-making for the sustainable use of southern African communal rangelands. Alice, Department of Livestock and Pasture Science, University of Fort Hare
Delmas RA, Marenco A, Tathy JP, Cros B, Baudet JGR (1991) Sources and sinks of methane in the African savanna. CH_4 emissions from biomass burning. Journal of Geophysical Research 96: 7287–7299
Desanker P, Magadza C (2000) Chapter 10 of draft IPCC WG2 third assessment. Cambridge, Cambridge University Press
Downie C (1964) Glaciations of Mount Kilimanjaro, North-East Tanganyika. Bulletin of the American Geological Society 75:1–16
Downing TE, Gitu KW, Crispen MK (eds) (1989) Coping with drought in Kenya. Boulder and London, Lynne Rienner
Duce RA, Liss PL, Merrill JT, Atlas EL, Buat-Menard P, Hicks BB, Miller JM, Prospero JM, Arimoto R, Church TM, Ellis W, Galloway JN, Hansen L, Jickels TD, Knap AH, Reinhardt KH, Scheider B, Soudine A, Tokos JJ, Tsunogai S, Wollast R, Zhou M (1991) The atmospheric input of trace species to the world ocean. Global Biogeochemical Cycles 5:193–259
Duncombe Rae CM (1991) Agulhas retroflection rings in the South Atlantic Ocean; an overview. South African Journal of Marine Science 11:327–344
Duncombe Rae CM, Shillington FA, Agenbag JJ, Taunton-Clark J, Gründlingh ML (1992) An Agulhas ring in the South East Atlantic Ocean and its interaction with the Benguela upwelling frontal system. Deep Sea Research 39:2009–2027
Du Plessis HM, Veelen M van (1991) Water quality: Salinisation and eutrophication time series and trends in South Africa. South African Journal of Science 87:11–16
East R (1984) Rainfall, soil nutrient states and biomass of large savanna animals. African Journal of Ecology 22:245–270
Ellery WN, Scholes RJ, Mentis MT (1991) An initial approach to predicting the sensitivity of the South African grassland biome to climatic change. South African Journal of Science 87:499–503
European Space Agency (1999) World fire atlas. *http://www.shark1.esrin.esa.it/FIRE/fire.html*
Fairbanks DHK, Scholes RJ (1999) South African country study on climate change. Vulnerability and adaptation assessment for plantation forestry. ENV-P-C 99013. Pretoria: Environmentek, CSIR
FAO (1995) Irrigation in Africa in figures. Water Reports 7. Rome, Food and Agriculture Organisation
Finney BP, Johnson TC (1991) Sedimentation in Lake Malawi (East Africa) during the past 10 000 years: a continuous palaeoclimate record from the southern tropics. Palaeogeography Palaeoclimatology and Palaeoecology 85:351–366
Finney BP, Scholz CA, Johnson TC, Trumbore S, Southon J (1996) Late Quaternary lake-level changes of Lake Malawi. In: Johnson TC, Odada E (eds) The limnology, climatology and palaeoclimatology of the East African Lakes. Toronto, Gordon and Beach, 495–508
Fishman J (1991) Probing planetary pollution from space. Environmental Science and Technology 25:612–621
Fishman J, Fakhruzzaman K, Cros B, Mganga D (1991) Identification of widespread pollution in the southern hemisphere deduced from satellite analyses. Science 252:1693–1696
Foley JA, Prentice IC, Ramankutty N, Levis S, Pollard D, Sitch S, Haxeltine A (1996) An integrated biosphere model of land surface processes, terrestrial carbon balance and vegetation dynamics. Global Biogeochemical Cycles 10:603–628
Freeman C (1984) Drought and agricultural decline in Bophuthatswana. In: South African Research Services (eds) South African review II. Johannesburg, Ravan Press, 284–289
Freiman MT, Piketh SJ, Tyson PD, Mittermaier MP (1999) The long-range transport of aerosols and trace gases over South Africa. IUGG 99, Birmingham, England, 19–30 July, 1999, A214
Garstang M, Fitzjarrald DR (1999) Observations of surface to atmosphere interactions in the tropics. New York, Oxford University Press
Garstang M, Tyson PD, Swap R, Edwards M, Kallberg P, Lindesay JA (1996) Horizontal and vertical transport of air over southern Africa. Journal of Geophysical Research 101:23721–23727
Garstang M, Ellery WN, McCarthy TS, Scholes MC, Scholes RJ, Swap RJ, Tyson PD (1998) The contribution of aerosol- and water-borne nutrients to the Okavango Delta ecosystem, Botswana. South African Journal of Science 94:223–229
Garzoli SL, Goñi GJ, Mariano AJ, Olson DB (1997) Monitoring the upper southeastern Atlantic transports using altimeter data. Journal of Marine Research 55:453–481
Gasse F, Ledee, V, Massault M, Fontes JC (1989) Water-level fluctuations in phase with oceanic changes during the last glaciation and deglaciation. Nature 342:57–59
Gatebe CK, Tyson PD, Piketh S, Annegarn H, Helas G (1999) A seasonal air transport climatology for Kenya. Journal of Geophysical Research 104:14237–14244
Global Change and Subsistence Rangelands in Southern Africa (1999) Second annual report to the European Commission. Cape Town, National Botanical Institute, 18–28 (See also *http:/www.cazs.bangor.ac.uk/rangeland*)
Goni GJ, Garzoli SL, Roubicek AJ, Olson DB, Brown OB (1997) Agulhas ring dynamics from TOPEX/POSEIDON satellite altimeter data. Journal of Marine Research 55:861–883
Gordon AL (1985) Indian-Atlantic transfer of thermocline water at the Agulhas retroflection. Science 227:1030–1033
Gordon AL (1986) Inter-ocean exchange of thermocline water. Journal of Geophysical Research 91:5037–5046
Gordon AL, Haxby WF (1990) Agulhas eddies invade the South Atlantic – evidence from GEOSAT altimeter and shipboard conductivity-temperature-depth survey. Journal of Geophysical Research 95:3117–3125
Gregory GL, Fuelberg HE, Longmore SP, Anderson BE, Collins JE, Blake DR (1996) Chemical characteristics of tropospheric air over the tropical South Atlantic Ocean: relationship to airmass source and trajectory history. Journal of Geophysical Research 101:23957–23972

Grove AT (1996) African river discharges and lake levels in the twentieth century. In: Johnson TC, Odada E (eds) The limnology, climatology and palaeoclimatology of the East African Lakes. Toronto, Gordon and Beach, 95–100

Gründlingh ML (1983) On the course of the Agulhas Current. South African Geographical Journal 65:49–57

Guenther A, Hewitt CN, Erickson D, Fall R, Geron C, Graedel T, Harley P, Klinger L, Lerdau M, McKay WA, Pierce T, Scholes B, Steinbrecher R, Tallamraju R, Taylor J, Zimmerman P (1995) A global model of natural volatile organic compound emissions. Journal of Geophysical Research 100:8873–8892

Guenther A, Otter L, Zimmerman P, Greenberg J, Scholes R, Scholes M (1996) Biogenic hydrocarbon emissions from southern African savannas. Journal of Geophysical Research 101:25859–25865

Hall M (1984) Prehistoric farming in the Mfolozi and Hluhluwe valleys of southeast Africa: an archaeobotanical survey. Journal of Archaeological Science 11:223–235

Hamilton JE (1997) Palaeoclimates of the East African Rift Valley region. Unpublished MSc dissertation, University of the Witwatersrand, 85 pp

Harris GW, Wienhold FG, Zenker T (1996) Airborne observations of strong biogenic NO_x emissions from the Namibian savanna at the end of the dry season. Journal of Geophysical Research 101:23707–23712

Harrison MSJ (1983) The Southern Oscillation, zonal equatorial circulation cells and South African rainfall. Preprints of the First International Conference on Southern Hemisphere Meteorology, 302–305

Hastenrath S (1984) The glaciers of Equatorial East Africa. Dordrecht D, Reidel

Hastenrath S (1985) Climate and the circulation of the tropics. Dordrecht D, Reidel

Hastenrath S (1991) Climate dynamics of the tropics. Dordrecht, Kluwer Academic Publishers

Held G, Gore BJ, Surridge AD, Tosen GR, Turner CR, Walmsley RD (1996) Air pollution and its impacts on the South African Highveld. Cleveland, Johannesburg, Environmental Scientific Association

Hemminga MA (ed) (1995) Interlinkages between Eastern-African coastal ecosystems. Yerseke: Netherlands Institute of Ecology, Centre for Estuarine and Coastal Ecology

Herman JR, Bhartia PB, Torres O, Hsu C, Seftor C, Celarier E (1997) Global distribution of UV-absorbing aerosols from Nimbus-7/TOMS data. Journal of Geophysical Research 102:16749–16759

Herold CE, Pitman WV, Bailey AK, Taviv I (1996) Lower Vet River water quality situation analysis with special Reference to the OFS goldfields. WRC Report No 523/1/96. Pretoria, Water Research Commission

Hewitson BC (1999) Deriving regional climate scenarios from general circulation models. WRC Report 751/1/99. Pretoria, Water Research Commission

Hinrichs K-U, Schneider RR, Muller PJ, Rullkotter J (2000) A biomarker perspective on palaeoproductivity variations in two Late Quaternary sediment sections in the Southeast Atlantic Ocean. Organic Geochemistry, 30:341–366

Hoffman MT (1995) Environmental history and the desertification of the Karoo, South Africa. Giornale Botanico Italiano 129: 261–273

Hoffman MT, Todd S, Ntshona Z, Turner S (1999) Land degradation in South Africa, Final Report. Pretoria: Department of Environmental Affairs and Tourism

Holmgren K, Karlén W, Shaw P (1995) Palaeoclimatic significance of the stable isotope composition and petrology of a late Pleistocene stalagmite from Botswana. Quaternary Research 43:320–328

Holmgren K, Karlén W, Lauritzen SE, Lee-Thorp JA, Partridge TC, Piketh S, Repinski P, Stevenson C, Svanered O, Tyson PD (1999) A 3 000-year high-resolution stalagmite-based record of palaeoclimate for north-eastern South Africa. Holocene 9:295–309

Howard WR (1985) Late Quaternary southern Indian Ocean circulation. South African Journal of Science 81:253–254

Hudson DA (1998) Antarctic sea-ice extent, southern hemisphere circulation and South African rainfall. Unpublished PhD thesis, University of Cape Town, 308 pp

Huffman TN (1989) Ceramics, settlements and Late Iron Age migrations. African Archaeological Review 7:155–182

Huffman TN (1996) Archaeological evidence for climatic change during the last 2 000 years in southern Africa. Quaternary International 33:55–60

Huffman TN, Herbert RK (1994) New perspectives on Eastern Bantu. Azania 29–30:1–10

Hulme M (1996) Climate change and Southern Africa. Report to the World Wildlife Foundation. Norwich, Climatic Research Unit, University of East Anglia

Hulme M, Doherty R, Ngara T, New M (2001) African climate change: 1900–2100. Climate Research 17:145–168

Hutchings L (1992) Fish harvesting in a variable productive environment – searching for rules or searching for exceptions? South African Journal of Marine Science 12:297–318

Hutchings L, Hampton I (1998) Variability in fish resources in the southern Benguela region In: International Symposium on Environmental variability in the Benguela Current. Swakopmund, Ministry of Fisheries and Marine Research

Hutchins DG, Hutton LG, Hutton SM, Jones CR, Loenhart EP (1976) A summary of the geology, seismicity, geomorphology and hydrogeology of the Okavango Delta. Botswana Geological Survey, Bulletin 7

Hutchings L, Pitcher GC, Probyn TA, Bailey GW (1995) The chemical and biological consequences of coastal upwelling. In: Summerhayes CP, Emeis K-C, Angel MV, Smith RL, Zeitzschel B (eds) Upwelling in the ocean: modern processes and ancient records. Chichester, John Wiley & Sons, 65–82

IIED (1997) The hidden harvest. The value of wild resources in agricultural systems – A summary. London, International Institute for Environment and Development

Iliffe J (1995) Africans, the history of a continent. Cambridge, Cambridge University Press

Imbrie J, Hays JD, Martinson DG, McIntyre A, Mix A, Morely JJ, Pisias NG, Prell W, Shackleton NJ (1984) The orbital theory of Pleistocene climate: support from a revised chronology of the marine $\delta^{18}O$ record. In: Berger A, Imbrie J, Hays J, Kukla G, Saltzman B (eds) Milankovitch and climate, Pt. 1. NATO ASI Series, 269–305. Dordrecht, Reidel

IPCC (1996) Climate change 1995. In: Houghton JT, Miera Filho LG, Callander BA, Harris N, Kattenburg A, Maskell K (eds) The science of climate change. Cambridge, Cambridge University Press

Jackson SP (1961) Climatological atlas of Africa. CCTA/CSA, Nairobi, 55 plates

Johnson TC, Odada E (eds) (1996) The limnology, climatology and Palaeoclimatology of the East African lakes. Toronto, Gordon and Beach

Johnson TC, Halfman JD, Showers WJ (1991) Palaeoclimate of the past 4 000 years at Lake Turkana, Kenya, based on the isotopic composition of authigenic calcite. Palaeogeography, Palaeoclimatology and Palaeoecology 85:189–198

Johnson TC, Scholz CA, Talbot MR, Kelts K, Ssemanda I, McGill JW (1996) Late Pleistocene desiccation of Lake Victoria and rapid evolution of cichlid fishes. Science 273:1091–1093

Johnson TC, Chan Y, Beuning KRM, Kelts K, Ngobi G, Verschuren D (1998) Biogenic silica profiles in Holocene cores from Lake Victoria: implications for lake level history and initiation of the Victoria Nile. In: Lehman JT (ed) Environmental change and response in East African Lakes. Dordrecht, Kluwer Academic Publishers, 75–88

Joubert AM (1995) Simulations of southern African climate by early general circulation models. South African Journal of Science 91:85–91

Joubert AM, Kohler MO (1996) Projected temperature increases over southern Africa due to increasing levels of greenhouse gases and sulphate aerosols. South African Journal of Science 92:524–526

Joubert AM, Mason SJ, Galpin JS (1996) Droughts over southern Africa in a doubled-CO_2 climate. International Journal of Climatology 16:149–158

Joubert AM, Katzfey JJ, McGregor JL, Nguyen KC (1999) Simulating mid-summer climate over southern Africa using a nested regional climate model. Journal of Geophysical Research 107:19015–19025

Jousaumme S, Taylor KE (1995) Status of the Palaeoclimate Modelling Intercomparison Project (PMIP). Proceedings of the First International AMIP Scientific Conference 92:425–430
Jury M, Rouault M, Weeks S, Schorman M (1997) Atmospheric boundary-layer fluxes and structure across a land-sea transition zone in south-eastern Africa. Boundary Layer Meteorology 83:311–330
Jury MR, Majodina M (1997) Preliminary climatology of southern Africa extreme weather 1973–1992. Theoretical and Applied Climatolology 56:103–112
Jury MR, Valentine HR, Lutjeharms JRE (1993) Influence of the Agulhas Current on summer rainfall on the southeast coast of South Africa. Journal of Applied Meteorology 32:1282–1287
Kendall RL (1960) An ecological history of the Lake Victoria basin. Ecology Monograph 39:121–176
Kienzle SW, Lorentz SA, Schulze RE (1997) Hydrology and Water Quality of the Mgeni Catchment. Pretoria, Water Research Commission
Kinsey B, Burger K, Willem Ginning J (1998) Coping with drought in Zimbabwe: Survey evidence on responses of rural households to risk. World Development 26:89–110
Kirst GJ, Schneider RR, Muller PJ, Storch I von, Wefer G (1999) Late Quaternary temperature variability in the Benguela Current system derived from alkenones. Quaternary Research 52:92–103
Konig W, Sausen R, Sielman F (1993) Objective identification of cylones in GCM simulations. Journal of Climate 6:2217–2231
Kostianoy AG, Lutjeharms JRE (1999) Atmospheric effects in the Angola-Benguela frontal zone. Journal of Geophysical Research 104:20963–20970
Kunz RP, Schulze RE, Scholes RJ (1995) An approach to modelling spatial changes of plant carbon: nitrogen ratios in southern Africa in relation to anticipated global climatic change. Journal of Biogeography 22:401–408
Lancaster N (1981) Palaeoenvironmental implications of fixed dune systems in southern Africa. Palaeogeography Palaeoclimatology and Palaeoecology 33:327–346
Lancaster N (2000) Eolian deposits. In: Partridge TC, Maud RR (eds) The Cenozoic of Southern Africa. New York, Oxford University Press, 73–87
Landman WA, Mason SJ (1999) Change in the association between Indian Ocean Sea-Surface temperatures and summer rainfall over South Africa and Namibia. International Journal of Climatology 19:1477–1492
Leach M, Mearns R (1988) Beyond the woodfuel crisis: People, land and trees in Africa. London, Earthscan Publications
Leach M, Mearns R (1996) The lie of the land: Challenging received wisdom on the African environment. Portsmouth, Heinemann
LeCanut P, Andreae MO, Harris GW, Wienhold FG, Zenker T (1996) Aerosol optical properties over southern Africa during SAFARI-92. In: Levine JS (ed) Biomass burning and global change. Cambridge, MA, MIT Press, 441–459
Lee-Thorp JA, Talma AS (2000) Stable light isotopes and environments in the southern African Quaternary and late Pliocene. In: Partridge TC, Maud RR (eds) The Cenozoic of Southern Africa. New York, Oxford University Press, 236–251
Lee-Thorp AM, Rouault M, Lutjeharms JRE (1998) Cumulus cloud formation above the Agulhas Current. South African Journal of Science 94:351–354
Lee-Thorp AM, Rouault M, Lutjeharms JRE (1999) Moisture uptake in the boundary layer above the Agulhas Current: a case study. Journal of Geophysical Research 104:1423–1430
Lemoine F (1998) Changements de l'hydologie de surface de l'Ocean Austral en relation avec les variations de la circulation thermohaline au cours des deux derniers cycles climatiques. PhD thesis, University of Paris VI, Paris
Levine JS, Winstead EL, Parsons DAB, Scholes MC, Scholes RJ, Cofer WR III, Cahoon DR Jr., Sebacher DI (1996) Biogenic soil emissions of nitric oxide (NO) and nitrous oxide (N_2O) from savannas in South Africa: The impact of wetting and burning. Journal of Geophysical Research 101:23689–23697
Levine JS, Parsons DAB, Zepp RG, Burke RA, Cahoon DR Jr., Cofer WR III, Miller WR, Scholes MC, Scholes RJ, Sebacher S, Winstead EL (1997) Southern African grasslands as a source of atmospheric gases. In: Van Wilgen BW, Andreae MO, Goldammer JG, Lindesay JA (eds) Fire in the Southern African savanna: Ecological and environmental perspectives. Johannesburg, Witwatersrand University Press, 135–160
Lipton M, Ellis F, Litpton M (1996) Land, labour and livelihoods in rural South Africa. Volume Two: KwaZulu-Natal and Northern Province. Durban, Indicator Press
Lindesay JA (1998) Present climates of southern Africa. In: Hobbs JE, Lindesay JA, Bridgman HA (eds) Climates of the southern continents, present, past and future. Chicester, Wiley, 6–52
Livingstone DA (1980) Environmental changes in Nile headwaters. In: Williams M, Faure H (eds) The Sahara and the Nile. Rotterdam, Balkema, 339–359
Lutjeharms JRE (1987) Meridional heat transport across the subtropical convergence by a warm eddy. Nature 331:251–253
Lutjeharms JRE, Ansorge I (2001) The Agulhas Return Current. Journal of Marine Systems 30:115–138
Lutjeharms JRE (1996) The exchange of water between the South Indian and the South Atlantic. In: Wefer G, Berger WH, Siedler G, Webb D (eds) The South Atlantic: Present and past circulation. Berlin, Springer-Verlag, 125–162
Lutjeharms JRE, Ballegooyen RC van (1984) Topographic control in the Agulhas Current system. Deep Sea Research 31:1321–1337
Lutjeharms JRE, Ballegooyen RC van (1988b) Anomalous upstream retroflection in the Agulhas Current. Science 240:1770–1772
Lutjeharms JRE, Cooper J (1996) Interbasin leakage through Agulhas Current filaments. Deep Sea Research 43:213–238
Lutjeharms JRE, Gordon AL (1987) Shedding of an Agulhas ring observed at sea. Nature 325:138–140
Lutjeharms JRE, Meeuwis JM (1987) The extent and variability of the South East Atlantic upwelling. South African Journal of Marine Science 5:51–62
Lutjeharms JRE, Roberts HR (1988) The Natal Pulse; an extreme transient on the Agulhas Current. Journal of Geophysical Research 93:631–645
Lutjeharms JRE, Ruijter WPM de (1996) The influence of the Agulhas Current on the adjacent coastal zone: possible impacts of climate change. Journal of Marine Systems 7:321–336
Lutjeharms JRE, Stockton PL (1987) Kinematics of the upwelling front off Southern Africa. South African Journal of Marine Science 5:35–49
Lutjeharms JRE, Valentine HR (1987) Water types and volumetric considerations of the South East Atlantic upwelling. South African Journal of Marine Science 5:63–71
Lutjeharms JRE, Valentine HR (1988) Eddies at the sub-tropical convergence south of Africa. Journal of Physical Oceanography 18:761–774
Lutjeharms JRE, Mey RD, Hunter IT (1986) Cloud lines over the Agulhas Current. South African Journal of Science 82:635–640
Lutjeharms JRE, Shillington FA, Duncombe Rae CM (1991) Observations of extreme upwelling filaments in the South East Atlantic Ocean. Science 253:774–776
Lutjeharms JRE, Monteiro PMS, Tyson PD, Obura D (2001) The oceans around southern Africa and regional effects of global change. South African Journal of Science 97:119–130
Lyle MW, Prahl FG, Sparrow MA (1992) Upwelling and productivity changes inferred from a temperature record in the central equatorial Pacific. Nature 355:812–815
Maenhaut W, Salma I, Garstang M, Meixner F (1993) Size-fractionated atmospheric aerosol composition and aerosol sources at Etosha, Namibia and Victoria Falls, Zimbabwe. Paper presented at the 1993 AGU Fall Meeting, San Francisco, CA
Maggs T (1984) The Iron Age south of the Zambezi. In: Klein RG (ed) Southern African prehistory and palaeoenvironments. Rotterdam, Balkema, 329–360
Mahaney WC (1990) Ice on the equator. Ellison Bay, Caxton
Maley J (1993) The climate and vegetational history of the equatorial regions of Africa during the Upper Quaternary. In: Shaw T, Sinclair P, Barsey A, Okpoko A (eds) The archaeology of Africa. London, Routledge, 43–52
Marufu L, Ludwig J, Andreae MO, Lelieveld J, Helas G (1999) Spatial and temporal variation in biofuel consumption rates and patterns in Zimbabwe: implications for atmospheric trace gas emissions. Biomass and Bioenergy 16311–16332

Marufu L, Dentener F, Lelieveld J, Andreae MO, Helas G (2000) Photochemistry of the African troposphere: The influence of biomass burning emissions. Journal of Geophysical Research 105:14513–14546

Mason SJ (1995) Sea-surface temperature – South African rainfall associations, 1910–1989. International Journal of Climatology 15:119–135

Mason SJ (1996) Rainfall trends over the Lowveld of South Africa. Climatic Change 32:35–54

Mason SJ, Tyson PD (2000) The occurrence and predictability of droughts over southern Africa. In: Whilhite DA (ed) Drought, Volume 1, a global assessment. London, Routledge, 113–134

Mason SJ, Waylen PR, Mimmack GM, Rajaratnam B, Harrison MJ (1999) Changes in extreme rainfall events in South Africa. Climatic Change 41:249–257

Mather GK, Dixon MJ, Jager JM de (1996) Assessing the potential for rain augmentation – the Nelspruit randomized convective cloud seeding experiment. Journal of Applied Meteorology 35:1465–1482

Mather GK, Terblanche DE, Steffens FE, Fletcher L (1997a) Results of the South African cloud seeding experiment using hygroscopic flares. Journal of Applied Meteorology 36:1433–1447

Mather GK, Terblanche DE, Steffens FE (1997b) The National Precipitation Research Programme, Final Report 1993–1996. WRC Report 726/1/97. Pretoria, Water Research Commission, 147pp

May H, Hoffman T, Marinus T (1998) The case for the communally-managed reserves in semi-arid Namaqualand, South Africa, Global Change and Subsistence Rangelands in Southern Africa, Proceedings of the 1st Joint Workshop held in Gaborone, Botswana, 13–17th October, 1997, 41–56

McCarthy TS, Metcalfe J (1990) Chemical sedimentation in the Okavango Delta, Botswana. Chemical Geology 89:157–178

McClanahan TR, Obura D (1995) Status of Kenyan coral reefs. Coastal Management 23:57–76

McClelland L (1996) Climatic variability and outlooks for maize production in South Africa. Unpublished MSc thesis, University of the Witwatersrand

McNae W (1968) A general account of the flora and fauna of the mangrove forests in the Indo-Pacific region. Advances in Marine Biology 6:73–270

Meigh JR, McKenzie AA, Austin BN, Bradford RB, Reynard NS (1998) Assessment of global water resources – phase II. Estimates for present and future water availability for Eastern and Southern Africa. DFID Report 98/4. Wallingford, Institute of Hydrology

Meixner FX, Fickinger T, Marufu L, Serça D, Nathaus FJ, Makina E, Makurumbira L, Andreae MO (1997) Preliminary results on nitrous oxide emission from a southern African savanna ecosystem. Nutrient Recycling Agroecosystems 48:123–138

Mentis MT, Grossman D, Hardy MB, O'Connor TG, O'Reagain PJ (1989) Paradigm shifts in South African range science, management and adminstration. South African Journal of Science 85:684–687

MICOA (1998) The biological diversity of Mozambique. In: Hatton J, Munguambe F (eds) Impacto: projectos e Estudos Ambientais. Maputo, Ministreio da Coordenacao do Ambiente

Midgley GF, O'Callaghan M (1993) Review of likely impacts of climate change on South African flora and vegetation. Report prepared for the southern African Nature Foundation and the World Wildlife Fund

Midgley GF, Stock WD, Juritz JM (1995) The effects of elevated CO_2 on Cape Fynbos species adapted to soils of different nutrient status: Nutrient- and CO_2-responsiveness. Journal of Biogeography 22:185–191

Midgley GF, Wand SJE, Pammenter NW (1999) Nutrient and genotypic constraints on CO_2 responsiveness: Photosynthetic regulation in congenerics of a nutrient-poor environment. Journal of Experimental Botany 50:533–542

Midgley GF, Bond WJ, Wand SJE, Roberts R (2000) Will Gullivers travel? Potential causes of changes in savanna tree success due to rising atmospheric CO_2. In: Ringrose S, Chanda R (eds) Towards sustainable management in the Kalahari Region. Gaborone: Directorate of Research and Development, University of Botswana

Mokitimi N, Mokuku C, Prasad G, Quinlan T, Letsela T (1998) Subsistence rangelands in Lesotho, global change and subsistence rangelands in Southern Africa, Proceedings of the 1st Joint Workshop held in Gaborone, Botswana, 13–17th October, 1997, 46–50

Monteiro PMS (1997) The oceanography, the biogeochemistry and the fluxes of carbon dioxide in the Benguela upwelling system, Unpublished PhD Thesis, University of Cape Town

Monteiro PMS, James AG, Sholto-Douglas AD, Field JG (1991) The $\delta^{13}C$ trophic position spectrum as a tool to define and quantify carbon pathways in marine food webs. Marine Ecology Progress Series 78:33–40

Monteiro PMS, Nelson G, Probyn TA, Field JG (2000) The spatial and temporal variability of particulate organic carbon (POC) export fluxes in the Benguela upwelling system: a box model approach. In preparation

Moody JL, Pszenny AAP, Grandy A, Keene WC, Galloway JN, Polian G (1991) Precipitation composition and its variability in the southern Indian Ocean: Amsterdam Island, 1980–1987. Journal of Geophysical Research 96:20769–20786

Mooney HA, Canadell J, Chapin FS, Ehleringer JR, Korner C, McMurtrie RE, Parton WJ, Pitelka LF, Schulze E-D (1999) Ecosystem physiology responses to global change. In: Walker BH, Steffen W, Canadell J, Ingram J (eds) The terrestrial biosphere and global change. Cambridge, Cambridge University Press, 190–228

Moorsom R, Franz J, Mupotola M (eds) (1995) Coping with aridity: drought impacts and preparedness in Namibia. Windhoek, Brandes and Apsel Verlag

Morris JW, Bezuidenhout JJ, Furniss PR (1982) Litter decomposition. In: Huntley BJ, Walker BH (eds) Ecology of tropical savannas. Berlin, Springer-Verlag, 535–553

Muller PJ, Cepek M, Ruhland G, Schneider RR (1997) Alkenone and coccolithophorid species changes in late Quaternary sediments from the Walvis Ridge: Implications for the alkenone temperature method. Palaeogeography, Palaeoclimatology and Palaeoecology 135:71–96

Murphy JM, Mitchell JBF (1995) Transient response of the Hadley centre coupled ocean-atmosphere model to increasing carbon dioxide. Part 2. Spatial and temporal structure of the response. Journal of Climate 8:57–80

NASA (20001) *http://www.earthobservatory.nasa.gov/*

Newell RE, Kidson JW, Vincent DG, Boer GJ (1972) The general circulation of the tropical atmosphere and interactions with extratropical latitudes, Vol. 1. Cambridge, Mass., MIT Press

Nicholson SE (1986) The nature of rainfall variability in Africa south of the equator. Journal of Climatology 6:515–530

Nicholson SE (1993) An overview of African rainfall fluctuations of the last decade. Journal of Climate 6:1463–1466

Nicholson SE, Entekhabi D (1986) The quasi-periodic behaviour of rainfall variability in Africa and its relationship to the Southern Oscillation. Archiv für Meteorologie, Geophysik und Bioklimatologie 34A:311–348

Niiler PP (1992) The ocean circulation. In: Trenberth K (ed) Climate system modelling. Cambridge, Cambridge University Press, 117–148

NOAA (1999) The ocean's role in climate variability and change and the resulting impacts on coasts. Natural Resources Forum 23:123–134

Odada EO (1996) Coastal change: Its implications for eastern Africa. In: Duursma E (ed) IOC Workshop Report No. 105. Paris, Unesco, 265–267

Odada E, Totolo O, Stafford-Smith M, Ingram J (1996) Global change and susbistence rangelands in southern Africa: the impact of livelihood variability and resource access on rural livelihoods. GCTE Working Document 20. Stockholm, IGBP

Ogallo LJ (1988) Relationships between seasonal rainfall in east Africa and the Southern Oscillation. Journal of Climatology 8:31–43

Osmaston HA (1989) Glaciers, glaciations and ELAs on the Ruwenzori. In: Mahaney WC (ed) Quaternary and environmental research on East African mountains. Rotterdam, Balkema, 31–104

Otter LB, Marufu L, Scholes MC (2001) Biogenic and biofuel sources of trace gases in southern Africa. South African Journal of Science 97:131–138

Owen RB, Crossley R, Johnson TC, Tweddle D, Kornfield I, Davison S, Eccles DH, Engstrom DE (1990) Major low levels of Lake Malawi and implications for speciation rates in cichlid fishes. Proceedings of the Royal Society of London B240:519–553

Parsons DAB, Scholes MC, Scholes RJ, Levine JS (1996) Biogenic NO emissions from savanna soils as a function of fire regime, soil type, soil nitrogen and water status. Journal of Geophysical Research 101:23683–23688

Partridge TC (1993) The evidence for Cainozoic aridification in southern Africa. Quaternary International 17:105–110

Partridge TC (1997) Cainozoic environmental change in southern Africa, with special emphasis on the last 200 000 years. Progress in Physical Geography 21:3–22

Partridge TC (1998) The sedimentary record and its implications for rainfall fluctuations in the past. In: Partridge TC (ed) Tswaing: Investigations into the origin and palaeoenvironments of the Pretoria Saltpan Crater. Memoir 85, South African Council for Geoscience, 127–142

Partridge TC, Avery M, Botha GA, Brink JS, Deacon J, Herbert RS, Maud RR, Scott L, Talma AS, Vogel JC (1990) Late Pleistocene and Holocene climatic changes in southern Africa. South African Journal of Science 86:302–306

Partridge TC, Demenocal PB, Lorentz SA, Paiker MJ, Vogel JC (1997) Orbital forcing of climate over South Africa: a 2 000-year rainfall record from the Pretoria Saltpan. Quaternary Science Review 16:1125–1133

Partridge TC, Scott L, Hamilton JE (1999) Synthetic reconstructions of southern African environments during the Last Glacial Maximum (21–18 kyr) and the Holocene Altithermal (8–6 kyr). Quaternary International 57/58:207–214

Perkins JS, Thomas DSG (1993) Spreading deserts of spatially confirmed environmental impacts? Land degradation and cattle ranching in the Kalahari Desert of Botswana. Land Degradation and Rehabilitation 4:179–194

Perks LA, Schulze RE (2000) Assessment of the impact of climate change on agriculture in South Africa. ACRUcons Report 34. Pietermaritzburg, School of Bioresources Engineering and Environmental Hydrology, University of Natal

Perks LA, Schulze RE, Kiker GA, Horan MJC, Maharaj M (2000) Preparation of climate data and information for application in impact studies of climate change over southern Africa. ACRUcons Report 32. Pietermaritzburg, School of Bioresources Engineering and Environmental Hydrology, University of Natal, Pietrmaritzburg

Pethick J (1991) Marshes, mangroves and sea level rise. Geography 76:1–5

Petit JR, et al. (1999) Climate and atmospheric history of the past 420 000 years from the Vostok ice core, Antarctica. Nature 399: 429–436

Pike LD, Rimmington GT (1965) Malawi, a geographical study. Oxford, Oxford University Press

Piketh SJ (2000) Transport of aerosols and trace gases over southern Africa. Unpublished PhD Thesis. Johannesburg, University of the Witwatersrand

Piketh SJ, Formenti P, Swap RJ, Anderson CA, Tyson PD, Annegarn HJ, Maenhaut W (1996) Identification of industrial aerosols at a remote site in South Africa. Proceedings of the AGU Fall Meeting, San Fransisco, CA. American Geophysical Union, A12B-7:77

Piketh SJ, Annegarn HJ, Tyson PD (1999a) Lower-tropospheric aerosol loadings over South Africa: The relative impacts of aeolian dust, industrial emissions and biomass burning. Journal of Geophysical Research 104:1597–1607

Piketh SJ, Freiman MT, Tyson PD, Annegarn HJ, Helas G (1999b) Transport of aerosol plumes in ribbon-like structures over southern Africa. IUGG 99, Birmingham, England, 19–30 July 1999, A215

Piketh SJ, Tyson PD, Steffen W (2000) Aeolian transport from southern Africa and iron fertilisation of marine biota in the South Indian Ocean, South African Journal of Science, 96:244–246

Pitcher GC, Boyd AJ (1996) Across-shelf and alongshore dinoflag-ellate distributions and the mechanisms of red tide formation within the southern Benguela upwelling system. In: Yasumoto T, Oshima Y, Fukuyo Y (eds) Harmful and toxic algal blooms. Intergovernmental Oceanographic Commission. Paris, UNESCO, 243–246

Pitcher GC, Cockroft A (1998) Low oxygen, rock lobster strandings and PSP. In: Wyatt T (ed) Harmful algal news. IOC – UNESCO 17:1–3

Pitcher GC, Calder D, Davis, G and Nelson G (1996) Harmful algal blooms and mussel farming in Saldanha Bay. Proceedings of the Aquaculture Association of Southern Africa 5:87–93

Pittock AB (1999) Coral reefs and environmental change: adaptation to what? American Zoologist 39:10–29

Pollard RT, Smythe-Wright D (1996) Understanding ocean circulation. UK-WOCE – The First Six Years, NERC

Poolman E (1999) Heavy rain events over South Africa. Paper presented at the 15th Annual Conference of the South African Society for Atmospheric Sciences, Richards Bay, 18–19 November 1999

Poth MA, Anderson IC, Miranda AC, Riggan PJ (1995) The magnitude and persistence of soil NO, N_2O, CH_4 and CO_2 fluxes from burned tropical savanna in Brazil. Biogeochemical Cycles 9:503–513

Potter CS, Matson PA, Vitousek PM, Davidson EA (1996) Process modelling of controls on nitrogen trace gas emissions from soils worldwide. Journal of Geophysical Research 101:1361–1377

Preston-Whyte RA, Tyson PD (1973) Note on pressure oscillations over South Africa. Monthly Weather Review 101:650–653

Quartly GD, Srokosz MA (1993) Seasonal variations in the region of the Agulhas retroflection: studies with Geosat and FRAM. Journal of Physical Oceanography 23:2107–2124

Rayner PJ, Law RM (1995) A comparison of modelled responses to prescribed CO_2 sources. CSIRO Australia, Division of Atmospheric Research, Technical Paper No 36

Reynard NS (1996) Changes in surface water availability. In: Hulme M (ed) Climate change and Southern Africa: An Exploration of Some Potential Impacts and Implications in the SADC Region. Norwich, Climatic Research Unit, University of East Anglia, 31–39

Richardson AJ, Verheye HM, Field JG (1998) Are pelagic fish food limited on their spawning grounds in South Africa? In: International workshop on zoo-ichthyoplankton monitoring in the Benguela Current ecosystem off Namibia, Ministry of Fisheries and Marine Research, Swakopmund, Namibia

Rind D (1998) Latitudinal temperature gradients and climatic change. Journal of Geophysical Research 103:5943–5971

Rind D, Overpeck J (1993) Hypothesized causes of decadal-to-century-scale climate variability: climate model results. Quaternary Science Review 12:357–374

Ringrose S, Matheson W (1991) A landsat analysis of range conditions in the Botswana Kalahari drought. International Journal of Remote Sensing 12:1023–1051

Ringrose S, Sefe F, Chanda R, Musisi-Mkambwe S (1996) Environmental change in the mid-Boteti area of north-central Botswana: Biophysical processes and human perceptions. Environmental Management 23:27–39

Robertson L, Langer J, Engardt M (1996) MATCH – Mesoscale atmospheric transport and chemistry modelling system, basic transport model description and control experiments with ^{222}Rn. Swedish Meteorological and Hydrological Institute, Report SMHI RMK, No. 80

Rorich RP, Turner CR (1994) Ambient monitoring network annual data report for 1993 and regional trend analysis. Report TRR/S94/059/rw, TRI, ESKOM, 139 pp

Rosqvist G (1990) Quaternary glaciations in Africa. Quaternary Science Review 9:281–297

Ross KE (1999) Long-term trends in water quality, the possible influence of global climate change, with reference to six contrasting drainage basins in South Africa. Unpublished paper. Johannesburg, Department of Geography and Environmental Studies, University of the Witwatersrand

Rouault M, Lutjeharms JRE (2001) Air-sea exchange over an Agulhas eddy at the Subtropical Convergence. Global Atmosphere and Ocean Systems 7:125–150

Rouault M, Lee-Thorp AM, Lutjeharms JRE (2000) Observations of the atmospheric boundary layer above the Agulhas Current during alongcurrent winds. Journal of Physical Oceanography 20:30–50

Rutherford MC, Westfall RH (1986) The biomes of southern Africa – an objective categorization. Memoirs of the Botanical Survey of South Africa 54:1–98

Rutherford MC, Midgley GF, Bond WJ (2000) South African country study on climate change, impacts and adaptation. Pretoria, Department of Environment Affairs and Tourism
Scarpe C (1986) Plant community structure in relation to grazing and environmental change along a north-south transect in the western Kahalari. Vegetatio 68:3–18
Schade GW, Hoffmann R-M, Crutzen PJ (1999) CO emissions from degraded plant matter: estimates of global source stenght. Tellus B 51:909–918
Scholes MC, Andreae MO (2000) Biogenic and pyrogenic emissions from Africa and their impact on the global atmosphere. Ambio 29:23–29
Scholes MC, Martin R, Scholes RJ, Parsons D, Winstead E (1997) NO and N_2O emissions from savanna soils following the first simulated rains of the season. Nutrient Cycling Agroecosystems 48:115–122
Scholes RJ (1993) Nutrient cycling in semi-arid grasslands and savannas: Its influence on pattern, productivity and stability. Proceedings of the XVII International Grassland Congress. Palmerston North, International Grasslands Society:1331–1334
Scholes RJ, Walker BH (1993) An African savanna: Synthesis of the Nylsvley Study. Cambridge, Cambridge University Press
Scholes RJ, Ward D, Justice CO (1996a) Emissions of trace gases and aerosol particles due to vegetation burning in southern-hemisphere Africa. Journal of Geophysical Research 101: 23677–23682
Scholes RJ, Kendall J, Justice CO (1996b) The quantity of biomass burned in southern Africa. Journal of Geophysical Research 101:23667–23676
Scholes RJ, Midgley GF, Wand SJE (1999) The vulnerability and adaptation of rangelands. South African Climate Change Country Studies Contract Report. Pretoria, Division of Water, Environment and Forest Technology, CSIR
Schouten MW, Ruijter WPM de, Leeuwen PJ van, Lutjeharms JRE (2000) Translation, decay and splitting of Agulhas rings in the south-eastern Atlantic Ocean. Journal of Geophysical Research 105:21913–21925
Schulze RE (1995) Hydrology and agrohydrology. Pretoria, Water Research Commission
Schulze RE (1997a) Impacts of global climate change in a hydrologically vulnerable region: Challenges to South African hydrologists. Progress in Physical Geography 21:113–136
Schulze RE (1997b) South African atlas of agrohydrology and agroclimatology. WRC Report TT82/96. Pretoria, Water Research Commission
Schulze RE (2000) Modelling hydrological responses to landuse and climatic change: A southern African perspective. Ambio 29:12–22
Schulze RE, Perks LA (2000) Assessment of the impact of climate change on hydrology and water resources in South Africa. Report to the South African Country Studies for Climate Change Programme, ACRUcons Report 33. Pietermaritzburg, School of Bioresources Engineering and Environmental Hydrology, University of Natal
Schulze RE, Kiker GA, Kunz RP (1995) Global climate change and agricultural productivity in southern Africa. Global Environmental Change 3:330–349
Scoones I (1995) Living with uncertaintity: New directions in pastoral development in Africa. London, Intermediate Technology Publications
Scoones I (1996) Crop production in a variable environment: a case study from southern Zimbabwe. Experimental Agriculture 32:291–303
Scott L, Thackeray JF (1987) Multivariate analysis of late Pleistocene and Holocene pollen spectra from Wonderkrater, Transvaal, South Africa. South African Journal of Science 83:93–98
Sea Trust (1997) A guide to the seashores of eastern Africa and the western Indian Ocean islands. Zanzibar, Western Indian Ocean Marine Science Association
Seely MK, Jacobson KM (1994) Desertification and Namibia: a perspective. Journal of African Zoology 108:21–36
Seiler W, Conrad R (1987) Contribution of tropical ecosystems to the global budgets of trace gases, especially CH_4, H_2, CO and N_2O. In: Dickinson RE (ed) The geophysiology of Amazonia. New York, John Wiley
Serça D, Delmas R, Le Roux X, Parsons DAB, Scholes MC, Abbadie L, Lensi R, Ronce O, Labroue L (1998) Comparison of nitrogen monoxide emissions from several African tropical ecosystems and influence of season and fire. Global Biogeochemical Cycles 12:637–651
Shackleton CM (1993) Are the communal grazing lands in need of saving? Development Southern Africa 10:65–78
Shackelton CM (1996) Potential stimulation of local rural economies by harvesting secondary products: a case study of the central eastern Transvaal lowveld. Ambio 25:33–38
Shackelton CM, Shakelton S (1997) The use and potential for commercialisation of Veld products in the Bushbuckridge Area, Nelspruit. Unpublished report, DANCED Community Forest Project. Nelspruit: Department of Water Affairs and Forestry
Shannon DA (2000) Climate-biosphere feedbacks over southern Africa. Unpublished PhD thesis, University of Cape Town
Shannon LV (1985) The Benguela ecosystem. 1. Evolution of the Benguela, physical features and processes. Oceanography and Marine Biology Annual Review 23:105–182
Shannon LV, Nelson G (1996) The Benguela: large scale features and processes and system variability. In: Wefer G, Berger WH, Siedler G, Webb D (eds) The South Atlantic: Present and past circulation. Berlin, Springer-Verlag, 163–210
Shannon LV, Agenbag JJ, Buys MEL (1986a) Large- and meso-scale features of the Angola-Benguela front. South African Journal of Marine Science 5:11–34
Shannon LV, Boyd AJ, Brundrit GB, Taunton-Clark J (1986b) On the existence of an El Niño-type phenomenon in the Benguela system. Journal of Marine Research 44:495–520
Shannon LV, Crawford RJM, Pollock D, Hutchings L, Boyd AJ, Taunton-Clark J, Badenhorst A, Melville-Smith R, Augustyn CJ, Cochrane KL, Hampton I, Nelson G, Japp SW, Tarr RJQ (1992) The 1980s – a decade of change in the Benguela ecosystem. South African Journal of Marine Science 12:271–296
Smayda T (1998) Ecological features of harmful algal blooms in coastal upwelling systems. In: International symposium and workshop on harmful algal blooms in the Benguela Current and other upwelling ecosystems. Swakopmund, Ministry of Fisheries and Marine Research
Smith JMB (1994) Crop, pasture and timber yield index. Cedara Report, N/A/94/4. Cedara, Natal Agricultural Research Institute
Snedaker SC (1995) Mangroves and climate change in the Florida and Caribbean region: Scenarios and hypotheses. Hydrobiologia 295:43–49
Sonzogni C, Bard E, Rostek F (1998) Tropical sea-surface temperatures during the Last Glacial period: a view based on alkenones in Indian Ocean sediments. Quaternary Science Review 17: 1185–1201
Stokes S, Thomas DSG, Shaw PA (1997a) New chronological evidence for the nature and timing of linear dune development in the southwest Kalahari. Geomorphology 20:81–93
Stokes S, Thomas DSG, Washington R (1997b) Multiple episodes of aridity in southern Africa since the last Interglacial. Nature 388:154–158
Stokes S, Washington R, Preston A (1998) Late quaternary evolution of the central and southern Kalahari: environmental responses to changing climatic conditions. In: Andrews P, Banham P (eds) Late Cenozoic Environments and Hominid Evolution; a tribute to Bill Bishop. Geological Society, London, 247–268
Stramma L, Lutjeharms JRE (1997) The flow field of the subtropical gyre of the South Indian Ocean. Journal of Geophysical Research 102:5513–5530
Sturman AS, Tyson PD, D'Abreton PC (1997) Transport of air from Africa and Australia to New Zealand. Journal of the Royal Society of New Zealand 27:485–498
Stute M, Talma AS (1998) Glacial temperatures and moisture transport regimes reconstructed from noble gas and $\delta^{18}O$, Stampriet aquifer, Namibia. In: Proceedings of the International Symposium on Isotope Techniques in the Study of Past and Current Environmental Changes in the Hydrosphere and Atmosphere. IAEA, Vienna, April 1997, 307
Subak S, Raskin P, Hippel D von (1993) National greenhouse gas accounts: current anthropogenic sources and sinks. Climatic Change 25:15–58

Summerhayes CP, Kroon D, Rosell-Mele A, Jordan RW, Schrader, H-J, Hearn R, Villanueva J, Grimalt JO, Eglington G (1995) Variability in the Benguela Current upwelling system over the past 70 000 years. Progress in Physical Oceanography 35:207–252
Swap RJ (1996) Transport and impact of southern African aerosols. Unpublished PhD dissertation. Charlottesville, University of Virginia
Takahashi T, et al. (1997) Global air-sea flux of CO_2: an estimate based on measurements of sea-air pCO_2 difference. *Proceedings of the National Academy of Science* 94:8292–8299
Taljaard JJ (1981) Upper air circulation, temperature and humidity over South Africa. Pretoria, South African Weather Bureau, Technical Paper, No. 10
Taljaard JF, Zunckel M (1992) Comparative study of pollutants in the Eastern Transvaal. Pretoria, EMATEK, CSIR
Talma AS, Vogel JC (1992) Late Quaternary palaeotemperatures derived from a speleothem from Cango Caves, Cape Province, South Africa. Quaternary Research 37:203–213
Thackeray JF (1987) Late Quaternary environmental changes inferred from small mammalian fauna, southern Africa. Climatic Change 10:1–21
Thackeray JF (1988) Molluscan fauna from Klasies River Mouth, South Africa. South African Archaeological Bulletin 43:27–32
Thackeray JF (1990) Temperature indices from late Quaternary sequences in South Africa: comparisons with the Vostok core. South African Geographical Journal 72:47–49
Thackeray JF (1992) Chronology of late Pleistocene deposits associated with *Homo sapiens* at Klasies River Mouth, South Africa. Palaeoecology of Africa 23:177–191
Thackeray JF, Avery DM (1990) A comparison between temperature indices for late Pleistocene sequences at Klasies River Mouth and Border Cave. Palaeoecology of Africa 21:311–316
Thomas DSG, Sporton D (1997) Understanding the dynamics of social and environmental variability. Applied Geography 17:11–27
Thompson BW (1965) The Climate of Africa. Nairobi, Oxford, 131 maps
Todd SW, Hoffman M (1998) Fenced in and nowhere to go, the story of Namqualand's communal rangelands. Veld and Flora 84:84–85
Todd SW, Hoffman MT (1999) A fence-line contrast reveals effects of heavy grazing on plant diversity and community composition in Namaqualand, South Africa. Plant Ecology 142:169–178
Tyson PD (1986) Climatic Change and Variability in Southern Africa. Cape town, Oxford University Press
Tyson PD (1997) Atmospheric transport of aerosols and trace gases over southern Africa. Progress in Physical Geography 21:79–101
Tyson PD, D'Abreton PC (1998) Transport and recirculation of aerosols off southern Africa: macroscale plume structure. Atmospheric Environment 32:1511–1524
Tyson PD, Partridge TC (2000) Evolution of Cenozoic climates. In: Partridge TC, Maud RR (eds) The Cenozoic of Southern Africa. New York, Oxford University Press
Tyson PD, Preston-Whyte RA (2000) The Weather and Climate of Southern Africa. Cape Town, Oxford University Press
Tyson PD, Du Toit WJF, Fuggle RF (1972) Temperature structure above cities: review and preliminary findings from the Johannesburg Urban Heat Island Project. Atmospheric Environment 6:533–542
Tyson PD, Dyer TGJ, Mametse MN (1975) Secular changes in South African rainfall: 1880 to 1972. Quarterly Journal of the Royal Meteorological Society 101:817–833
Tyson PD, Garstang M, Swap R, Kållberg P, Edwards M (1996a) An air transport climatology for subtropical southern Africa. International Journal of Climatology 16:265–291
Tyson PD, Garstang M, Swap R (1996b) Large-scale recirculation of air over southern Africa. Journal of Applied Meteorology 35:2218–2236
Tyson PD, Gasse F, Bergonzini L, D'Abreton PC (1997) Aerosols, atmospheric transmissivity and hydrological modelling of climatic change over Africa south of the equator. International Journal of Climatology 17:1651–1665
Tyson PD, Mason SJ, Jones MQW, Cooper GRJ (1998) Global warming in South Africa: evidence from geothermal profiles. Geophysical Research Letters, 25:2711–2713
Tyson PD, Odada EO, Partridge TC (2001) Late-Quaternary and Holocene environmental change in Southern Africa. South African Journal of Science 97:139–149
Tyson PD, Lee-Thorp J, Holmgren K, Thackeray JF (2002) Changing gradients of climate change in southern Africa during the past millennium: their implications for population movements, Climatic Change 52:129–135
UNDP (1999) Human development report. Oxford, Oxford University Press
UNEP (1999) Global environment outlook 2000. London, Earthscan Publications
Valentine TR (1993) Drought, transfer entitlements, income distribution: the Botswana experience. World Development 21: 109–126
Van Ballegooyen RC, Gründlingh ML, Lutjeharms JRE (1994) Eddy fluxes of heat and salt from the southwest Indian Ocean into the southeast Atlantic Ocean: a case study. Journal of Geophysical Research 99:14053–14070
Van Katwijk MM, Meier NF, Van Loon R, Van Hove EM, Giesen WBJT, Van der Velde G, Den Hartog C (1993) Sabaki River sediment load and coral stress: Correlation between sediments and condition of the Malindi-Watamu Reefs in Kenya (Indian Ocean). Marine Biology 117:675–683
Van Leeuwen PJ, Ruijter WPM de, Lutjeharms JRE (2000) Natal Pulses and the formation of Agulhas rings. Journal of Geophysical Research 105:6425–6436
Verheye HM (2000) Decadal-scale trends across several marine trophic levels in the southern Benguela upwelling system off South Africa. Ambio 29:30–34
Verheye HM, Richardson AJ (1998) Long term increase in crustacean zooplankton abundance in the southern Benguela upwelling region (1951–1996) bottom-up or top-down control? In: International Workshop on Zoo-ichthyoplankton Monitoring in the Benguela Current Ecosystem off Namibia. Ministry of Fisheries and Marine Research, Swakopmund, Namibia
Verheye HM, Hutchings L, Huggett JA, Painting SJ (1992) Mesozzoplankton dynamics in the Benguela ecosystem with emphasis on herbivorous copepods. South African Journal of Marine Science 12:561–584
Verschuren D, Laird KR, Cumming BF (2000) Rainfall and drought in equatorial east Africa during the past 1100 years. Nature 403:410–414
Vogel C, Laing M, Monnik C (2000) Drought in South Africa, with special reference to the 1980–1994 period. In: Whilhite DA (ed) Drought, Vol. 1, A Global Assessment. London, Routledge, 348–366
Vogel CH (1995) People and drought in South Africa: reaction and mitigation. In: Binns T (ed) People and the environment in Africa. London, Wiley and Sons
Vogel CH (2000) Usable science: an assessment of long-term seasonal forecasts amongst farmers in rural areas of South Africa. The South African Geographical Journal 82:107–116
Vogel JC (1995) The temporal distribution of radiocarbon dates for the Iron Age of southern Africa. South African Archaeological Bulletin 50:106–109
Von Kotze A, Holloway A (1996) Reducing risk: Participatory learning activities for disaster mitigation in Southern Africa. Durban, Department of Adult and Community Education, Natal University
Vrba ES, Denton GH, Partridge TC, Burckle LH (eds) (1995) Palaeoclimate and evolution with emphasis on human origins. New Haven, Yale University Press
Waldron HN, Brundrit GB, Ballegooyen RC van (1998a) Sea level fluctuation and its potential as an upwelling index in the northern Benguela. International symposium on environmental variability in the Benguela Current. Swakopmund, Ministry of Fisheries and Marine Research
Waldron HN, Probyn TA, Brundrit GB (1998b) Carbon pathways and export associated with the southern Benguela upwelling system: a re-appraisal. South African Journal of Marine Science 19:113–118
Walker ND (1990) Links between South African summer rainfall and temperature variability of the Agulhas and Benguela Current Systems. Journal of Geophysical Research 95:3297–3319

Wand SJE, Midgley GF, Jones MH, Curtis PS (1999a) Responses of wild C4 and C3 grass (*Poaceae*) species to elevated atmospheric CO_2 concentration: A test of current theories and perceptions. Global Change Biology 5:723–741

Wand SJE, Midgley GF, Stock WD (1999b) Predicted responses of C4 grass-dominated southern African rangelands to rising atmospheric CO_2 concentrations. Proceedings of the VI International Rangeland Congress 2:922–923

Ward D, Ngairorue BT, Kathena J, Samuels R, Ofran Y (1998) Land degradation is not a necessary outcome of communal pastoralism in arid Namibia. Journal of Arid Environments 40:357–371

Watson AJ (1999) The global survey of pCO_2. IGPB Newsletter 37:6–7

Webster JB (1980) Drought, migration and chronology in the Lake Malawi littoral. Transafrican Journal of History 9:70–90

Weijer W, Ruijter WPM de, Dijkstra HA, Leeuwen PJ van (1999) Impact of interbasin exchange on the Atlantic overturning circulation. Journal of Physical Oceanography 29:2266–2284

Weijer W, Ruijter WPM de, Sterl A, Drijfhout SS (2000) Response of the Atlantic overturning circulation to the South Atlantic sources of buoyance. Journal of Physical Oceanography 29: 2266–2284

Weijer W, Ruijter WPM de, Dijkstra HA (2001) Stability of the competition between the Bering Strait freshwater flux and Agulhas heat and salt sources. Journal of Physical Oceanography 31: 2385-2402

Wells RB, Snyman GM, Held G, Dos Santos A (1987) Air pollution on the Eastern Transvaal Highveld. Foundation for Research Development Report. CSIR Atmos/87/23. Pretoria, Atmospheric Sciences Division, NPRL, CSIR

White WB, Petersen R (1996) An Antarctic circumpolar wave in surface pressure, wind and temperature and sea ice extent. Nature 380:699–702

Wilkinson CR (1996) Global change and coral reefs: Impacts on reefs, economies and human cultures. Global Change Biology 2:547–558

World Bank (1998) The World Bank and climate change: Africa (*http://www.world bank.org.html/extdr/climchng/afrclim.htm*)

WMO (1990) The role of the World Meteorological Organization in the International Decade for Natural Disaster Reduction. WMO, No. 745. Geneva, World Meteorological Organization

Wunsch C (1994) Tracer inverse problems. In: Anderson DLT, Wildebrand J (eds) Ocean circulation models: Combining dynamics and data. Hingham, MA, Kluwer

Yienger JJ, Levy H II (1995) Empirical model of global soil-biogenic NO_x emissions. Journal of Geophysical Research 100: 11447–11464

Zeidler J, Seely MK, Hanrahan S, Scholes M (1999) An index of biological integrity for habitat assessment by Namibian farmers, Proceedings VI International Rangeland Meeting, Townsville

Zepp RG, Miller WL, Burke RA, Parsons DAB, Scholes MC (1996) Effects of moisture and burning on soil-atmosphere exchange of trace gas carbon gases in a southern African savanna. Journal of Geophysical Research 101:23699–23706

Zunckel M, Robertson L, Tyson PD, Rodhe H (1999) Dry deposition of sulphur over Southern Africa. Atmospheric Environment 34:2797–2808

Chapter 3

Global Change and Biogeochemical Cycles: the South Asia Region

A. P. Mitra · M. Dileep Kumar · K. Rupa Kumar · Y. P. Abrol · Naveen Kalra · M. Velayutham · S. W. A. Naqvi[1]

3.1 Introduction

South Asia includes the Indian subcontinent (India, Pakistan, Bangladesh and Nepal), as well as Sri Lanka, Maldives and Mauritius (Fig. 3.1). The region is one of the most densely populated in the world, with present population densities of 100–500 persons km^{-2}. Although total land area comprises only about 3% of the world's land masses, the 1990 population was 21.3% of the global total and by 2025 is expected to rise to ~24%. The urban population in the region is increasing rapidly. In 1980 the percentage of the population living in urban areas was 23% for India, 28% for Pakistan and 11% for Bangladesh. By 2000 estimates had increased to 38%, 37% and 21% respectively. Several of the world's most polluted cities are found in South Asia. Calcutta, Delhi, Mumbai, Karachi and Dhaka are examples of megacities that produce unacceptably high emissions of health-endangering gaseous and particulate matter into the atmosphere. Once considered only a local problem, such urban pollution is now recognized to have regional and even global implications.

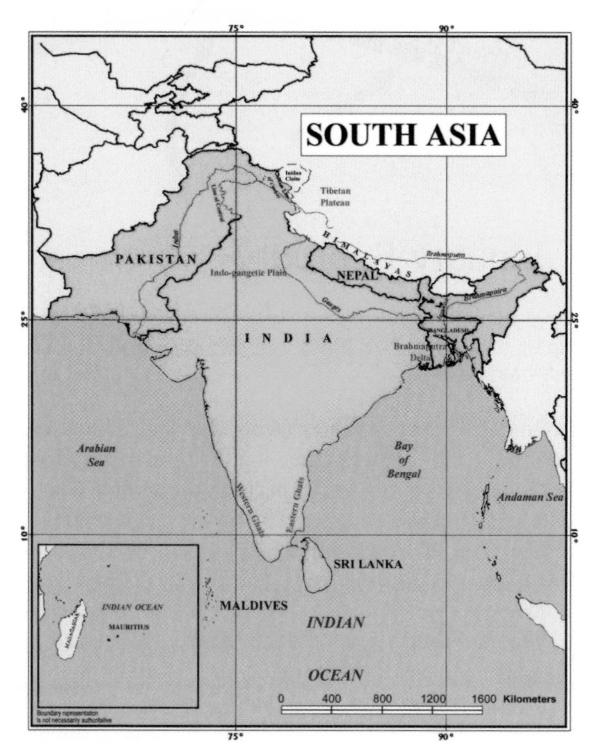

Fig. 3.1. The South Asia region

One of the major driving forces for global change is the rapid increase in the greenhouse gas content of the atmosphere, in which the role of South Asia is still small, but rapidly rising. Atmospheric composition changes over South Asia are due to both long-lived greenhouse gases and short-lived gases like CO, NO_x and SO_2, as well as particulate material. In South Asia atmospheric trace gases and aerosols are projected to increase at rates several times higher than those of the developed world. For example, SO_2, NO_x and suspended particulate matter emissions from fossil fuel sources alone are likely to increase at an average annual growth rate of 3%, 5% and 2% respectively by 2025 (Shukla 2000). Per capita CO_2 emission from South Asia of 0.47 Tg in 1995 is projected to increase to 0.90 Tg by 2025. Methane emission in India from various sectors was 18 Tg in 1990 and is projected to increase to 26 Tg by 2025 (ALGAS 1998).

Although the total regional emission of greenhouse gases (in CO_2-equivalents) is only 3% of the global total, its annual growth rate is high at 6% for India, 8% for Bangladesh and 10% for Pakistan. Emissions from agricultural sources are equivalent to 50% of the emissions from energy-use and industry for India. By comparison, agricultural emissions for Pakistan reach 62%, in Bangladesh 126% and in Sri Lanka 325% of their energy/industry equivalents. For the subcontinent as a whole the ratio is 55%. Per capita emissions at present are low, as exemplified by 0.2 TC (tons of carbon) per person for India in comparison to the global average of 1.2 TC per person and 5 TC for the US. Emission growth rates for the region are high. In the case of CO_2, the increase was 23% from 1990 to 1994, i.e. an annual rate of 6% compared to the global growth rate of 1%. The projected scenario for 2020 is for over a five-fold increase in CO_2

[1] *Contributing authors:* C. Sharma, S. Bhattacharya, G. Beig, T. K. Mandal, J. Ratnasiri, P. S. Swathi, M. M. Sarin, G. B. Pant, K. Krishna Kumar, R. G. Ashrit, G. S. Mandal, L. P. Devekota, L. Zubair, A. M. Chaudhury, C. Qamar-ur-Zaman, A. Muhammad, P. K. Agrawal, M. Aggrawal, S. Gadgil, S. Kesh, R. K. Mall, M. Pal, L. S. Rathore, A. Sastri, D. C. Uprety, N. C. Gautam, V. Subramanyam, S. R. Singh, M. Lal, P. R. Shukla.

emissions from the energy sector alone and a six-fold increase in total CO_2 equivalent emissions (ALGAS 1998).

Regional atmospheric pollutants are of three distinct categories: long-lived gases (CO_2, CH_4, N_2O), short-lived gases (CO, NO_x, SO_2), and particulate materials of different sizes and composition, including climatically important soot carbon. All contribute to radiative forcing (directly or indirectly) and have regional impacts on food security, health and water balance. The interface between air quality and regional emissions has become blurred, with urban pollution rapidly becoming a regional and extra-regional problem.

Several aspects of the oceans surrounding the South Asian landmass have extra-regional, if not global implications. Apart from their dominant role in the unique monsoon regime, the two arms of the northern Indian Ocean – the Arabian Sea and the Bay of Bengal – behave differently as ocean basins. The Arabian Sea experiences evaporation in excess of precipitation and runoff. On the other hand, the Bay of Bengal receives extensive rainfall and substantial runoff from a number of rivers. Both are regions of pronounced oxygen deficiency. The Arabian Sea is one of only three sites in the open ocean where denitrification occurs in the water column on a large scale. Both the Arabian Sea and the Bay of Bengal are subject to major human interference: increasing population and rapid industrialization of the coastal areas, increasing use of fertilizers and agricultural chemicals and damming of rivers are inducing large-scale modification of biogeochemical cycles. The effects of these may extend beyond the coastal zone to the open sea.

The Arabian Sea is a net emitter of CO_2 (Sarma et al. 1998). The net flux to the atmosphere is estimated to be as large as 80 Tg C yr^{-1}. In contrast, the Bay of Bengal is a CO_2 sink, with an uptake of around 20 Tg C yr^{-1}. In the Arabian Sea, the emission rate of CH_4 is several times higher than the global average, but is not large enough to be a major determinant of the global CH_4 balance.

South Asia is characterized by a tropical monsoon climate. Differences in rainfall, rather than temperature, are of primary significance in defining the climatic regimes within the region. The most important feature is the seasonal alteration of atmospheric flow patterns associated with the monsoon (Fig. 3.2). Two monsoon systems operate in the region: the southwest or summer monsoon and the northeast or the winter monsoon. The summer monsoon accounts for 70–90% of annual rainfall over most of South Asia, except over Sri Lanka and Maldives where the northeast monsoon is dominant. Considerable monsoon variability occurs on both space and time scales. There is also a clear association between El Niño events and weak monsoons. Over the period 1871–1988, 11 of 21 drought years were El Niño years. During the 90-year period between 1901 and 1990, rainfall was deficient in all 7 strong El Niño cases (Pant and Parthasarathy 1981; Pant and Rupa Kumar 1997).

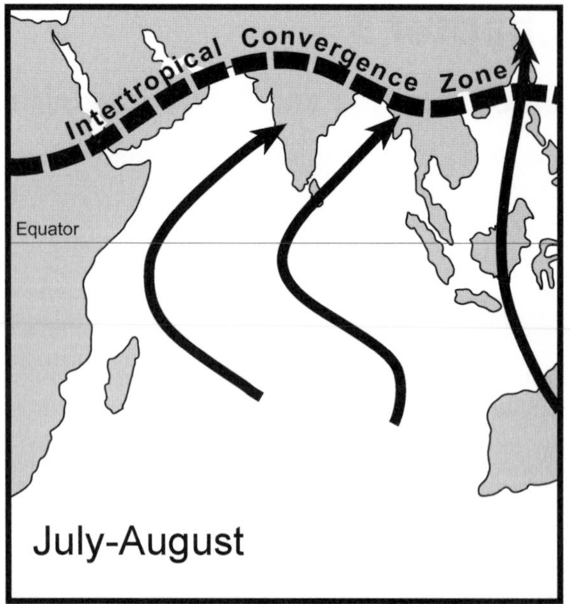

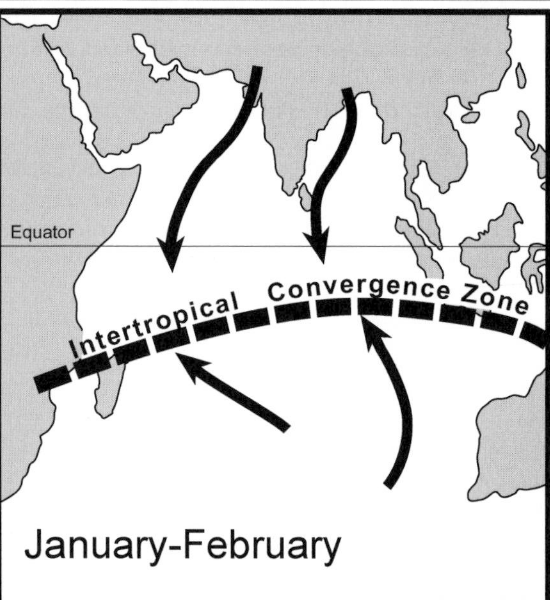

Fig. 3.2. Seasonal surface airflow over the region in the summer and winter monsoons

One of the major issues facing the South Asian region in the 21st century is meeting the food demands of its increasing population, especially in the face of climate and land use changes. The South Asian region is largely agrarian and is highly dependent on the summer monsoon. Knowledge of future variability of the monsoon system with global change is of great concern for regional food security. There is an urgent need to assess the current agricultural conditions and practices within the region, and the potential impacts of global change on them, especially the effect of climatic change on the regional agricultural production. Land use changes may be equally important for agriculture, especially

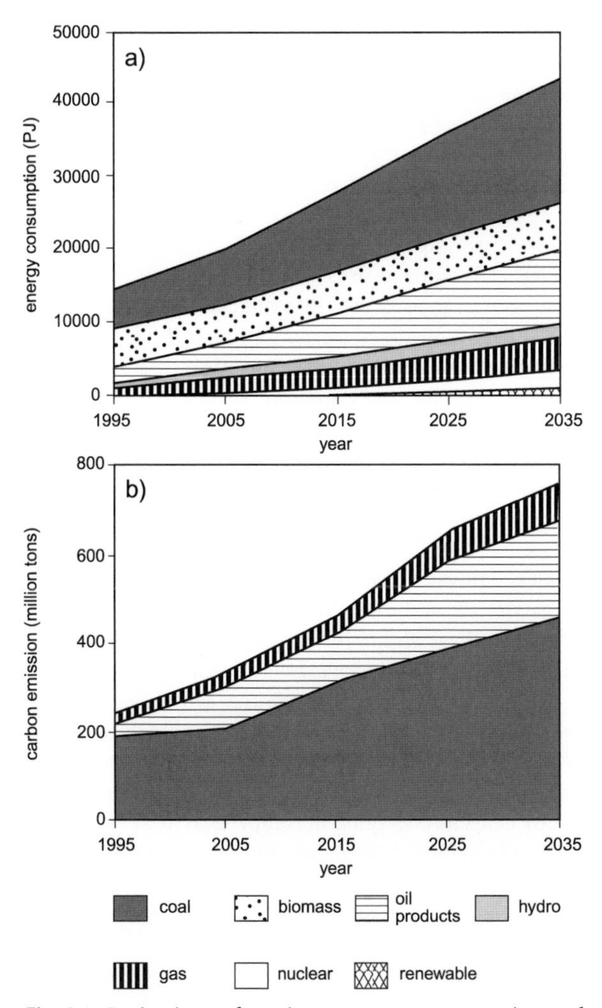

Fig. 3.3. Projections of **a** primary energy consumption and **b** carbon emissions over India (Shukla 2000)

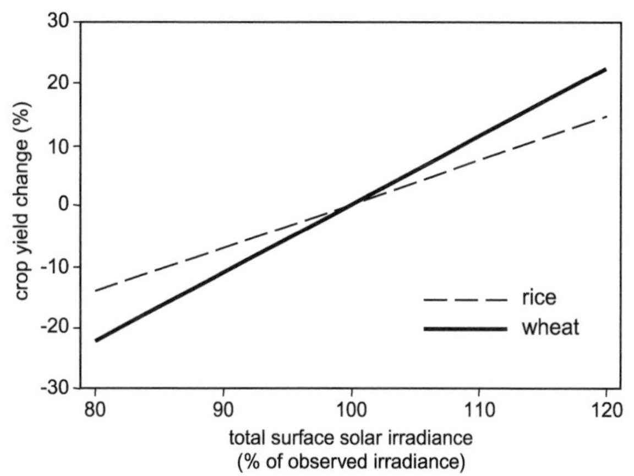

Fig. 3.4. The sensitivity of rice and wheat crop yields to changes in solar irradiance (Chemeidies et al. 1999)

within the Indo-Gangetic Plain. Driven by the urgency to increase production of food grains, measures taken in the past four decades have brought about significant changes in land use and associated biophysical and socio-economic processes. These changes profoundly influence the ability to ensure continued food security for the region's growing population.

Increasing population and economic liberalization has lead to a rapid growth in industry with consequent higher energy consumption and emissions of greenhouse gases and particulate pollutants. Energy consumption projections in India alone show an increase from 14.3 EJ in 1995 to 35.7 EJ by 2035 (Fig. 3.3) (Shukla 2000). Carbon emissions are projected to rise to about 800 Mt by 2035. The increasing population in the region is also putting pressure on forest areas through conversion to agriculture land, encroachment for settlements and the harvesting of forest biomass for fuel. Modelling suggests a decline in forest cover for all the South Asian countries by 2020, except in India, where forest cover is expected to remain more or less same due to various actions taken by government.

Earth system interactions at various scales are complex and not always well understood; in many cases the feedback systems cannot be quantified adequately at present. For example, the changes in CH_4 emissions from paddy fields, under CO_2 enrichment or with increasing ozone and aerosol loading, are unknown. The relative contributions to anthropogenic haze from rural biomass and fossil fuel burning have yet to be quantified adequately. INDOEX results have shown that aerosol loading over Indian Ocean reduces the solar heating of the northern Indian Ocean by about 15% (~25 $W m^{-2}$) and enhances the heating of the boundary layer by about 0.4 K day^{-1} (~12 $W m^{-2}$) (Lelieveld et al. 2001). This may have significant impacts on monsoonal circulation, cloudiness, the hydrological cycle and the functioning of the regional ecosystems (Ramanathan 2000, pers. comn.).

The impacts of global changes at regional level occur at different time scales, varying from centuries to decades for the impact of long-lived greenhouse gases to months to weeks for aerosols and ozone. The effects of regional aerosol loadings on agricultural production in South Asia may be larger than presently thought. The effect of aerosol-induced sunlight reduction on agricultural yields in China amounts to losses of 10% or more (Fig. 3.4) (Chemeides et al. 1999). For India, changes in the receipt of sunlight have appreciable effects that must be added to those of changes in temperature, precipitation and CO_2 concentration. Undoubtedly changing atmospheric conditions over South Asia will have increasing impacts on the region in future.

3.2 Changing Atmospheric Composition over South Asia

The atmosphere is the key medium connecting the lithosphere and biosphere and is exposed to both natural as well as anthropogenic perturbations. Under normal conditions, several *in-situ* mechanisms are capable of accommodating small perturbations. However, in the last few decades anthropogenic influences have increased to the point that they are now capable of overriding natural corrective mechanisms and are now driving atmospheric changes.

The rapid change in atmospheric composition over South Asia, derived principally from emissions from energy, agriculture, forestry and land-use, together with waste disposal, comprises three categories of pollutants: long-lived gases (e.g. CO_2, CH_4 and N_2O), short-lived gases (e.g. CO, NO_x and SO_2) and particulates of different sizes and composition (Table 3.1). All of these pollutant species are increasing rapidly, with annual growth rates of 4–6%. They no longer constitute local urban air pollution problems affecting individual cities, but have regional and extra-regional consequences for climate, agriculture, health and the hydrological cycle.

3.2.1 Greenhouse Gases

Although more than 20% of the global population lives in South Asia, greenhouse gas emissions are low as a proportion of global emissions: approximately 2.7% for CO_2 from all sources and 7% for CH_4 from all sources (Table 3.2). In terms of CO_2-to-C, the total CH_4 emission from the region is about 142 Tg yr^{-1} of the total CO_2 emission of 198 Tg yr^{-1}. For fossil fuels alone, the CO_2 contribution is 3% (173 Tg yr^{-1}), but is expected to rise to 11–12% (690–800 Tg yr^{-1}) by the year 2025 under baseline scenarios.

Greenhouse gas emissions are not a serious problem in the region at present, notwithstanding land-use changes. However, carbon emissions from the Indian energy sector are expected to increase steadily, causing the Indian share of greenhouse gas emissions amongst the ALGAS countries in CO_2 equivalent terms to increase from 20% in 1990 to 30% in 2020. Data for 1990, energy and agricultural sources produce most CO_2 and CH_4 for India (Fig. 3.5), with net emissions projected to increase in future. Uptake and emissions from the forestry sector nearly balance each other.

3.2.2 Biosphere-Atmosphere Interactions

In CO_2 equivalent values, agricultural trace gas emissions are 55% of the combined energy and industry sectors. Spatial variability is considerable: for India the figure is 50%, for Bangladesh 126% and for Sri Lanka 325%.

Rice cultivation is a major source of CH_4 emission in the global inventory. While South Asia has nearly one-third of the total global paddy harvest area, ALGAS emission estimates suggest that only 6 Tg yr^{-1} are emitted from paddy fields in the region in comparison to 23 Tg yr^{-1} for the whole ALGAS-ASIA region and 30–60 Tg yr^{-1} for glo-

Table 3.1. Emissions of trace gases and other pollutants in 1990 (Mitra 2000). *EC* denotes elemental carbon and *BC* black carbon

Region	Emission (Tg yr^{-1})							
	CO_2	CH_4	N_2O	CO	NO_x	SO_2	EC	BC
India	585	18.5	0.26	57	3	3.5	0.45	1.9
South Asia	715	24	0.26	69	4			
China	1770	25 – 33	0.2 – 0.5		22		1.46	2.23
China and India	2355	47	0.61		25		1.91	4.13
USA	4521	27	0.4	83	21	22		
Global	26400	375	5.7	1800	103	75	6.4	10.1

Table 3.2. CO_2 and CH_4 emissions for 1990 from the South Asian Region (ALGAS 1998)

Country/region	CO_2-to-C emission (Tg yr^{-1})			CH_4 emission (Tg yr^{-1})				
	Fossil fuel	Forest clearing	Total	Paddy	Animals	Wasteland	Total	CO_2-to-C equivalent (Tg)
India	146	14	160	4	7.6	3.5	18.7	107
Pakistan	20	2.7	23	0.5	1.6	0.2	2.7	15.5
Bangladesh	6	5.5	11	0.5	0.5	0.04	1.7	9.7
Sri Lanka	1.1	1.5	3.7	0.5	0.09	0.1	0.8	4.5
Nepal	0.3		0.3	0.4	0.4		1	5.6
Total	173	25	198	6	10	4	25	142.3
% of world			2.7				7	

bal paddy fields. Continuous CH_4 field measurements suggest that the annual methane emission from rice paddy fields in the region may actually be about 4 Tg yr^{-1} (Gupta and Mitra 1999).

Forests are also net emitters of carbon in the region, with South Asia emissions exceeding uptake owing to the forest degradation that has occurred in the region. In the case of Sri Lanka, the forest cover decreased from 70% in 1900 to 22% in 1985. In India, periodic satellite assessments have shown that following an earlier decline, the area under forests has stabilized since 1982 (Ravindranath and Hall 1994) and has shown an increase in recent years.

3.2.3 Aerosols

Aerosols are recognized increasingly as a major factor affecting the climate at local, regional and global scales. In South and East Asia, atmospheric aerosols are of particular importance for global change. Accelerated rates of economic growth and development in the region have a direct bearing on the aerosol loading in the atmosphere over the region. Atmospheric SO_2 from anthropogenic sources is no longer negligible. The RAINS-ASIA model indicates that SO_2 emissions for 1990 were 4 470 Gg for India, 614 Gg for Pakistan, 122 Gg for Nepal, 118 Gg for Bangladesh and 42 Gg for Sri Lanka (Carmichael et al. 1998). Modelling further reveals that the S-loading in the atmosphere is likely to increase exponentially during the next few decades, increasing the risk of acidification everywhere, but particularly in India. Acid rain is not yet a problem in India (Table 3.3), owing to the presence of neutralizing alkaline dust in the atmosphere (Das et al. 1999). Similarly, Pakistan does not have any significant acid rain related problems at present (Bhatti 1998).

Aerosol sources in the region include natural dust from surface soils, particulates from industrial and urban emissions, chemical transformation of gaseous products to solids in the atmosphere and vehicular emissions. In Bangladesh, about 27% of the yearly average mass concentration of fine (PM2.5) aerosol particulate matter at urban sites is from vehicular emissions. During the dry season it may be as high as 41%. The world's highest levels of lead in PM2.5 particles have been recorded in Dhaka (Biswas et al. 1999).

The extent to which the regional aerosols are transported over and out of South Asia has been assessed prior to and during INDOEX. Even though the atmos-

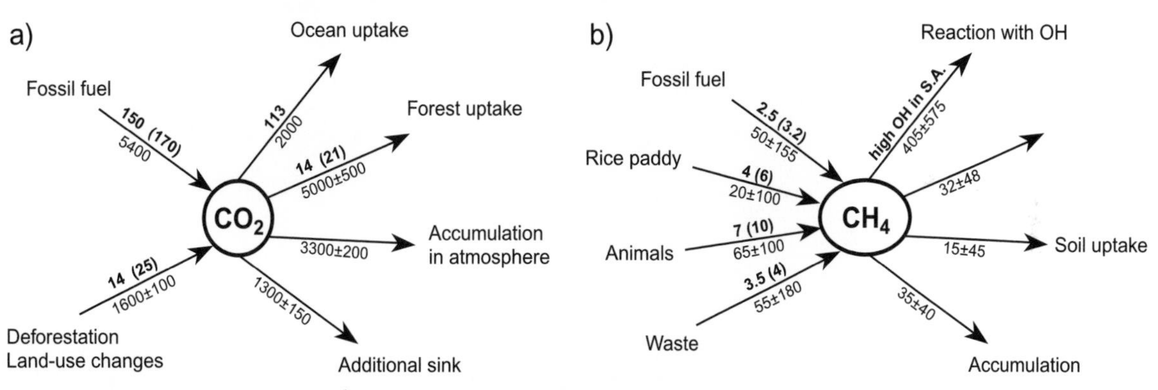

Fig. 3.5. South Asian budgets (Tg C yr^{-1}) for 1990 for **a** CO_2 and **b** CH_4

Table 3.3. pH observations for some stations in South Asia (Das et al. 1999)

Station	Number of observation	pH variation	Acid rain trend	Observations
Allahabad	56	6.35 – 9.00	Decreasing	00
Jodhpur	58	6.35 – 8.70	Decreasing	00
Kodaikanal	62	5.18 – 8.70	Decreasing	03
Minicoy	66	5.52 – 8.90	Decreasing	01
Mohanbari	42	5.50 – 8.30	No trend	02
Nagpur	8	5.55 – 7.35	Increasing	01
Port Blair	48	5.30 – 8.50	Decreasing	05
Pune	76	5.65 – 8.90	Decreasing	00
Srinagar	88	6.15 – 8.40	Decreasing	00
Visakhapatnam	77	5.92 – 8.20	Decreasing	00
Khulana (BD)		6.5 – 7.3		
Horton Plains (SL)		5.37 – 7.47		

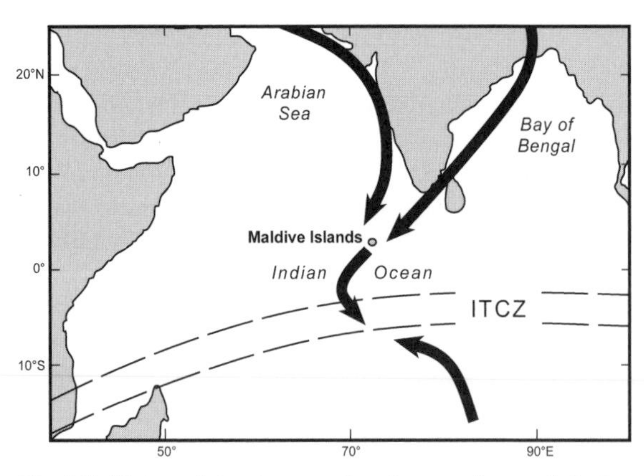

Fig. 3.6. Major winter transport pathways observed during INDOEX in winter 1999 (Lelieveld et al. 2001)

pheric circulation may have been anomalous during the main phase of INDOEX and consequently aerosol loadings may have been higher than normal, measurements of aerosol characteristics during INDOEX observational programmes show a large hemispherical difference within the region, with high values of aerosol optical depth north of the equator. It is clear that long-distance trace gas and aerosol transport from the subcontinent over the Arabian Sea, Bay of Bengal and northern Indian Ocean is significant (Sikka 2000, pers. comn.; Lelieveld et al. 2001) (Fig. 3.6). Often the transport is clearly evident in satellite photographs (Fig. 3.7). Submicron size particles (0.1 to 0.4 micron radius) appear to contribute most to the measured optical depth. Low-level transport of aerosols takes 6–7 days to reach the

Fig. 3.7. An example of aerosol and trace gas transport over and out of South Asia (NASA 2001)

equatorial inter-tropical convergence zone where the air from the northern hemisphere converges with relatively pristine air from the southern hemisphere. Anthropogenic particles may form more than 85% of the total aerosol loading observed over the northern-hemisphere tropical Indian Ocean (Lelieveld et al. 2001).

INDOEX has revealed the presence of an extensive and highly scattering and absorbing winter haze layer over much of the northern Indian Ocean. Approximately 29% of the total aerosol loading is from non-sea salt sulphates and ammonia, 11% from soot, 17% is sea salt, 15% dust, 8% ash and 20% organic materials (Satheesh et al. 1998). The haze layer can reduce monthly solar radiation at the ocean surface substantially. The main species responsible for surface forcing and atmospheric heating is soot originating from fossil fuel and biomass burning over India and its neighbours. In the 500 nm optical band alone, the aerosol optical depth over the Indian Ocean region for January to March 1999 (Fig. 3.8) was responsible for a diurnally averaged aerosol radiative forcing of about $-5.2\ W\,m^{-2}$ (Jayaraman 1999). Total regional radiative forcing may be 10–40 $W\,m^{-2}$. Radiative forcing by aerosols at the top of atmosphere, in the atmosphere and at surface over coastal India, the adjacent polluted ocean and clean ocean beyond are very different (Table 3.4).

Modelled CO transport from the South Asian region, resulting from large-scale biomass burning, shows a plume moving offshore to the southwest during the winter monsoon, crossing the Maldive Islands before moving into the southern hemisphere (Lelieveld et al. 2001) (Fig. 3.9). Soundings from the Maldives through the low-level transport plume reveal maximum concentrations of CO, CH_3CN (methyl cyanide), SO_2, and O_3 at altitudes

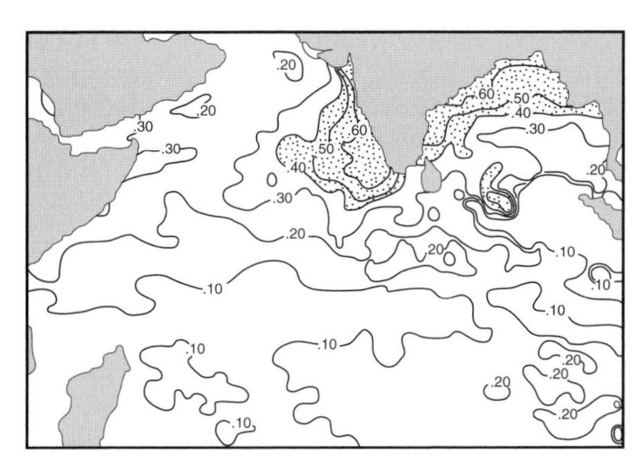

Fig. 3.8. Aerosol optical depth during March 1999 over the INDOEX region (Rajeev and Ramanathan 1999, pers. comn.)

of 2–3 km, with a second peak at 3–4 km. Aerosol scattering also shows a maximum at around 2–3 km (Fig. 3.9) (Lelieveld et al. 2001). Transported aerosols and trace gases are trapped in a stable layer that develops over the Indian Ocean between 1–3 km in the monsoonal outflow from South Asia during winter.

Modelling indicates that at present in-situ low-level ozone formation in the haze layer is not significant due to limitation of NO_x (Beig 2001, pers. comn.). The situation may change as NO_x emissions increase with increasing use of vehicular road transport. IPCC moderate-growth scenarios indicate that ozone smog formation in the South Asia which is likely to increase markedly in the coming decades may also have strong influence on the Northern hemispheric atmosphere by adding to the background ozone level (Lelieveld et al. 1999).

3.2.4 Ozone Changes

Columnar ozone is naturally low in equatorial regions generally and so in South Asia. The major regional problem is not stratospheric ozone depletion, but increasing tropospheric ozone concentrations resulting from increases in the emissions of the ozone precursor gases CO and NO_x into the atmosphere from the surface. Occasionally stratospheric ozone penetrates downward through breaks in the tropopause at heights of around 10 km over the Indian Ocean region to add to lower-tropospheric concentrations (Mandal et. al. 1999).

The consequences of tropospheric ozone changes on agriculture and health are important for South Asia. At Lahore, Pakistan where 6-hourly mean ozone concentrations sometimes reach 60 ppb, wheat and rice yields decline significantly as a consequence (Magg et al. 1995). Reductions in two successive seasons ranged from 33% to 46% in wheat and from 37% to 51% in rice.

3.2.5 Anthropogenic Links

The extent to which anthropogenic activities contribute to the trace gas and aerosol loading of the atmosphere in India is exemplified by the contribution made by vehicular emissions to urban air pollution in Mumbai and Calcutta. During a strike by transport workers in Calcutta in May 2000, with buses and taxis absent from the streets, 24-hour average near-surface concentrations of particulate matter decreased by 30–50% at different localities in comparison to non-strike days in the same

Table 3.4. Radiative forcing ($W\,m^{-2}$) in the INDOEX region (Jayaraman 1999)

	Coastal India	Polluted ocean	Clean ocean
Top of atmosphere	+7	+4.8	+1.2
Atmosphere	+20	+7.9	+0.08
Surface	−27	−12.7	−2

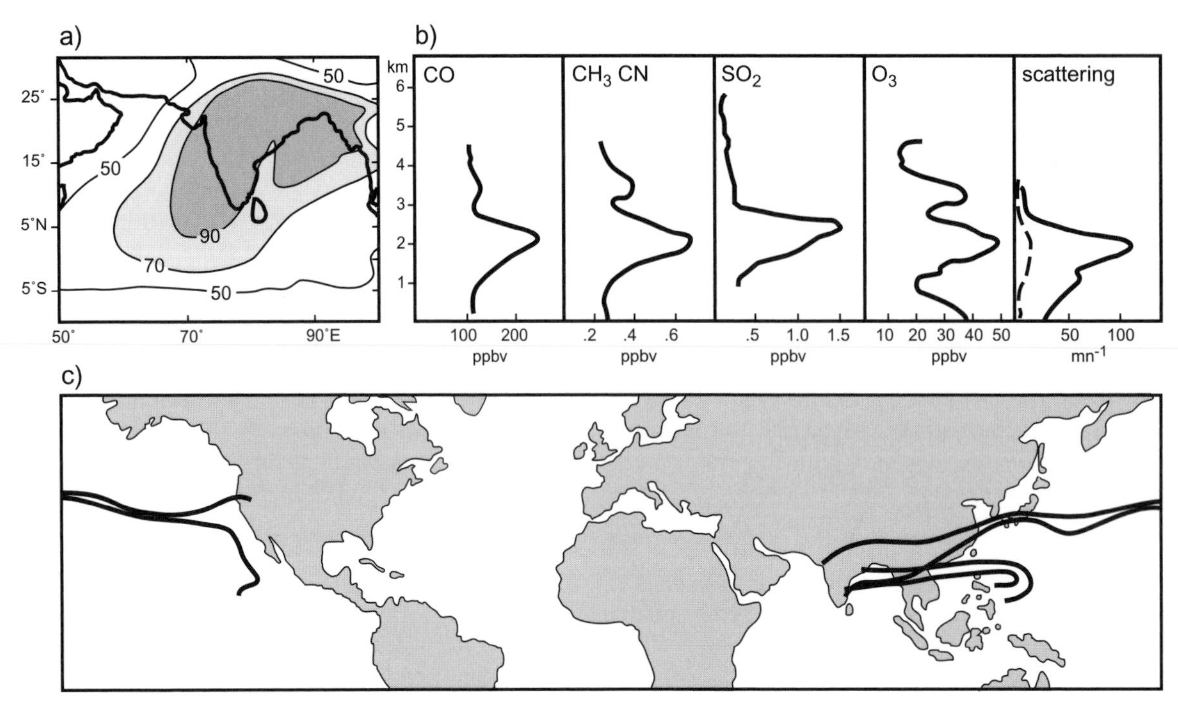

Fig. 3.9. **a** Modelled near-surface mean CO (ppbv) from biomass burning over South Asia over the Indian Ocean during February 1999, **b** CO, CH_3CN, SO_2, and O_3 profiles and aerosol scattering (*solid line*) and absorption (*broken line*) observed over the Maldives on 13 March 1999 (Lelieveld et al. 2001) and **c** trajectories of aerosol and trace gas transport from South Asia (courtesy T. N. Krishnamurti)

month. Concentrations of SO_2 decreased by 20% and NO_2 by 44% (Hasan 2001, pers. comn.). A similar finding was observed during a 24-hour strike in Mumbai in 1999. Particulate concentrations diminished by 30%, SO_2 by 48% and NO_2 by 36%. In both cases, such changes exceed those normally associated with variability produced by changing meteorological and other conditions.

3.2.6 Links between the Lower and Upper Atmosphere

Surface emissions affect not only the lower troposphere, but also extend to the upper regions of atmosphere affecting the chemistry and radiation balance at various levels. With doubling of CO_2, mean lapse rates may change to give a climatologically different temperature structure of the atmosphere. Daily radiosonde data for the troposphere and stratosphere (Rupa Kumar et al. 1987; Rupa Kumar and Kothawale 2001, pers. comn.) and temperature profiles determined from weekly rocket flights form Thumba to heights of 70 km (Golitsyn et al. 1996) reveal that warming in the lower atmosphere has taken place over India from 1963 to 1997, whereas the stratosphere and lower mesosphere have undergone cooling (Fig. 3.10). Modelling suggests other changes may also have occurred. An increase in water vapor and decrease in NO in the mesosphere and thermosphere is most probable; ion composition is also likely to change in the middle atmosphere with ramifications for a number of atmospheric processes (Beig and Mitra 1997a,b). The models also suggest the mesosphere may become wetter and cooler.

Changing vertical temperature structure of the atmosphere over South Asia is likely to affect neighbouring regions as meridional temperature gradients at different heights adjust and regional, and possibly even larger-scale, circulation patterns change accordingly. Lelieveld et al. (1999) conclude that South Asian pollution emissions will significantly contribute to the changes in atmospheric composition in other regions of the northern hemisphere and to climate forcing in regions as far as Europe in future.

3.3 The South Asian Monsoon

The most important feature in the meteorology of South Asia is the seasonal reversal of atmospheric flow patterns associated with the monsoons, owing to the seasonally modulated excess heating of the Asian land mass in summer and excess cooling in winter in comparison to the waters of the adjacent oceans. Orographic effects play an important role. While the South Asian southwest summer monsoon in general terms is a consequence of the thermal differences between the land and the sea, it is more immediately due to the seasonal shifting of thermally produced planetary belts of pressure and winds under continental influences. The upper-air circulation, under the influence of the Central Asian

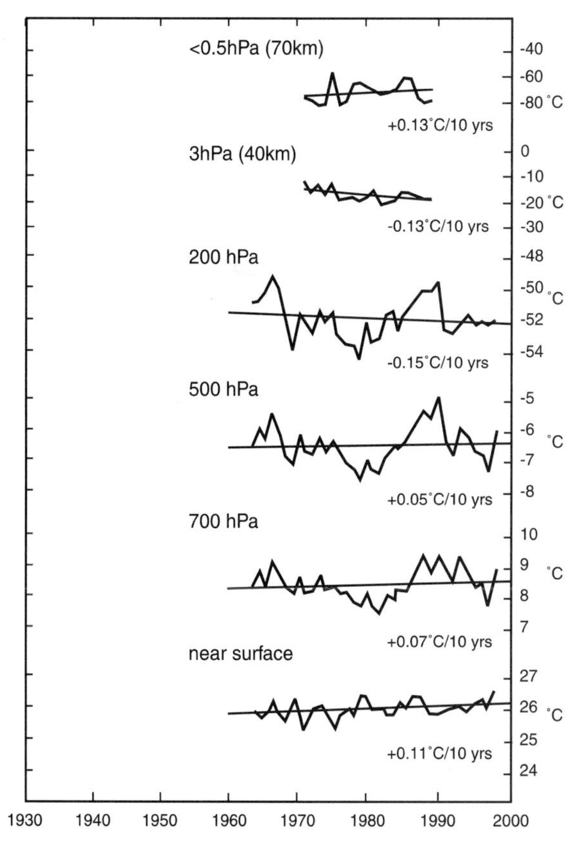

Fig. 3.10. Temperature trends from near surface to an altitude of 70 km, 1963–1997 (updated from Golitsyn et al. 1996)

highlands, also undergoes complex seasonal changes, which form an essential part of the establishment of the monsoon over South Asia. The mountains and high plateaus of Central Asia favour the summer monsoon not only by deflecting the upper westerlies northward, but also by constituting an important heat source in the upper troposphere. Intense heating of the Tibetan plateau deepens the low pressure over northern India, creating favourable conditions for the southwest summer monsoon to develop into a powerful air stream (Pant and Rupa Kumar 1997). During winter, the development of an extensive surface high pressure area over the cold continent of Central Asia and northern South Asia, and of low pressure over the Indian Ocean, facilitates the flow of air from the northeast towards the Indian Ocean at lower levels. The northeast monsoon brings dry condition to most of South Asia, but winter rains to Sri Lanka and southeastern parts of India. Northern parts of South Asia receive additional precipitation from westerly disturbances and southern parts, especially Sri Lanka, from weather associated with the Inter-tropical Convergence Zone (ITCZ). For most of the South Asian region, however, the southwest summer monsoon is the major rain-bearing circulation bringing approximately 80% of the total annual rainfall to most of the region. South Asia depends critically on monsoon rainfall.

The monsoon system operates via connections between atmosphere, land and ocean systems, through fluxes of heat, moisture, and momentum between them and the loss of heat to outer space (Webster et al. 1998). Besides possessing the largest annual amplitude of any tropical or subtropical climate feature, the monsoons also exhibit considerable variability on a wide range of time scales. The monsoon system is potentially sensitive to changes in radiative forcing resulting from changing concentrations of long-lived greenhouse gases and aerosols, as well as through changes in boundary conditions such as sea surface temperature and land surface conditions such as snow and vegetation cover. While the terrestrial ecosystems are known to be sensitive to monsoon variability, modelling reveals that large-scale, sustained modification of ecosystems and vegetation cover has significant effects on the monsoon (Wei and Fu 1998).

Monsoon rainfall, though a dependable source of water for South Asia with a coefficient of variation of only 10%, displays a variety of spatial and temporal variations (Pant and Rupa Kumar 1997). The annual rainfall increases from west to east across the Indian subcontinent by almost three orders of magnitude. The simultaneous occurrence of catastrophic floods in some areas and devastating droughts in others are a common feature of the climate of the region. Such extreme events also show much temporal variability on inter-annual, decadal and longer time scales. How spatial and temporal variability may alter with global change is a question of importance, but unclear. Much clearer is how the monsoon has changed in the past.

3.3.1 Palaeomonsoon Changes

The climates of Asia appear to be affected significantly by the extent and height of the Himalayan Mountains and the Tibetan plateau (Zhisheng et al. 2001). Records of aeolian sediments from China and marine sediments from the Indian and North Pacific oceans identify three stages of evolution of the monsoon. The first occurred 9–8 Myr ago when enhanced aridity prevailed in the Asian interior at the time the Indian and east Asian monsoons began to develop. The second stage began 3.6–2.6 Myr ago, when the east Asian summer and winter monsoons began to intensify, together with increased dust transport to the North Pacific Ocean. Finally, increased variability and possible weakening of the Indian east Asian summer monsoons and continued strengthening of the east Asian winter monsoons appears to have been initiated about 2.6 Myr ago. The results from numerical climate modelling support the argument that the stages in evolution of Asian monsoons are linked to phases of Himalayan-Tibetan plateau uplift and to Northern Hemisphere glaciation.

More recently, it is believed that four monsoon maxima occurred during interglacial conditions over the past 150 000 years, owing to changes in orbital forcing and solar radiation and changes in surface boundary conditions (Prell and Kutzbach 1987). GCM modelling reveals that with interglacial orbital configurations, increasing northern hemisphere radiation strengthens seasonal land-ocean temperature gradients and the monsoon circulation over South Asia (Dong et al. 1996). In general the models have shown that cold periods in the past are characterized by strengthened winter-like circulation, whereas warm periods are associated with strong summer monsoon flow. Observations support the model findings. Oxygen isotopes from Arabian sea sediment cores indicate lower salinity and weak upwelling during the last glacial maximum and reverse conditions during interglacial phases (Cullen 1981; Duplessy 1982). Lacustrine pollen sequences from northwest India suggest a cold and dry period with weak monsoon activity during the last glacial maximum at around 18 ka (Singh et al. 1974). Increased moisture from about 10 ka to 5 ka has been inferred from mesic pollen and high lake levels in northwest India, Kashmir, Nepal and the southwest of the Tibetan plateau (Singh et al. 1974; Swain et al. 1983; Wasson et al. 1984) and from increases in monsoon-transported pollen found in Gulf of Aden and Arabian Sea ocean sediment cores (Van Campo et al. 1982; Van Campo 1986).

Multidisciplinary evidence from the northwestern parts of South Asia shows that the Holocene climate from 10–4.5 ka was warm and humid and associated with vegetation rich in grass, poor in halophytes and experiencing a relatively high frequency of floods as the intensity of the summer monsoon increased after 18 ka (Pant and Maliekal 1987). Sediment cores from the Arabian Sea indicate that the continental heat low was strongest around 8 ka, when the southwest monsoon probably reached its northernmost position (Sirocko et al. 1993). The period thereafter, until about 5 ka, was marked by fluctuating precipitation and lake levels (Swain et al. 1983; Prasad et al. 1997). The period 4.5–3.5 ka experienced fewer extremes, but increased rainfall, swamp vegetation and mesophytic dominance. Evidence of human interference increased in later times. Around 3.5 ka, a trend towards aridity set in over the entire northwest, with a total absence of preserved pollen in Rajasthan lakes and increased sand mobility. Pollen data from the Central Himalayan region suggests that that the period 4–3.5 ka represents the weakest phase of the monsoon during the Holocene (Phadtare 2000). After 3.5 ka the present climate, with its frequent droughts and seasonal extremes, became established and the current vegetation was fixed (Pant and Maliekal 1987). Recent high-resolution studies clearly suggest that short-term oscillations in the Indian monsoon occurred within the Holocene, with the possibility of tropical soil hydrology and continental vegetation playing a role in driving high-frequency variations (Overpeck et al. 1996; Thamban et al. 2000).

Unlike the situation for the northwestern parts of South Asia, the palaeoclimatic evidence for changing conditions in the past over the peninsular Indian region is meagre. Palynological studies of marine cores from the west coast of India reveal major vegetational changes around 3.5 ka (Caratini et al. 1991). Marine microorganisms from the same west coast region suggest periods of higher precipitation at around 280, 840, 1 610 and 2 030 B.P., separated by three significantly dry episodes at around 420, 910 and 1 680 B.P. (Nigam et al. 1992; Nigam 1993).

There are indications that the northern, desert-margins-edge of the monsoon, in the northwest part of South Asia, underwent wide fluctuations leading to the appearance *and disappearance of human civilizations in the region. The Harappan* civilization of the Indus valley flourished during the period 2300 to 1700 B.C., the *Painted-Gray-Ware culture* between 700 and 300 B.C., and the *Rangamahal culture* between A.D. 100 and 200. Ramaswamy (1968), commenting on the good monsoon regimes in northwest India, has postulated that during the period 2000–500 B.C. deep troughs in the upper westerlies may have extended into Pakistan more frequently than now, causing monsoon depressions to curve to the north or northeast, leading to active monsoon conditions over the entire Indus valley.

The best high-resolution evidence for climate change in South Asia over the last millennium comes from the high-altitude (7 200 m) ice-core record from Dasoupu, Tibet (Thompson et al. 2000). Changing dust and chloride concentrations in the core give a precise record of decadally changing South Asian monsoon intensities from A.D. 1000–1440. From 1440 onwards the resolution is annual. The greatest recorded failure of the monsoon and occurrence of drought in the last thousand years was from 1790 to 1796. Less severe monsoon failures took place in the 1640s, 1590s, 1530s, 1330s, 1280s and 1230s events. During the drought of 1790–1796 at least 600 000 people died of starvation in just one region of northern India in 1792 alone. The consequences on the whole region over the entire drought would have been catastrophic. The twentieth century increase in anthropogenic activity in India and Nepal, downwind from the Dasoupu site, is recorded by a doubling of chloride concentrations and a four-fold increase in dust. Like other cores from the Tibetan Plateau, the Dasoupu data suggest a large-scale warming trend that appears to be amplified at higher elevations in the high mountain terrain (Thompson et al. 2000).

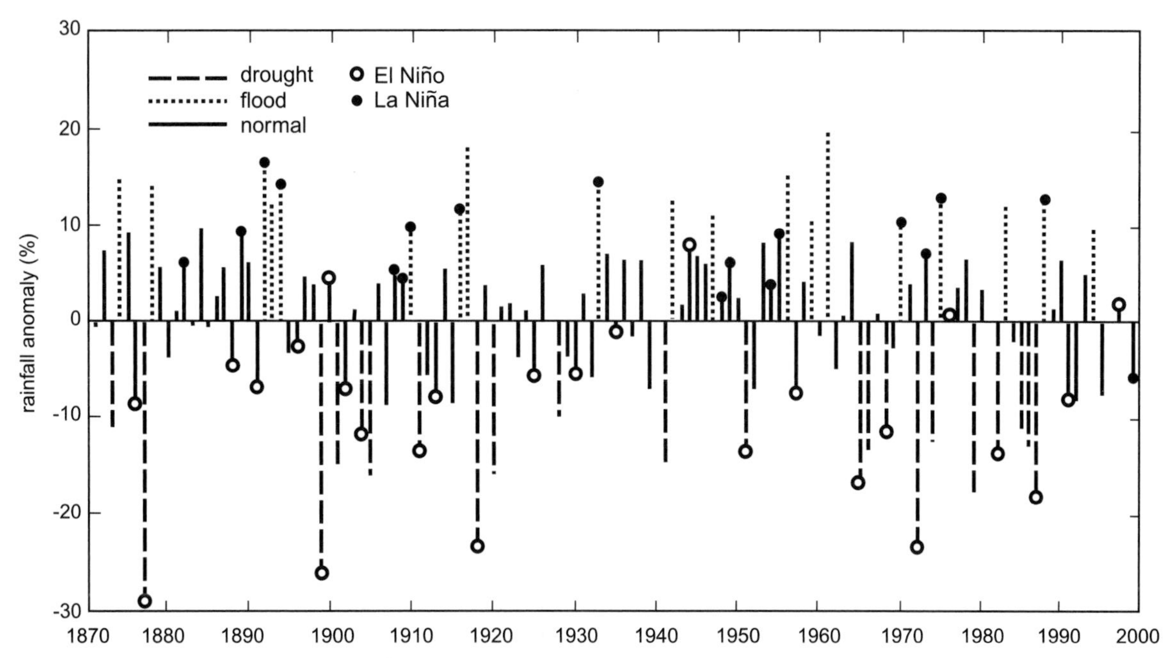

Fig. 3.11. All-India summer monsoon rainfall, 1871–1999 (updated from Parthasarathy et al. 1987)

3.3.2 Monsoon Changes during the Period of Meteorological Record

A network of stations, with at least one rain gauge per district having a continuous and homogeneous rainfall record, is available for India since 1871 (Parthasarathy et al. 1987). Reliable means for areas larger than districts can be extended back in time to 1844 using fewer stations; in some cases it is possible to go back to 1813, but with diminished reliability (Sontakke et al. 1993). From these records it is possible to assess the variability of all-India rainfall and the monsoon during the period of meteorological record (Fig. 3.11). Inter-annual variability is the predominant mode of variability of monsoon rainfall. A feature of anomalous monsoon situations is the spatial coherence of seasonal rainfall anomalies over large areas of South Asia. The effect of droughts is accentuated by the higher coefficient of variability over arid regions (Parthasarathy 1984) and their frequent occurrence in consecutive years (Chowdhury et al. 1989).

Whereas no significant trends are to be found in the all-India rainfall over the period of record (Mooley and Parthasarathy 1984), on smaller spatial scales significant areas experiencing long-term rainfall changes have been reported (Rupa Kumar et al. 1992). By grouping together the contiguous districts showing significant and near-significant trends, it is possible to show that increasing monsoon rainfall (+10 to +12% per century) is occurring along the west coast, north Andhra Pradesh and in northwest India, while decreases in rainfall (−6 to −8% per century) are being observed over east Madhya Pradesh and adjoining areas, northeast India and parts of Gujarat and Kerala (Fig. 3.12).

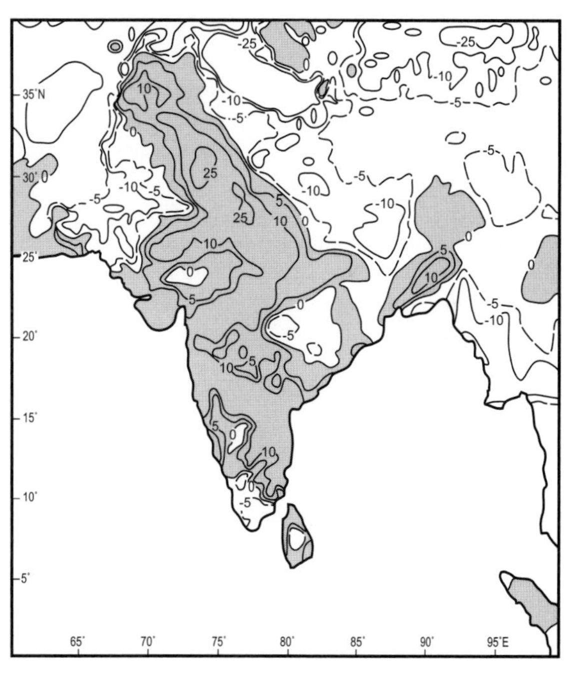

Fig. 3.12. Spatial variation in linear trends (positive and negative percentage changes per century) in monsoon rainfall over South Asia, 1901–1990) (Rupa Kumar et al. 1992)

Inter-annual variability dominates the rainfall spectrum of South Asia. For more than a century a clear association between a weak monsoon, a large negative Southern Oscillation Index and El Niño events has been apparent and conversely between a strong monsoon,

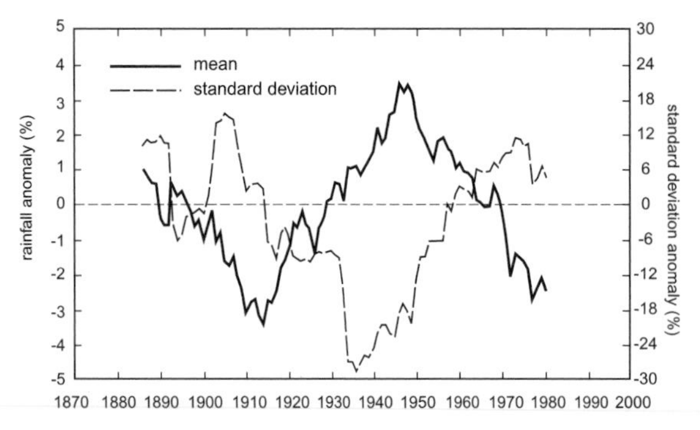

Fig. 3.13. Epochal variation of all-India summer monsoon rainfall (Mooley and Parthasarathy 1984)

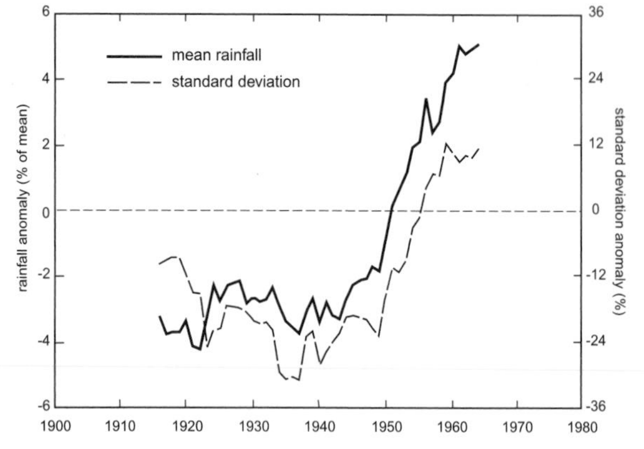

Fig. 3.14. Anomalies and standard deviations of Bangladesh rainfall (Kripalani et al. 1996; Pant and Rupa Kumar 1997)

large positive SOI and the absence of El Niño events (Sikka 1980; Pant and Parthasarathy 1981). During the period 1871–1999, 11 of 21 drought years were El Niño years, 18 cases of normal rainfall years occurred with El Niños and only 2 above-normal years (in excess of 110% of normal) in 18 were El Niño years. Factors other than ENSO also contribute to the inter-annual variability of the monsoon rainfall (Krishna Kumar et al. 1995; Anderson 1999).

Quasi-biennial periodicities in the range of 2 to 3 years have been found to be significant in many monsoon rainfall time series. Other quasi-periodicities ranging between 6 and 14 years have also been detected, but with less spatial coherence (Parthasarathy 1984). These periodicities were more prominent during the second half of the twentieth century than in the first (Mooley and Parthasarathy 1984) and may be due to interactions between the quasi-biennial oscillation of stratosphere (Bhalme et al. 1987) and ENSO in the tropics (Bhalme and Jadhav 1984), or may be relate to other factors.

On multi-decadal scales, several major monsoon changes have occurred since the late nineteenth century in South Asia (Fig. 3.13). The periods 1895–1932 and 1965–1987 were characterized by frequent droughts, whereas 1872–1894 and 1933–1964 were practically drought free (Mooley and Parthasarathy 1984). This low frequency signal is present not only in summer monsoon seasonal rainfall totals, but also in other monsoon parameters, such as onset, break monsoon days and number of storms and depressions (Pant et al. 1988). Recent work confirms the significance of multi-decadal changes in the monsoon record (Parthasarathy et al. 1991).

Much of the above is based in the analysis of Indian rainfall. All the other countries of the region experience the South Asia monsoon, but it often has different effects in different parts of the region, as has been pointed out previously for Sri Lanka. In Nepal precipitation decreases from east to west as the monsoon flow advances from Bay of Bengal and proceeds northwestward, due mainly to mechanical topographical forcing. Thus, the Eastern Himalayas (Assam) have about eight months of rainy season (March-October) with a fairly active pre-monsoon season, while the Central Himalayas (Nepal) have only a four-month rainy season. In the Western Himalayas summer monsoon is active only in July and August (Mani 1981). Regional mean precipitation time series for Nepal show no long-term trends, but significant variability on the annual and decadal time scales. The Nepal monsoon record correlates well with ENSO.

In contrast, by virtue of its location, Pakistan receives both summer and winter precipitation. Summer rainfall is mostly from the southwest monsoon, whereas winter rainfall is derived from western disturbances. Significant inter-annual variability occurs in the rainfall of both seasons. Over the southeastern coastal belt and southeastern deserts of Pakistan an upward trend in rainfall is apparent (Qamar-uz-Zaman et al. 1998) and is spatially coherent with the trends reported by Rupa Kumar et al. (1992). Over Bangladesh, monsoon rainfall has increased since 1950 (Kripalani et al. 1996; Pant and Rupa Kumar 1997) (Fig. 3.14).

3.3.3 The South Asian Monsoon and Global Warming

South Asia, like many regions of the world, experienced clear surface warming during the twentieth century. Seasonal and annual surface air temperatures over India show a significant warming of 0.4 °C per century from 1901 onwards (Hingane et al. 1985, updated). The warming is mainly attributed to the post-monsoon and winter seasons. West coast, interior peninsula, north central and northeast regions of India all show pronounced warming in the mean annual temperatures (Fig. 3.15).

However, contrary to the findings in other regions (Karl et al. 1993), the surface heating is manifested through increases in maximum temperatures, not in minima (Srivastava et al. 1992; Rupa Kumar et al. 1994;

Lal et al. 1996). The increase in the all-India mean temperatures is almost solely attributed to the increase in maximum temperatures (0.6° per century significant at 1% level); there is no significant trend in the minimum temperatures (Fig. 3.16). Consequently, a general increase in the diurnal range of temperatures has occurred.

3.3.4 Climate Sensitivity of the Monsoon

Increases in atmospheric concentrations of greenhouse gases and aerosols and the associated changes in radiative forcing may lead to significant changes in the circulation patterns and the hydrological cycle. The sensitivity of the monsoon circulation to such changes is a matter of great concern in South Asia. The fact that the monsoon is sensitive to planetary scale anomalies like ENSO and major volcanic eruptions is well established (Mukherjee et al. 1985; Handler 1986). Despite its complexity on regional and smaller scales (Pant and Rupa Kumar 1997), many of the large-scale features of the monsoon circulation are now captured by current-generation atmospheric GCMs (Stephenson et al. 1998). The GCM simulated monsoon response to increased amounts of CO_2 is complex and strongly model-dependent.

Several modelling studies have found that the simulated Asian summer monsoon becomes more intense with doubling of CO_2 in the atmosphere. Five mixed-layer GCMs suggest that wetter summer conditions are likely to occur over both South and Southeast Asia (Zhao and Kellogg 1988). Using an early generation coupled ocean-atmosphere model, Meehl and Washington (1993) obtained the same result and Bhaskaran et al. (1995) showed in a transient coupled experiment with a gradual increase of the CO_2 concentration a northward shift and an intensification of the monsoon rainfall occurred. Considerable uncertainty exists as to what actually may happen, however, as contrary results pointing to the delayed onset of the summer monsoon with a possible future net reduction in rainfall have been found in other studies (Lal et al. 1999; Hassel and Jones 1999).

When sulphate aerosols are incorporated in the models, a net reduction in area-averaged monsoon precipitation appears likely (Lal et al. 1995; Bhaskaran and Mitchell 1998). Other studies suggest that the monsoon sensitivity to CO_2 doubling is not only related to changes in the horizontal transport of water vapor, but also to changes in the precipitation efficiency, which depends on soil moisture (Lal et al. 1995; Kitoh et al. 1997; Douville et al. 2000; Stephenson et al. 2000; Rupa Kumar and Ashrit 2001). The treatment of land surface hydrology in the GCMs is a critical factor in determining monsoon sensitivity to regional and global change. Simulation of mean seasonal precipitation of the monsoon remains a formidable challenge and many modelling uncertainties remain to be resolved.

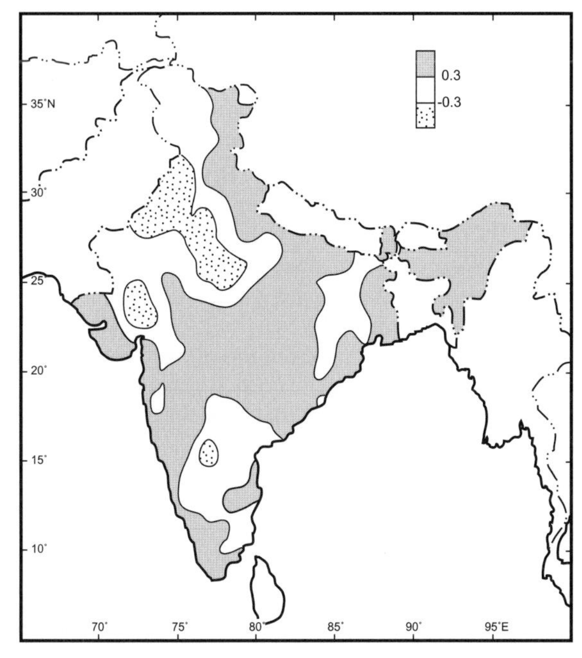

Fig. 3.15. Spatial variation of mean temperature trends (°C per century) over India, 1901–1990 (Hingane et al. 1985, updated)

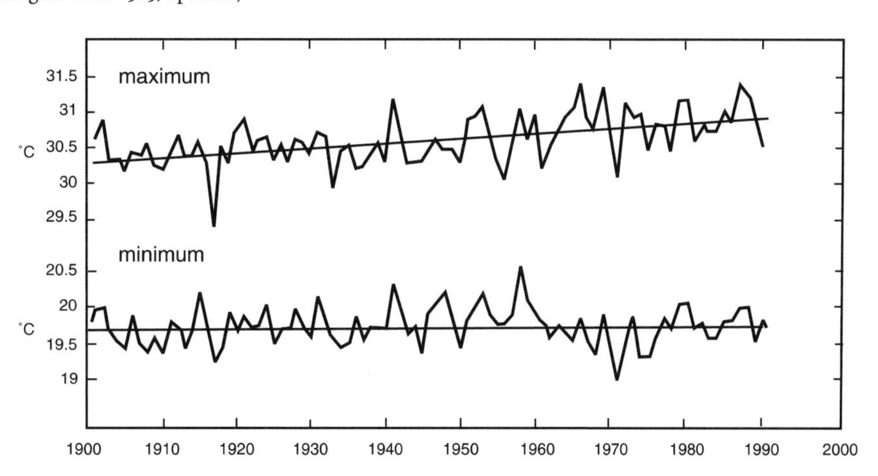

Fig. 3.16. Trends in all-India maximum and minimum temperatures (Rupa Kumar et al. 1994)

Box 3.1. Extremes in the Indian Ocean

For the tropical Pacific and Atlantic Oceans, internal modes of variability that lead to climatic oscillations have been recognized, but in the Indian Ocean region a similar ocean-atmosphere interaction causing inter-annual climate variability has not been found. However, recent analyses of observational data (Saji et al. 1999; Webster et al. 1999) indicate, for the first time, the existence of a dipole mode in the Indian Ocean with a pattern of internal variability with anomalously low sea surface temperatures (SSTs) off Sumatra and high SSTs in the western Indian Ocean, having strong coupling through precipitation and ocean dynamics. This air-sea interaction process is unique and inherent in the Indian Ocean, and has been shown to be independent of the ENSO phenomenon. While the dipole mode appears to have a biennial character, as does the monsoon, how it is related to the monsoon rainfall is unclear and further research is needed to determine its precise role in monsoon variability.

3.3.5 Weakening of ENSO-Monsoon Relationships

Regional precursors of the monsoon are modulated by the ENSO (Parthasarathy et al. 1991). The relationship between Bombay pre-monsoon pressure and monsoon rainfall becomes dominant when the ENSO variance in Bombay pressure is high and ceases to exist when the ENSO variance is small. During the period 1951–1990 the atmosphere-land-ocean systems seem to have been strongly coupled and ENSO-related monsoon variability has been pronounced. It is not clear why the coupling was weaker before 1950 and again after 1990. Two possible reasons for the break-down of the inverse relationship between ENSO and the Indian summer monsoon in the 1990s have been proposed (Krishna Kumar et al. 1999) (Fig. 3.17). A southeast shift in the Walker circulation anomalies associated with ENSO events may lead to a reduced subsidence over the Indian region, thus favouring normal or weakened monsoon conditions. Additionally, increased surface temperatures over Eurasia in winter and spring (that are the result of the continental warming trend) favour a stronger monsoon owing to an enhanced land-ocean thermal gradient, are counteracting strong ENSO events which would otherwise lead to the weakening of the system. An alternative idea, suggested by GCMs that simulate a stronger warming over land than over sea, but do not simulate a consequentially stronger monsoon, is that the weakening of ENSO-monsoon correlation may be explained by an increase in precipitable water as a result of global warming, rather than by an increased land-sea thermal gradient (Douville et al. 2000).

3.4 Extreme Climatic Events

3.4.1 Cyclonic Storms, Monsoon Depressions and Low Pressure Systems

Approximately a century of well-documented data is available on cyclonic storms, monsoon depressions and low-pressure systems for South Asia. The annual number of cyclonic storms of different intensities over the Bay of Bengal and Arabian Sea over the last 100 years shows no long-term trends, exhibits a Poisson distribution and appears to be random (Mooley 1981; Mooley and Mohile 1984). The number, duration and westward displacement of low-pressure systems during the monsoon season is likewise random, but shows a Gaussian distribution (Mooley and Shukla 1989).

In the north Indian Ocean around 16 cyclonic disturbances occur each year of which about 6 develop in to cyclonic storms. The annual number of severe cyclonic storms with hurricane force winds averaged 1.3 over the period 1891–1990. From 1965–1990 the number was 2.3. No clear variability pattern appears to be associated with the occurrence of tropical cyclones. While the total frequency of cyclonic storms that form over the Bay of Bengal has remained almost constant over the period 1887–1997, an increase in the frequency of severe cyclonic storms appears to have taken place in recent decades (Fig. 3.18). Whether this is real or a product of recently

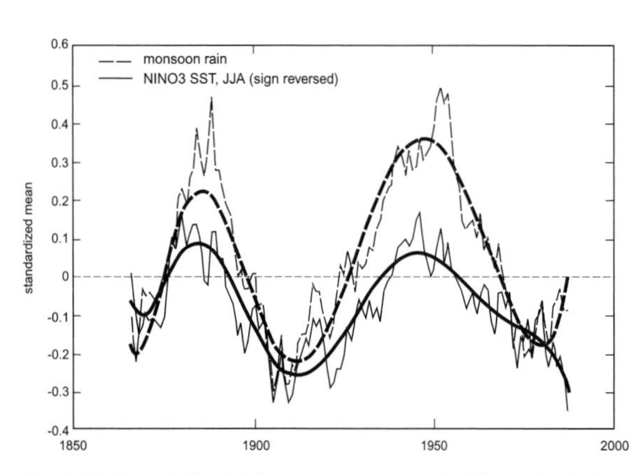

Fig. 3.17. The relationship between monsoon rainfall over South Asia and NINO3 sea surface temperatures (Krishna Kumar et al. 1999)

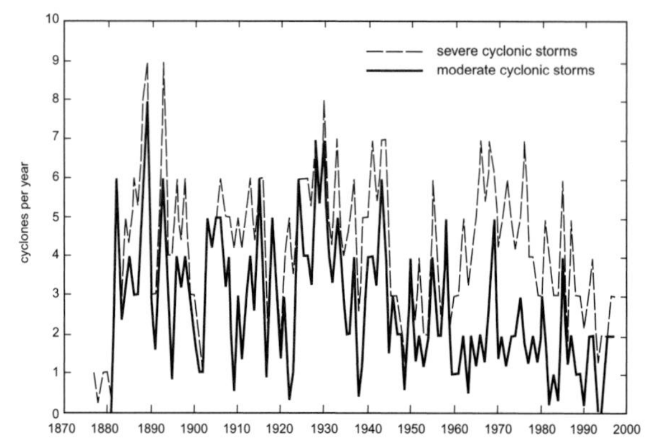

Fig. 3.18. Variation in cyclone frequency over the Bay of Bengal, 1880–1997 (Rupa Kumar and Mohile 2001, pers. comn.)

enhanced monitoring technology is not clear (Mooley and Mohile 1984). A slight decreasing trend in the frequency of cyclonic disturbances and tropical cyclones may be apparent during the monsoon season (Mandal 1992). High sea surface temperature is a necessary, though not sufficient, condition for the formation and growth of tropical cyclones. Over the Indian Ocean, Bay of Bengal and the Arabian Sea significant and consistent warming of the sea surface has occurred during the twentieth century (Fig. 3.19). Sensitivity experiments with a numerical model suggest that cyclone intensity may increase with the increasing sea surface temperatures (Bhaskar Rao 1997) (Fig. 3.20). Were such an increase to occur, it would exacerbate flooding in Bangla-

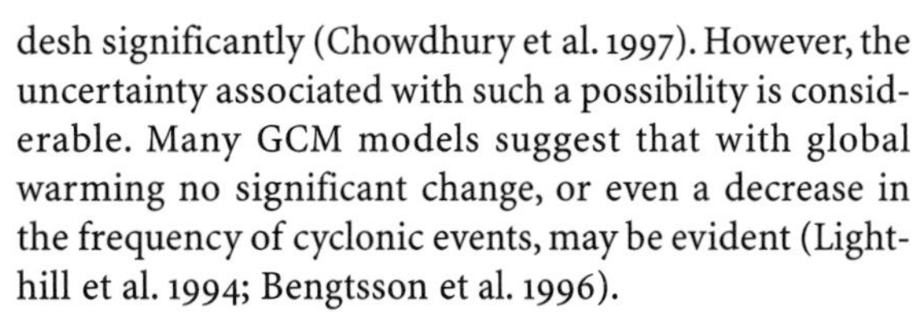
desh significantly (Chowdhury et al. 1997). However, the uncertainty associated with such a possibility is considerable. Many GCM models suggest that with global warming no significant change, or even a decrease in the frequency of cyclonic events, may be evident (Lighthill et al. 1994; Bengtsson et al. 1996).

Analysis of records from Bangladesh, India and Sri Lanka show increasing sea levels at all locations (Warrick and Ahmad 1996; Weerakkody 1996; Shanker and Shetye 1999). Continued rise of sea level with global warming has not only severe implications for the vulnerable populations in the coastal margins of the region, especially Bangladesh, but also for the biogeochemistry of the ocean basins of South Asia.

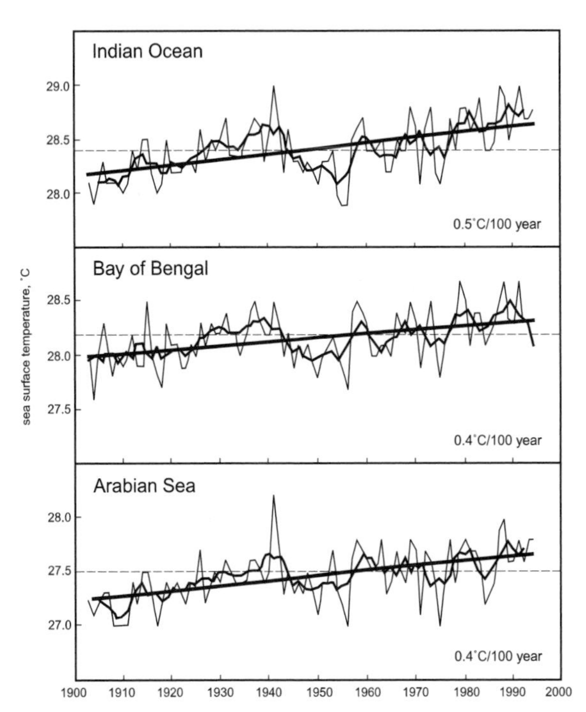

Fig. 3.19. Sea surface temperature trends (°C per century) over the Indian Ocean, Bay of Bengal and Arabian Sea (Rupa Kumar and Kothawale 2001, pers. comn.)

3.5 The Oceans around South Asia

3.5.1 Carbon-Nitrogen Cycling in the Aquatic Systems

The seasonal monsoon shift in wind over South Asia affects not only the land, but also the oceans of the region. Over the ocean, the strong summer, southwest monsoon winds cause extensive upwelling of nutrient-rich sub-surface waters off the coasts of Somalia, Oman and southwest India, as well as in the open Arabian Sea. This supports a high rate of biological production. By contrast, upwelling is much weaker in the Bay of Bengal due to freshwater influx. High production in the Arabian Sea causes mid-depth O_2 depletion, which is the most severe occurring anywhere in the world oceans. This leads to denitrification (Sen Gupta et al. 1976; Naqvi 1987) and changes in N_2O cycling to the west of South Asia.

The Arabian Sea and the Bay of Bengal are important not only because of the unusual biogeochemical processes operating in their waters, but also because of their susceptibility to human interference. With rapid industrialization of the coastal areas, land-use change,

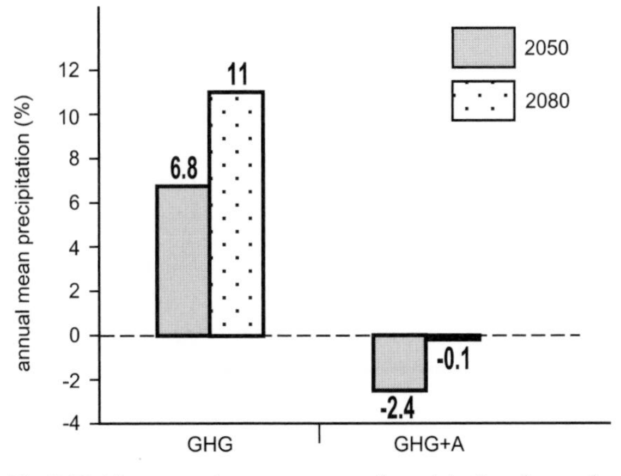

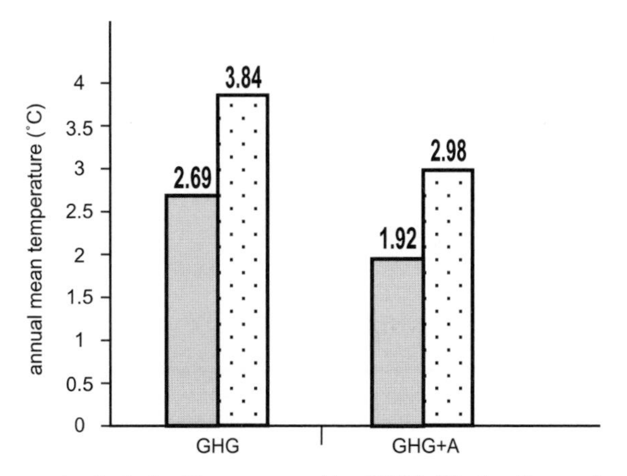

Fig. 3.20. Mean annual temperatures and precipitation changes by 2050 and 2080 derived from an ensemble of GCMs (Bhaskar Rao 1997)

ever-increasing use of fertilizers and other chemicals, human waste effluents and damming of rivers, the coastal zones of these seas are experiencing, and will increasingly continue to experience, large-scale biogeochemical modification, which may extend to the open sea. It will be useful to examine the sorts of changes taking place.

3.5.2 Carbon

Carbon dioxide exchange in the coastal zone and beyond is modulated by both natural and anthropogenic factors. An example of a natural process is given by the Mandovi-Zuari estuarine system in Goa, India. Significant CO_2 ejection (0–200 mmol $m^{-2} d^{-1}$) results from soil-water interactions. Protons are exchanged between humus rich particulate matter (drained by rivers into the estuary) and water due to ion-exchange reactions during the soil-water interactions. The proton released into water lowers the pH thereby enhancing the pCO_2 in water. In the estuarine system of Goa, the pCO_2 reached as high as 2 500 µatm, that is about 6–7 times that in atmosphere. The system is not heavily polluted and present high emissions are caused by natural reactions in a tropical estuary. Geological settings and nature of catchment soils vary considerably from estuary to estuary in the region and the carbon dynamics of each will respond accordingly.

3.5.2.1 *Horizontal Fluxes into the North Indian Ocean*

The transport of carbon by rivers constitutes one of the important components of the global carbon cycle. The Godavari River is the largest Indian peninsular river. Here dissolved inorganic carbon levels vary between 26–53 mg C l^{-1}. Assuming an average river water inorganic carbon content of 18 mg C l^{-1} for all rivers discharging into the into the Bay of Bengal and Arabian Sea and using the run-off values of ~1 600 and ~300 $km^3 yr^1$ into the adjacent seas respectively, inorganic carbon transports amount to 29 Tg yr^{-1} into the Bay of Bengal and 2.6 Tg yr^{-1} into the Arabian Sea. The organic carbon loads (based on data from Ittekkot et al. (1985) and Ittekkot and Arain (1986)) equal 3.2 Tg yr^{-1} into the Bay and 2.2 Tg yr^{-1} into the Arabian Sea. The Persian Gulf and the Red Sea in turn are estimated to receive 11.7 Tg C yr^{-1} and 42.8 Tg C yr^{-1} from the Arabian Sea (Somasundar et al. 1990).

Annual precipitation of 8 000 km^3 over the Bay of Bengal and 6 100 km^3 over the Arabian Sea imply an atmospheric scavenging of 30 and 26 Tg C yr^{-1}, respectively. The ^{234}Th disequilibrium between that in water column (dissolved and particulate) and that estimated from ^{238}U yielded a scavenging residence time of ~30 days at a removal rate of ~3 400 dpm $m^{-2} d^{-1}$ for ^{234}Th (Sarin et al. 1996). The activity ratios of ^{234}Th and ^{238}U in the mixed layers of the Arabian Sea varied from 0.4 to 0.8. This suggests intense scavenging of Th by sinking particles. Based on the C to ^{234}Th ratios in sinking particles and ^{234}Th deficiencies in surface layer (100 m) the fluxes of carbon from surface to deep Arabian Sea were found to be in the range of 320 to 1 150 mg C $m^{-2} d^{-1}$. These sinking fluxes are higher than the column productivity, possibly due to a differential cycling of sinking carbon components and ^{234}Th in the Arabian Sea surface layers. From the deepest trap collections (Nair et al. 1989; Ittekkot et al. 1991) sedimentation of carbon is estimated to be 12 Tg yr^{-1} and 2 Tg yr^{-1}, in the Bay of Bengal and Arabian Sea, respectively.

3.5.2.2 *Air-Sea Fluxes*

In the Arabian Sea, supply of CO_2 through upwelling of deep waters greatly overwhelms biological removal. The partial pressure of CO_2 (pCO_2) in surface waters within the three upwelling zones is among the highest (~700 µatm) reported anywhere from the oceans (Sarma 1999). Even outside the upwelling zone, the surface water pCO_2 is always higher than that in the atmosphere, leading to a net flux of CO_2 from the ocean to the atmosphere. Carbon dioxide emissions total 45 Tg C yr^{-1} from the eastern and central Arabian Sea, which cover

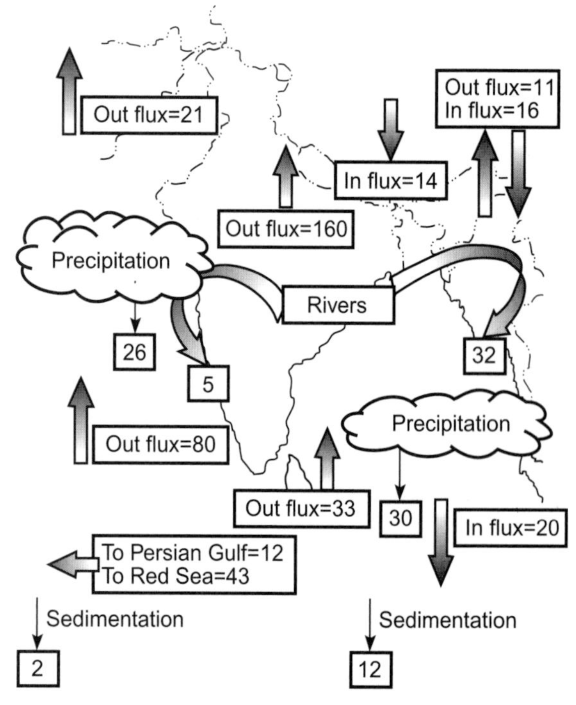

Fig. 3.21. Air-sea fluxes of carbon over the South Asian region

approximately 0.25 of the total area (Sarma et al. 1998). The Arabian Sea has been estimated to emit ~80 Tg yr^{-1} of CO_2 to the atmosphere (George et al. 1994). On the other hand, in the Bay of Bengal, which experiences near-estuarine conditions, the strong thermohaline stratification suppresses the pumping of CO_2 from the deep sea to the surface layer (Fig. 3.21). As the euphotic zone is substantially fertilized through river discharge and atmospheric inputs of nutrients (a significant fraction of which is of anthropogenic origin), there is a net removal of CO_2 through biological activity making the Bay of Bengal a unique seasonal sink of atmospheric CO_2 of an estimated strength of 20 Tg C yr^{-1} (Kumar et al. 1996). Higher $\delta^{18}O$ in sediments of the Andaman Sea during the last deglaciation has been attributed to the possibility of reduced river flow from the Irrawaddy and Salween Rivers (Ahmed et al. 2000), which can act as a weaker fresh water lens. Therefore, it is possible that the fluctuating river run-off in the past had a decisive effect on the air-to-sea CO_2 fluxes in the Bay of Bengal. Greenhouse warming could effectively increase run-off to the North Indian Ocean, through either melting ice in the Himalayas or increasing rainfall through atmospheric disturbances, leading to CO_2 absorption by the sea. Overall the external fluxes of carbon appear to be positively balanced in the Bay of Bengal (~70 Tg yr^{-1}) while that in the Arabian Sea is negatively balanced (106 Tg yr^{1}). The steady state carbon flow in these bodies is, however, maintained by water exchanges with the southern Indian Ocean.

Box 3.2. Importance of the Arabian Sea

The uniqueness of the Arabian Sea, which is comprised of highly variable physical and biogeochemical regimes, attracted several countries to undertake multi-disciplinary expeditions (between 1992 and 1997) to study the carbon flows in this dynamic region. Areas studied include the southwest monsoon upwelling and winter convective regimes. The experiments have established that the Arabian Sea acts as a CO_2 source to the atmosphere, although the extent varies in space and time. The results reveal that dynamics plays an important role in making nutrients available, which can potentially alter the trophic structure and particulate fluxes to the deep sea. Further, higher carbon demand by bacteria in the oxygen minimum zone appears to have been met by transparent exopolymer particles (Kumar et al. 1998). Intense regeneration processes that occur both at the sea surface and in the ocean-interior result in very little carbon sedimentation. Based on these measurements and the results of modelling studies, the budgets of carbon for the upper 100 m and 1000 m of the central and eastern Arabian Sea have been estimated. The transport of carbon by circulation is dominant over biological material fluxes. Carbon in the surface waters (100 m) of central and eastern Arabian Sea appears to be in a near steady state since the imbalance of ~0.11 Gt yr^{-1} is within 10% of the total fluxes (Table 3.5) and within the range of computational uncertainty. The Arabian Sea is an important and perennial source to CO_2 in atmosphere and results from North Indian Ocean are in excellent agreement with the physical-biogeochemical modelling results that depict emission from the Arabian Sea, but absorption by the Bay of Bengal (Swathi et al. 2000).

3.5.2.3 *Anthropogenic Carbon Dioxide in the Indian Ocean*

The invasion of anthropogenic carbon dioxide into the Indian Ocean has been established only recently. Anthropogenic CO_2 is higher in surface waters of the Bay of Bengal than of the Arabian Sea, with the Bay of Bengal surface water having ~50 µmol kg^{-1} while the Arabian Sea and Red Sea surface waters only contain ~42 and 47 µmol kg^{-1}, respectively. The penetration of the human-induced carbon is deepest in the Gulf of Aden followed by the Arabian Sea and Bay of Bengal (Goyet et al. 1999). Vertically integrated anthropogenic carbon is high at 30° to 40° S, but lower in the northern Indian Ocean. The accumulation of anthropogenic CO_2 to the north of 35° S between 1978 and 1995 is 4.1 Pg C, which amounts to an invasion rate of 0.24 Pg C yr^{-1} in the Indian Ocean (Sabine et al. 1999).

3.5.3 Methane

The Arabian Sea is an area where the emission rate of CH_4 is several times higher than the global average, but not large enough to affect the global CH_4 budget. The estimated flux of ~0.04 Tg C yr^{-1} is merely 0.01% of CH_4 emission rate to the atmosphere from all sources (Owens et al. 1991; Patra et al. 1998). The pronounced sub-surface maximum in CH_4 concentration observed in most oceanic areas is also found in the Arabian Sea. However, it seems to be intensified by an acute O_2 deficiency and may be formed *in situ* due to CH_4 production within anoxic interiors of particles (Jayaku-

Table 3.5. Carbon budgets in the upper 100 m and 1000 m of the water column in the central and eastern Arabian Sea

Process	Carbon (10^3 Tg yr^{-1})
Upper 100 m	
Sea-to-air flux	–0.032
Lateral influx	+0.569
Dissolved sinking flux	–0.568
Particulate sinking flux	–0.074
NET	–0.105
Upper 1 000 m	
Sea-to-air flux	–0.032
Lateral influx	+0.569
Lateral outflux	–0.795
Upward flux (1 000 m)	+0.212
Particulate sinking flux	–0.003
NET	–0.049

mar et al. 2000). Methane saturation in surface waters of the Arabian Sea is between 110% and 256%, giving rise to an average emission rate of 1.64 ±0.68 μmol $m^{-2} d^{-1}$ (Jayakumar 1999a). Close to the river mouths, CH_4 concentrations seem to be high. For example, in the coastal waters off Goa emission rates varying from 20.75 ±16 to 136.4 ±32 μmol $m^{-2} d^{-1}$ have been observed (Jayakumar 1999a). Large near-bottom anomalies, such as those observed in areas of known hydrocarbon seepage, are not seen in the inner shelf region, where the presence of gas charged sediments is indicated by geophysical data (Karisiddaiah and Veerayya 1996), indicating that the sedimentary supply does not dominantly control CH_4 distribution in coastal waters. Instead, the available data strongly suggest that the coastal wetlands, particularly mangrove swamps, generate large amounts of CH_4, and probably make an important contribution through freshwater discharge to CH_4 cycling in the coastal region (Jayakumar et al. 2000).

3.5.4 Nitrogen

3.5.4.1 *Open-Ocean Denitrification*

The waters of the northern Indian Ocean show severe O_2 deficiency at intermediate depths. Between approximately 150 and 1 000 m in the northern and central parts of the Arabian Sea O_2 levels are lower than 5 μM (~0.1 ml l^{-1} corresponds to ~2% of surface-water concentrations). The distribution of nitrite (NO_2^-), an intermediate product that accumulates in the water column during denitrification, reveals that the entire O_2-deficient layer is not strongly reducing. An abrupt build-up of NO_2^- occurs when the O_2 concentrations fall below 1 μM (~0.02 ml l^{-1}). Such a strong dependence of redox processes on minor changes in ambient O_2 levels implies an extremely sensitive ecosystem with a delicately poised biogeochemical balance, which may be expected to react rapidly to anthropogenic perturbations. Changes in ambient O_2 levels have occurred in the past. The rapid increase in $\delta^{13}C$ that took place during the early deglaciation has been attributed to fluctuations in North Atlantic Deep Water production (Naqvi et al. 1994a). The same may have occurred in the north Indian Ocean. Evaluations through nitrate-deficit and electron-transport calculations yielded a remarkably similar denitrification rate of around 30 Tg N yr^{-1} (Naqvi 1987). This is approximately one-third of the estimated global water column denitrification rate. Intense denitrification is found reflected in higher values of $\delta^{15}N$ in sediments (Schafer and Ittekkot 1996). Lower $\delta^{15}N$ were found associated with glacial events implying that non-denitrifying waters prevailed at these times (Altabet et al. 1995).

3.5.4.2 *Coastal Anoxia and Denitrification*

In addition to the perennial denitrifying layer found in the central Arabian Sea, reducing conditions also develop in near-bottom waters of the inner continental shelf along the central and southwestern coasts of India. However, this is a seasonal phenomenon confined to the southwest monsoon when moderate to strong upwelling occurs along the Indian coast. Observations during four southwest monsoons since 1995 reveal that cold, nutrient-rich and oxygen-poor waters are invariably found within ~10 m of the surface all along the coast south of Goa. However, the upwelled water is frequently overlain by a layer of low salinity (10 m thick) resulting from freshwater runoff and local precipitation. This leads to very strong thermohaline stratification preventing the upwelled water from reaching the sea surface. Where the stratification is broken by turbulence, very high nutrient concentrations (up to 16 μM nitrate) are observed. This triggers very high rates of primary production (up to 6.33 g C $m^{-2} d^{-1}$).

The water upwelling onto the inner shelf is probably derived from the poleward undercurrent, which has a slightly elevated oxygen content. However, intense photosynthetic production creates a very high oxygen demand below the thermocline. Accordingly, the sub-surface waters in the inner shelf become severely depleted in oxygen and reducing with the passage of time. Denitrification may at times be extreme leading to complete removal of NO_3^- after which nitrate-free subsurface waters may produce a slight odor of H_2S.

Periodic observations off Goa provide a record of the evolution of suboxic/anoxic conditions off the sub-continent. Reducing conditions develop during the peak southwest monsoon season (July–August) and gradually intensify thereafter causing sulphate reduction. Anoxia events in the inner shelf are closely related to the occurrence of the freshwater lid, underlying the control of terrestrial processes on near shore biogeochemistry. These appear to play a crucial role in determining the living resources of region. Their response to future climate change is expected to have significant socio-economic implications.

3.5.4.3 *Nitrous Oxide*

Large super-saturations of N_2O have been observed in the surface waters of the Arabian Sea, especially within the three upwelling zones, resulting in a very high rate of N_2O emission (Naqvi and Noronha 1991; Naqvi et al. 1998). The gas accumulates in high concentrations when O_2 levels fall to <0.5 ml l^{-1}, but do not reach suboxia. However, in strongly reducing environments, such as

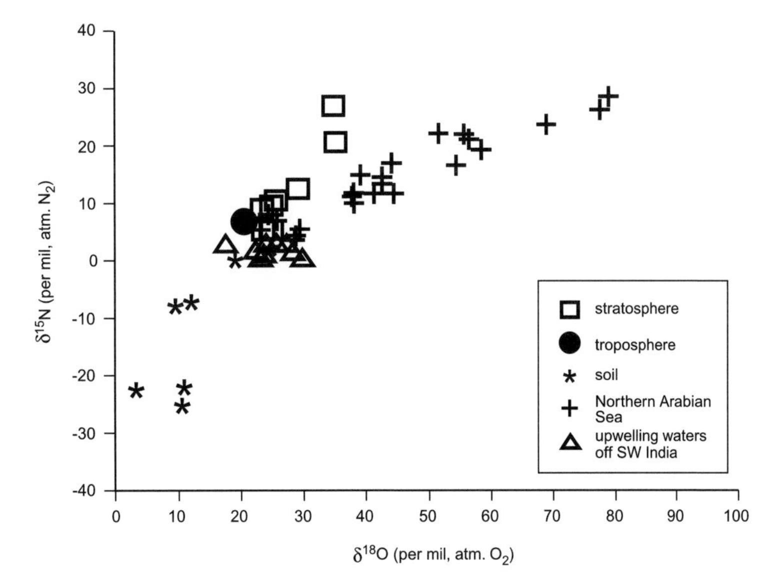

Fig. 3.22. Variation of $\delta^{18}O$ and $\delta^{15}N$ over South Asia (Naqvi et al. 1998)

those found within the core of the denitrifying zone, N_2O is itself reduced to N_2 by bacteria. The Arabian Sea contains sites located close to each other, vertically as well as horizontally, which serve both as strong sources and sinks of N_2O leading to a turnover which may be the most rapid on the planet (Naqvi et al. 1998).

The highest isotopic values ($\delta^{15}N$ and $\delta^{18}O$) and the lowest values (less than that in troposphere) of N_2O in the geochemical reservoirs (except soils) occur in the denitrifying intermediate waters of the Arabian Sea and in the upwelled waters of the southwest coast of India, respectively (Fig. 3.22). The rate of N_2O turnover is particularly high in the coastal upwelling zone off southwestern India during the southwest monsoon, where some of the highest concentrations of N_2O as well as the steepest vertical gradients in the world have been observed. Saturation of 128–300% is observed over the open Arabian Sea; in the Indian coastal waters during the southwest monsoon season it may be 1 364%. By comparison, in the Bay of Bengal the surface saturation is only 89–214% and consequently the sea-to-air flux of N_2O to the east of India is substantially smaller due to the strong thermohaline stratification. Sea-to-air fluxes from the Arabian Sea are greater (0.4–0.9 Tg yr^{-1}) than from the Bay of Bengal (0.03–0.08 Tg yr^{-1}). Moreover, due to the absence of bulk water-column denitrification, there is no water-column sink of N_2O. Since the mid-depth O_2 levels are most conducive for N_2O build-up, the Bay of Bengal region serves a large oceanic reservoir of N_2O (Naqvi et al. 1994b).

In the Arabian Sea, a steep decrease in $\delta^{15}N$ is observed in the denitrifying layer, a trend that is exactly opposite to that seen outside the suboxic zone. Although this feature can result from an input of unaltered NO_3^- horizontally advected from outside the suboxic zone, this is unlikely. A more plausible source of isotopically light

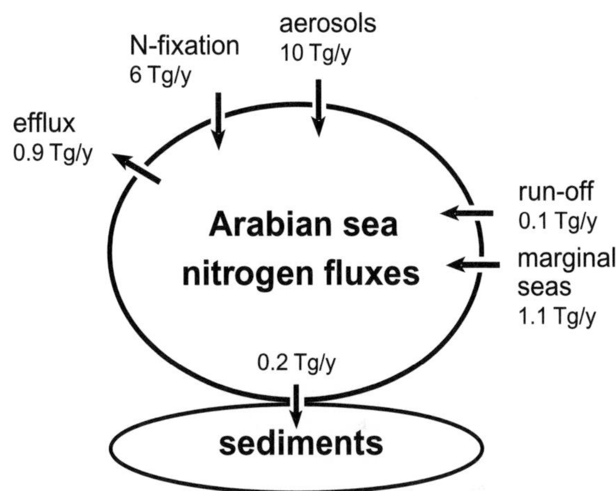

Fig. 3.23. The nitrogen budget for the Arabian Sea

NO_3^- in the surface layer may be nitrogen fixation, the extent of which may be estimated from the isotope mass balance. It is possible that 40% of the NO_3^- present at the bottom of the surface mixed layer might be derived through fixation. The estimated nitrogen fixation rate for the Arabian Sea (~6 Tg N yr^{-1}) is roughly a quarter of the water-column denitrification rate (Brandes et al. 1998). Thus, in spite of a high rate of nitrogen fixation, the Arabian Sea still serves as a net sink of combined nitrogen. From the external fluxes alone (Fig. 3.23), the nitrogen budget appears to be positively balanced.

3.5.5 Consequences of Global Change in South Asian Seas

Global warming is likely to lead to increased rainfall from intensified monsoons and possibly from increased cyclone activity, increased ice-melt water from the Himalayas and an increase in soil erosion. Such conditions favor large-scale transportation of water, suspended sediments and nutrients into the neighboring seas. Within the Bay of Bengal, this would result in salinity stratification of the surface water and promotion of biological production leading to higher rate of carbon sequestration. Enhanced biological production may favor an increase in commercial fish stock and have positive socio-economic impact. In contrast, the impact in the Arabian Sea would be an increase in occurrence of severe anoxia following the southwest monsoon, leading to a diminution in the oceanic living resources and negative socio-economic impact.

The biogeochemistry and biodiversity in the oceanic zones of South Asia, as manifested in coastal processes, are sensitive to climate change as well as land use-related activities, primarily agriculture, which contributes a significant fraction of nutrient inputs to the coastal margins.

The biogeochemical environments of the Bay of Bengal and Arabian Sea are very different. Consequently to ameliorate undesirable anthropogenic effects in the coastal zones and oceans off South Asia will require different strategies in future for eastern and western oceanic regions. Approaches to land use, river management, agriculture and coastal zone management will have to be adapted to what is required by each environment.

3.6 Agriculture and Global Change in the Region

The economy of the South Asian region is largely agrarian and is heavily dependent on the summer monsoon. Any changes in the variability of the monsoon in future will have significant impacts on the food security of the region. Pests and diseases affecting agriculture production will also be affected by the changing environmental conditions. Before future changes can be considered, it is necessary to consider present agricultural conditions in the region.

3.6.1 Current Agricultural Conditions and Practices

3.6.1.1 *India*

The agricultural sector represents 35% of India's Gross National Product (GNP) at present and will long continue to occupy an important place in the national economy. It sustains the livelihood of nearly 75% of the population. To meet the ever-increasing demand of more than one billion people, food grain production has increased since the mid twentieth century from 51 Mt to around 200 Mt. This has been made possible by increasing the area under cultivation to about 140 million hectares and by enhancing yield and productivity. The increase in area has been achieved by conversion of traditional grazing land of low productivity into irrigated agriculture lands. The availability of high yielding crop varieties, along with increased use of fertilizers, irrigation and pesticides has largely contributed to increased food grain production in the last three decades. The increase has come in relatively well-endowed areas with few soil and/or climate-associated constraints limiting production and has led to an increase in net income of the farmers in the Indian states of Punjab, Haryana, Western Uttar Pradesh, West Bengal, parts of Andhra Pradesh, Tamil Nadu and Rajasthan. Despite the increase in grain production over the last few decades (Table 3.6), the impact of monsoon variability has remained large throughout the period.

Future challenges include increasing food grain production with only a limited increase in cultivated area, a decline in farm size, a marginal increase in irrigation and a depleting soil resource base. While steps need to be taken to improve management technologies for well-endowed areas, the emphasis needs to be placed in future on regimes which have climate and soil constraints and account for over 65% of the cropped area (Abrol 1996; Gadgil et al. 1999a).

Table 3.6. Area, production and average yield of crops in India (1972–1995) (TEDDY 1998/99)

Crop	1975–76			1996–97		
	Area (Mha)	Production (Mt)	Average yield (t ha^{-1})	Area (Mha)	Production (Mt)	Average yield (t ha^{-1})
Rice	39.5	48.7	1.23	43.3	81.3	1.88
Wheat	18.2	28.8	1.41	25.9	69.3	2.67
Jute[a] and Mesta[a]	1.1	5.9	1.16	1.1	11.0	1.84
Sugarcane	2.6	140.6	50.9	4.2	277.3	67.0
Pulses	22.5	13.0	0.53	23.2	14.4	0.62
Oil seeds	16.6	10.6	0.63	26.8	25.0	0.93
Food grains	124.3	121.0	0.94	124.5	199.3	1.6
Cotton[b]	7.6	6.0	0.14	9.1	14.3	0.27

[a] In bales, 1 bale is 180 kg.
[b] In bales, 1 bale is 170 kg.

3.6.1.2 *Pakistan*

Agriculture is the largest sector of the Pakistan's economy and contributes about 25% to GDP. The predominantly rural population directly depends on agriculture for its livelihood. Major industries of the country are also based on agriculture, and agricultural products (raw and processed) constitute 80% of the national product. Wheat is the major crop and main staple food-grain. Cotton is the most important cash crop and is the basis for the country's extensive textile industry. Rice and sugarcane are other major crops besides fruits, vegetables, fodder and coarse grains.

Pakistan is largely an arid to semi-arid country; except for the mountainous areas in the northeastern part, the rest of the country, especially the provinces of Sindh and Baluchistan, are mostly arid. The effects of climate change on hydrology are therefore of great importance to Pakistan. Extensive irrigation systems have been developed to help assure water supply for crop production, but a major problem is the large extent of salinity and water logging in vast tracts of irrigated areas.

In Pakistan, the production of major crops like wheat, rice, cotton and sugarcane needs to be doubled by the year 2020 in order to meet the requirements of the country's growing population. It is believed that this increase in production can be achieved through improving irrigation efficiency and enhancing productivity, including better use of weather and climate forecasts. It is anticipated that the improved agronomic practices coupled with improved irrigation efficiency can lead the country towards achieving and continuing self-sufficiency in food production (UNEP 1998).

3.6.1.3 *Bangladesh*

About 74% of the effective land area of Bangladesh is under crop production. Currently, the agriculture sector employs about 65% of the labor force. More than 100 different crops are presently grown. Rice is the staple food and by far the most important crop (grown in all three growing seasons of the year and covering 78% of the total cropped area) although wheat is becoming increasingly important.

Cropping intensity in Bangladesh increased from 151% to 179% between 1972 and 1995 while the total agricultural land and the net cropped area decreased (Table 3.7). Rice is the main cereal crop and contributes nearly 94% of the total production; wheat contributes ~6% (BBS 1993). Rice production has increased from 9.9 Mt in 1972/73 to 18.34 Mt in 1992/93. The area under rice production has increased at an average annual rate of 0.3%, while the production has increased at nearly 4.3%.

3.6.1.4 *Nepal*

There are four physiographic regions in Nepal: the Terai, the Siwalik, the Middle Mountains and the High Mountains. The Terai is the northward extension of the Indo-Gangetic Plains and comprises 14.3% of the total landmass. Though the percentage of area of the Terai is small, more than 46% of the population resides in this area. The Terai is also the granary of the country and supplies food to the people living in the northern mountain areas. Nearly 82% of the total cultivated area in the Terai is under rice production, but wheat is becoming an increasingly important crop. The area and yield of different crops in the Terai are shown in Table 3.8.

The economic performance of Nepal depends primarily on the success and failure of its agricultural crops, particularly in the Terai. A major concern of the country is increasing the level of agricultural production in a sustained way without degradation of environmental resources. Agricultural practices are still traditional and the use of modern inputs in very limited. For example, improved seed varieties are used by only 37% of the total rice holdings, 46% of the wheat holdings and 18% of the maize holdings. Similarly, chemical fertilizers are used in 53% of the area under rice, 70% under wheat and 15% under maize (CBS 1994).

Table 3.7. Area, production and average yield of crops in Bangladesh (BBS 1972–1995)

Crop	1972–73			1994–95		
	Area (Mha)	Production (Mt)	Average yield (t ha^{-1})	Area (Mha)	Production (Mt)	Average yield (t ha^{-1})
Rice	9.65	9.9	1.02	9.92	16.83	1.69
Wheat	0.12	0.09	0.75	0.65	1.3	2.0
Jute	–	–	–	0.57	5.9	10.35
Sugarcane	–	–	–	0.19	7.9	41.57
Pulses	0.31	0.21	0.67	0.83	0.6	0.72
Oil seeds	0.30	0.23	0.76	0.50	0.45	0.9
Spices	0.15	0.30	2.0	–	–	–
Fruits	0.14	1.21	8.64	–	1.51	–
Potato	–	–	–	0.17	1.82	10.70

Table 3.8. Area and yield of different crops in the Terai, Nepal (CBS 1996)

Crop	1981		1994		Change	
	Area (ha)	Yield (t ha^{-1})	Area (ha)	Yield (t ha^{-1})	Area (ha)	Yield (t ha^{-1})
Paddy	1.033	1.90	0.989	2.12	−0.04	+0.22
Wheat	0.252	1.33	0.349	1.59	+0.09	+0.26
Maize	0.131	1.64	0.169	1.92	+0.03	+0.28
Millet	0.018	0.96	0.017	1.00	−0.001	+0.04
Barley	0.007	0.83	0.002	0.93	−0.004	+0.10

Table 3.9. Some estimates of global change effects on agriculture

Source	Parameter	Nature of effect	Approximate magnitude
Long-lived greenhouse gases	i) Increase in temp.	NEGATIVE	10% in yield at 2 °C
	ii) Changes in ppt.	POSITIVE	10^6 tons food grain per 100 mm change
Aerosols	i) Reduction of sunlight	NEGATIVE	10% in the haze region
	ii) Acidic properties	NEGATIVE	Uncertain
	iii) UV effects	Uncertain	Uncertain
CO_2 enrichment	CO_2 enrichment	POSITIVE	10–15% change in yield for 2 × CO_2
Ozone changes	i) Direct effect on plants from tropospheric ozone	NEGATIVE	Uncertain
	ii) UV-B effect	NEGATIVE	Little

3.6.2 The Future

Since cropping patterns have changed markedly in the last three decades, future farming strategies that are appropriate for the climate variability of agroclimatic zones in the region will have to be determined by the use of state of the art crop models in conjunction with climate data (Gadgil et al. 1999a,b). Many of the net impacts changes in the greenhouse gas and aerosol composition of the atmosphere will have on agriculture yields are not yet fully known, but some of their estimates may be made (Table 3.9). Large uncertainties exist.

3.6.2.1 *Changes in Monsoon Rainfall*

Grain yields in South Asia show high positive correlations with monsoon rainfall. Yields in the poor monsoon year of 1974 were 20% below the good monsoon year of 1975 (Gadgil 1996) (Fig. 3.24). Indian monsoon rainfall has in general been stable in historical times with extreme events being a part of the pattern of natural variability. These extremes, droughts and floods, have a great impact on the economic health of the region. Periods of more frequent droughts (1895–1932 and 1933–1987) and relatively drought-free periods (1872–1894 and 1933–1961) have occurred in the past. How drought and flood frequencies may change in the future and how alterations in the monsoons may be introduced or accentuated by global change are issues of great concern in agriculture.

Global warming might have played a role in the recent weakening of the ENSO-monsoon relationship, ensuring a normal monsoon even with severe El Niño conditions. As has been pointed out earlier, several GCM models suggest a greater intensity of the summer monsoon and rainfall in future; others show the opposite. In any event, the projected differences are less than existing inter-annual variability. Given the considerable uncertainties associated with GCM model outputs for South Asia, the development of regional climate modelling for South Asia, possibly using an approach similar to that being used in East Asia (see Chap. 5) is an urgent need.

3.6.2.2 *Effects of Changing Temperature*

The warming of 0.4 °C per century over the last hundred years or so in India, due mainly to an increase in maximum temperature, has had an adverse effect on grain yields. In South Asia, the productivity of wheat and other crops declines markedly at high temperatures (Fig. 3.25). The growing season of wheat is limited by high temperature at sowing and maturation phases. Rain-fed wheat depends upon soil moisture remaining after the monsoon recedes in September. High maximum and minimum temperature affect seedling estab-

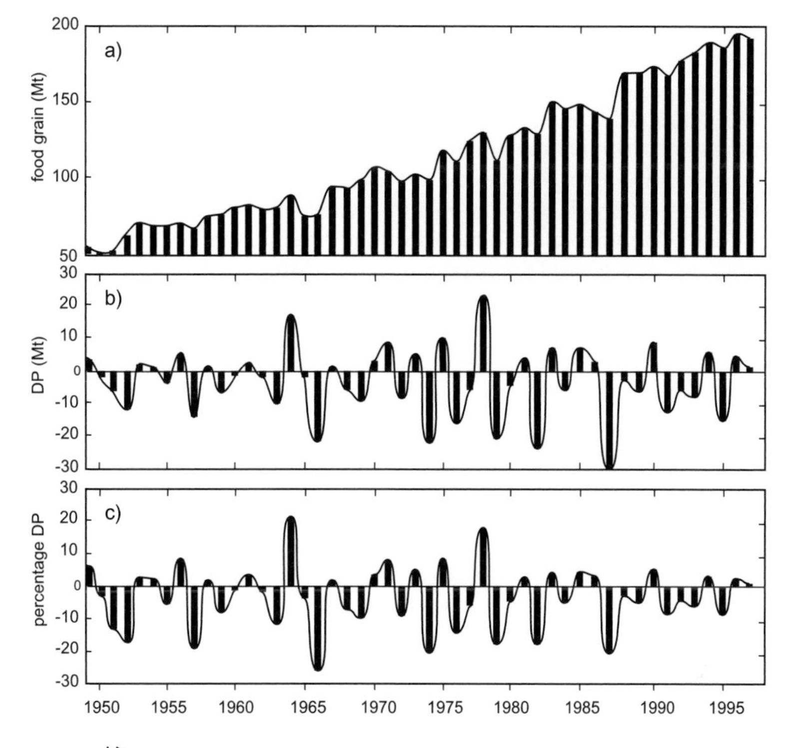

Fig. 3.24.
All-India food grain production, together with differences and percentage differences in that production, 1949–1996) (Gadgil 1996)

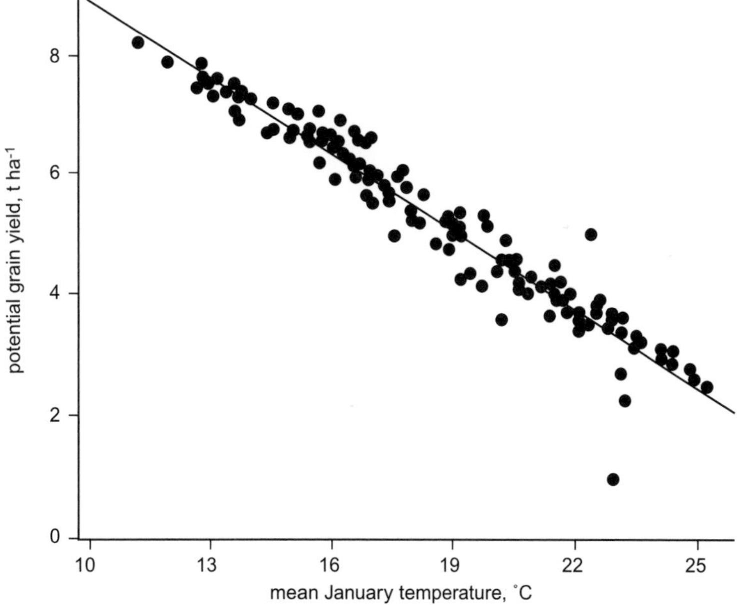

Fig. 3.25.
Sensitivity of grain yields to January temperature changes (Abrol et al. 1996)

lishment, accelerate early vegetative development, reduce canopy cover, tilling, spike size and yield. Hence, sowing is typically delayed until after mid-October when the seedbeds have cooled, even though much of the moisture may be lost. High temperatures in the later half of February (maximum: 25 °C/minimum: 10 °C), March (30 °C/13 °C) and April (30 °C/20 °C) reduce the number of viable florets and grain filling duration. High temperature stress reduces yield of wheat sown in December/January, which is necessitated in some regions because of a multiple-cropping system (Abrol et al. 1996).

Rice crops tolerate a maximum air temperature of around 30 °C during most of the growing period; higher temperatures may be sustained for short periods during breaks in or early withdrawal of the monsoon. Sustained low temperatures limit rice growth and yield. This is especially true at higher latitudes (sub-Himalayan rice growing areas) during the seedling and vegetative phase in the spring season and during the grain-filling phase in the wet season. Temperature also influences the growth rate and productivity of rice crops. Generally, the growth rate of the rice crop increases linearly with temperature in the range of 22–31 °C. Higher temperatures have a negative effect on rice growth and productivity. At flowering and during grain filling, high temperature reduces yield by causing spikelet sterility and shortening the duration of the grain-filling phase. The situation is similar for sorghum and pearl millet when exposed to extreme high temperatures in Rajasthan, India. After sowing, air and soil temperatures often exceed 40 °C and mid-day soil temperatures of 60 °C are common.

3.6.2.3 *Modelling Change*

It is important for agriculture to be able to assess the likelihood of future changes in frequency of extreme events and to ascertain whether the impact of increased rainfall and CO_2 will negate the impact of increased temperature and evapotranspiration (Gadgil 1996). This is especially so since many crops are highly sensitive to small changes in climate. For instance, under normal management conditions in North India, every half degree rise in temperature results in a decrease in the growing period for wheat by 7 days, which in turn reduces yield by 0.45 t ha^{-1} (Aggarawal 1991; Abrol and Ingram 1996).

A number of studies in India have been aimed at understanding the effects of atmospheric CO_2 and associated climatic change on crop yields (Abrol et al. 1991; Sinha and Swaminathan 1991; Aggarwal and Sinha 1993; Aggarwal and Kalra 1994; Gangadhar Rao and Sinha 1994; Lal et al. 1998; Saseendran et al. 2000). Some crop models suggest that the integrated impact of higher ambient surface temperatures with higher CO_2 concentrations may be negative, since the projected 3 °C rise in surface temperature more than cancels the increased CO_2 fertilization effect (Sinha and Swaminathan 1991). Strong interactions between changing components of climate and rice and wheat yields have been demonstrated (Lal et al. 1997) (Fig. 3.26). A similar result has been shown for soya bean yields in Central India (Lal et al. 1999).

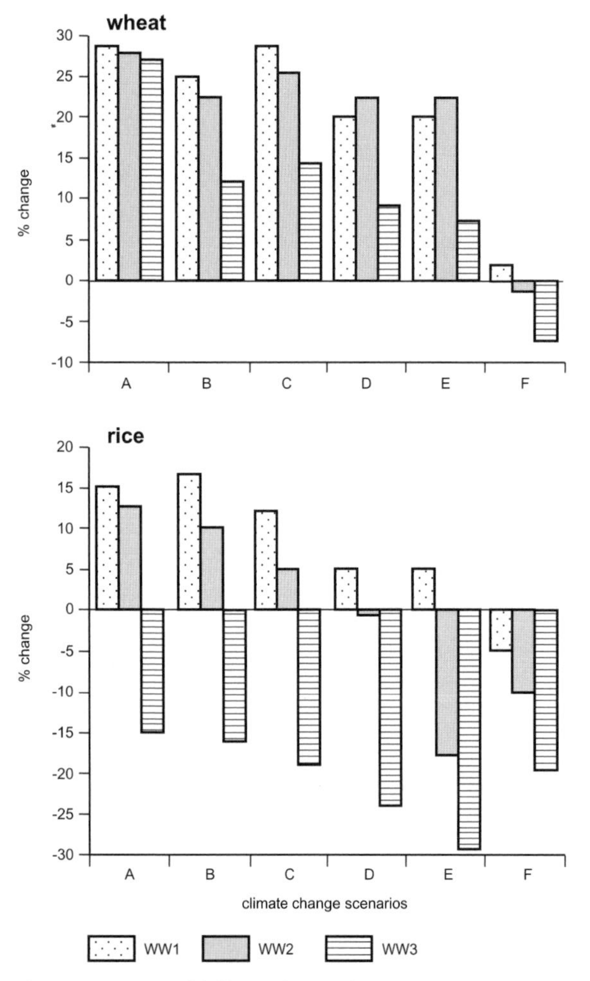

Fig. 3.26. Impacts of different climate change scenarios on wheat and rice production in northwest India (Lal et al. 1999). *WW1* denotes a business as usual irrigation scenario, *WW2* a moderate conservation scenario, *WW3* a highly conservative scenario

Use of the WTGROWS model suggests that with an enhanced CO_2 concentration of 425 ppm, and with no rise in temperature, irrigated and rainfed wheat yields would increase significantly in most parts of India (Aggarwal and Kalra 1994). With present-day CO_2 levels and an increase of 2 °C in temperature potential yields are reduced at most places by as much as 18% in tropical locations and by a smaller percentage in sub-tropical environments. When temperature increases by 2 °C and CO_2 to 425 ppm irrigated wheat yields are projected to increase slightly for latitudes greater than 27° N, but are reduced elsewhere. In several areas rainfed yields are projected to increase by as much as ~28%.

Sensitivity analysis in India with the CERES model on both wheat and rice give similar results (Lal et al. 1998). With doubling of CO_2 levels alone, yields of wheat and rice increase by 28% and 15% respectively. However a simultaneous increase in mean annual temperature of 3 °C in the case of wheat and 2 °C in the case of rice negates the CO_2 fertilization effects. While the adverse impacts of likely water shortage on wheat crops would be minimized to a certain extent under elevated CO_2 levels, they would largely be maintained for the rice crops resulting in about 20% net decline in the rice yields. Similar analyses using the CERES-rice model for Kerala State likewise show an increasing CO_2 fertilization effect leads to yield increases and enhanced water use efficiency (Saseendran et al. 2000). However, for every accompanying 1 °C increase in temperature yields decline by about 6%.

3.6.2.4 *Effects of Changing UV-B Levels*

The current increase in chlorofluorocarbons and other trace gases is linked with the depletion of stratospheric

Box 3.3. Free Air Carbon Dioxide Enrichment (FACE)

The exponential rise in carbon dioxide concentration of the atmosphere is now a global reality. A doubling of pre-industrial levels of CO_2 is anticipated in the 21st century. This is expected to affect the carbon balance in the biosphere and photosynthetic carbon assimilation in plants, effectively influencing the productivity of crop plants. Several studies have been conducted on the possible effects of elevated CO_2 on plant processes, growth and productivity due to differences in response at different stages of growth, crop species and other growth limiting factors, such as nutrition, moisture, pollutants, etc. For most such studies, open top chamber (OTC) facilities have been employed at various locations in South Asia. However, such facilities suffer from the confounding effects of growth chambers such as decreased light intensity, unnatural wind flow, elevated temperature and disturbed soil patterns.

The Free Air Carbon Dioxide Enrichment (FACE) Facility is a response to the need for realistic field data on carbon fixation under controlled, elevated CO_2 levels, without the confounding effects of growth chambers. The FACE system with advanced technologies provides new opportunities to answer questions ranging from molecular biology to plant canopy dynamics to global carbon balance including the effects on rising CO_2 on crop ecosystems, rhizosphere, microbial growth, soil system, plant insect relation, biogeochemistry of soils, etc.

The South Asia Regional Free Air Carbon Dioxide Enrichment programme includes one large and one medium size FACE facility at the Indian Agricultural Research Institute, New Delhi (Fig. 3.27). The South Asian FACE efforts have focused on the effects of elevated CO_2 on rice because it is the staple diet of the people of South Asia and its productivity has to be maintained even under elevated CO_2 levels to meet the growing demands of the region.

The salient findings of the experiments conducted so far on rice and other crops are:

- All the yield components in rice, including panicle number (effective tillers), filled grains per panicle and grain weight, responded positively to enhanced carbon dioxide levels.
- Carbon dioxide enrichment not only reduced the adverse effect of low levels of sulphur dioxide but the extra-carbon supplied by carbon dioxide enrichment took the advantage of air-borne sulphur as a nutritional supplement as shown by a maximum increment in growth and yield.
- At elevated levels of carbon dioxide, berseem (*Trifolium alexandrium*) responded by an increase in vegetative biomass so that early harvesting was possible. However, there was an increase in C:N ratio, making it necessary to increase nitrogen input or identify genotypes with improved nitrogen fixing ability in order to maintain optimum fodder quality.
- Enhanced carbon dioxide concentration effected the carbon dioxide assimilates and their partitioning within the source leaf and transport to the sink in mungbean and wheat.
- Carbon dioxide elevation partially compensates for the negative effect of moisture stress in *Brassica* plants and may possibly help them grow in the drier habitat than they are currently grown.
- *Brassica* spp. responded differently to elevated carbon dioxide levels. The species that are susceptible to moisture stress (*B. campestris* and *B. nigra*) responded significantly to higher carbon dioxide as compared to *B. juncea* and *B. carinata.*
- Parents (*B. campestris* and *B. oxyrrhina*) and their hybrids responded differentially to increased carbon dioxide. The variability in response in terms of photosynthesis and leaf/root ratio could be attributed to their sink potential (Uprety et al. 2000).

Fig. 3.27. The medium-sized FACE facility at the Indian Agricultural Research Institute, New Delhi

ozone layers and a consequent increase in ultraviolet-B (UV-B) rays reaching the earth surface. These rays are capable of disrupting proteins, nucleic acids and other important plant constituents. Although ozone depletion may lead to greater relative increments of effective UV-B at temperate and high altitudes than in the tropics, the resulting UV-B levels are already intense in the tropics because of low ozone content and any further change could be serious.

Most of the reports concerning the effect of UV-B radiation in crop plants have focused on temperate regions, with only a few studies investigating the response of tropical vegetation to enhanced levels of UV-B irradiation. It is necessary to establish whether tropical plants have an inherent resistance to UV-B given that tropical regions receive a higher flux of UV-B because of natural latitudinal variation in the thickness of the stratospheric ozone layer and the angle of incoming solar radiation.

Investigations on effects of ambient, as well as enhanced levels of UV-B on tropical legumes, cereals (C_3 and C_4) and Brassicas have been carried out. These have shown that the ambient UV-B level (22.8 $\mu W\ cm^{-2}\ \mu m^{-1}$) around Delhi has little effect on maize it inhibits optimal crop growth in mungbean (Pal et al. 1997). With enhanced UV-B, corresponding to 15 and 20% depletion of ozone, physiological functioning and biomass production in tropical legumes in inhibited, markedly so in some cases. In this respect C_3 crops are more susceptible than C_4 (Singh et al. 1994). The combined effects of future increases in temperature, radiation and CO_2 levels have been simulated for the Delhi area (Fig. 3.28).

3.6.2.5 *Changing Atmospheric Aerosols Concentrations*

Given that low-level haze is such a pervasive feature of the climate over South Asia, the effect of changing aerosol concentrations in the haze layer on radiation levels available for plant utilization in agriculture is an issue of concern. Significant diminution of incident solar ra-

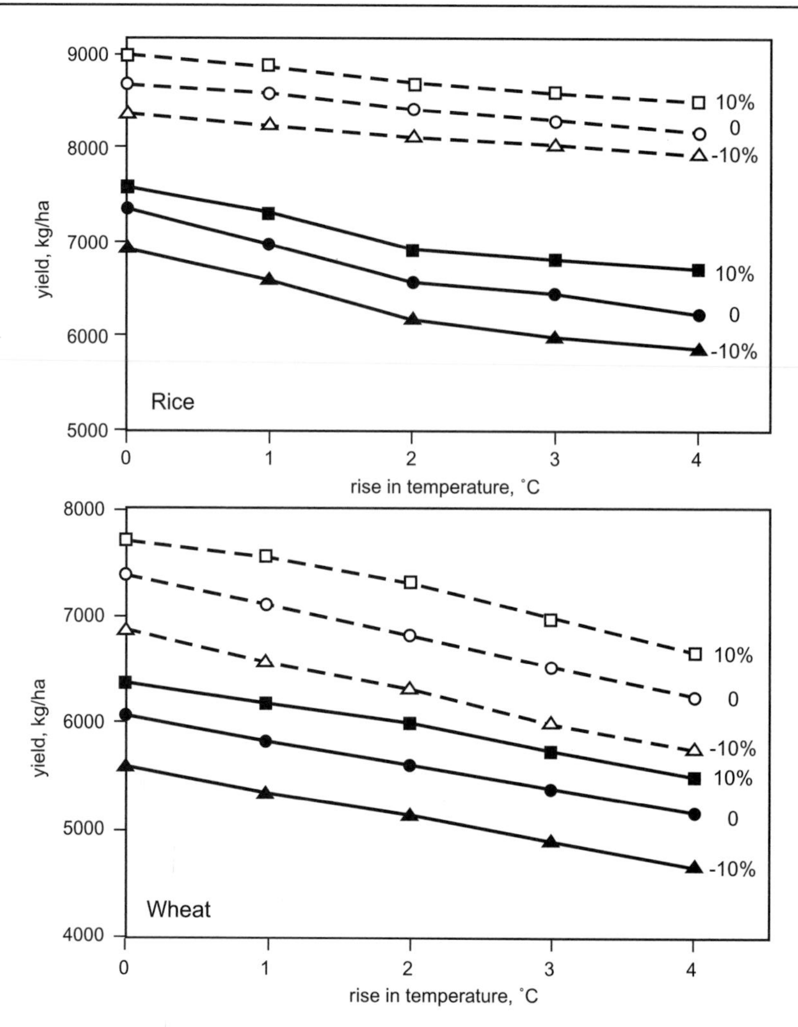

Fig. 3.28.
Climate change effects on rice and wheat yields in New Delhi environment (*bold lines* denote CO_2 levels of 350 ppmv; *dotted lines* those of 700 ppmv)

diation probably results in reduced photosynthesis and crop yield (Chemeides et al. 1999). Accumulation of dust and other aerosol particles on leaves may further reduce photosynthesis and plant growth. The effect of acid deposition is likewise a problem in some areas for regional ecosystems and may also be for agriculture. However, the precise magnitude and role of these effects remains to be determined for the region.

3.6.2.6 *Global Change and the Effects of Pests and Diseases*

As elsewhere, in South Asia the incidence of pest and diseases affecting crop yields is most severe in tropical regions, owing the warm moist climate, multiple cropping systems and availability of alternate pests throughout the year (Ramana Rao et al. 1994; Rao and Rao 1996; Abrol and Gadgil 1999). Pests and disease have marked deleterious effects on crop yields. How climate will change in future will determine the extent to which the consequences of pests and diseases will increase or decrease in times to come.

Changes in temperature, precipitation, humidity and extreme events will greatly influence the spread of pests and fungal disease. The impact of change on insect feeding behaviour associated with the direct physiological effects on host plant metabolism and chemical composition e.g. higher C:N ratios and accumulation of secondary compounds, such as phenols which are associated with allelopathy among plants and host plant resistance to insects and pathogens, may be significant (Abrol and Sharma 1996). Furthermore, modification in the susceptibility of pathogen attack because of biochemical or structural changes in the host and variations in host plant stress from higher temperatures or drought associated with global warming may increase vulnerability to pest attacks and diseases. Changes in the geographical distributions of agricultural insect pests, increased over-wintering, stimulation in rate of early season growth, reduction in growth, reduction in generation time, increase in number of generations per year, lengthening of development period, alteration in crop/pest synchrony, upsetting of natural control of predator and disease and increased invasion by migratory exotic pests will all occur with global change (Abrol and Sharma 1996).

Box 3.4. Use of Climate Information to Reduce Pest and Disease Effects: Groundnuts in South Asia

The Southern region of India is characterized by regular mono-cropping of groundnuts in vast areas with often the same variety (*viz.* TMV-2). Groundnuts, an important oilseed crop, are cultivated on about 8 million ha of which over 80% is rain-fed. Variation in yield arises to a large extent from the variation in the total rainfall during the growing-season. The crop variety is generally sown in July and harvested towards the end of about 120 days. The productivity is also critically dependent on the incidence of the pests and diseases. The organisms are always present at low levels of intensity and can multiply when the weather conditions are favourable and the plant susceptible to attack. For example, dry spells promote the incidence of *leaf miner* attacks, whereas wet spells promote *crown rot*, late *tikka* diseases and *collar rot*. Losses in yields of pod and straw of groundnut by pest diseases are variable depending on the intensity of attack and the crop growth during such attacks.

The approach to determine the incidence of dry and wet spells, and various crop phenological stage which are critical for determining yields, has been to determine probabilities, from 1901–1990 rainfall at locations in South India, of wet and dry spells at critical stages of growth in respect to the incidence of pests and diseases affecting the crop (Gadgil 1995). The numerical PNUTGRO model for growth, development and yield of groundnuts, has been validated for the prevalent groundnut variety (Singh et al. 1994) and used to determine the optimum sowing window (22 June to 17 August) for cultivating rainfed groundnuts (Gadgil 1995; Gadgil et al. 1999a,b). The sowing window is the period for climatologically minimizing the risk of failure. Analysis of crop phenological stages in relation to incidence of moisture stress reveales that the growth stage with maximum impacts on yield is the pod-filling stage and that the incidence of locally triggered pests and diseases, such as *leaf miner* and late leaf spot (*tikka*), is least when sowing is postponed to after mid-July.

The development of such models is useful, not only for assessing impacts of future change, but also for agricultural decision-support systems, such as those required for tailoring of crop varieties and operations to be optimal for the climate variability of a farming region. Costs and benefits of different farming strategies may be estimated and decisions regarding pesticide applications to control pests and diseases may be taken objectively (Gadgil et al. 1999b).

3.7 Worrisome Trends in the Bread-Basket Region of South Asia

The Indo-Gangetic Plain is among the most extensive fluvial plains of the world. It extends as an eco-region over 4 countries, namely Bangladesh, India, Nepal and Pakistan. In the case of India, the Indo-Gangetic Plain constitutes 21% of the total land area and houses over 40% of the country's population. It accounts for 26% of the net cropped area, 31% of the gross cropped area and contributes 51% to the national food grain output. The crop intensity has witnessed an increase of about 50% in this region from the 1960–1963 period to 1990–1993 period (Abrol 1998).

This region is well endowed with natural resources, deep productive soils, plentiful surface and ground water resources and a climate favorable for double, and in some cases even triple-cropping in a year. A detailed account of Indo-Gangetic soils and their properties is available (Velayutham et al. 2000). The carbon stock of this region has been analysed by Velayutham et al. (1999) and Bhattacharyya et al. (2000).

This region has been a major contributor to the green-revolution-era agriculture production increases in South Asia over the last 3 decades,. Rice-wheat cropping systems are the main production system in this bread-basket region of South Asia. Trends in rice and wheat yields in 2 selected sub-regions of the Indian part of the Indo-Gangetic Plain (Punjab and Bihar) illustrate growth in the region as a whole (Fig. 3.29) (Abrol 1998). While overall agricultural production has increased during the green-revolution-era, recently yields have begun to level off. There has also been a declining yield at constant input levels in long-term field experiments. Increasingly the realization has set in that increasing input levels are needed to maintain previous levels. While the exact reasons for the failure to maintain previous yields are not known with certainty, it seems likely that changes in soil structure and chemistry, through compaction and continued inputs of fertilizers and other nutrients, contribute to changes rendering soils less productive. Use of irrigation water has further exacerbated the situation; in some locations, declining water tables and surface salinity are also contributing to declining productivity.

3.7.1 Water in the Indo-Gangetic Plain

The river Ganges is the main source of irrigation for agriculture and other occupations on the Indo-Gangetic Plain (Table 3.10). By far the largest single usage is that of 145 km^3 annually for irrigation (Table 3.10).

The states most involved with water from the Indo-Gangetic Plain are Haryana, Punjab, Uttar Pradesh and West Bengal. Their ground water resources and the irrigation potential are given in Table 3.11.

Introduction of the rice-wheat cropping system in the alluvial plain has enhanced the consumptive water requirements beyond what can be met by precipitation and surface water resources. This has resulted in water tables declining differentially in the region and the replacement of crops, such as changes from cotton, maize and citrus to paddy and eucalyptus in the southwestern parts of the plain.

Rainfed agro-ecosystems occupy the major cropped areas in the Indo-Gangetic Plain. Out of the 142 Mha net sown area in the country, nearly two-thirds is under dryland, rainfed crops. Nearly 67 Mha of rainfed croplands have an annual rainfall of 500–1500 mm. The productivity and stability of rainfed crops is low and the agro-ecosystem is characterized by frequent moisture stress

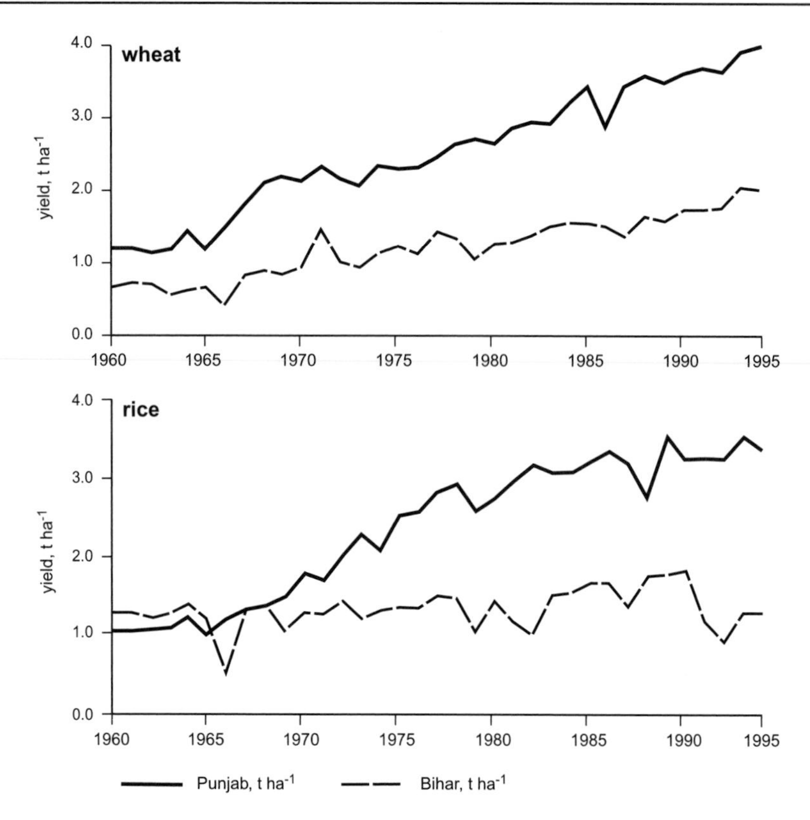

Fig. 3.29. Trends in wheat and rice yields in the Punjab and Bihar areas, India since 1960 (Abrol 1998)

Table 3.10. River Ganges water resources (Gupta et al. 2000)

Ganges Basin	
Total replenishable ground water resources (Mha m yr^{-1})	170.99
Provision for domestic industrial and other uses (Mha m yr^{-1})	26.03
Available ground water resources for irrigation (Mha m yr^{-1})	144.96
Net draft (Mha m yr^{-1})	48.59
Balance ground water potential available for exploitation (Mha m yr^{-1})	96.37
Level of ground water development (%)	33.52

Table 3.11. Groundwater resource and irrigation potential of the major states of Indo-Gangetic Plain (Gupta et al. 2000)

	Haryana	Punjab	Uttar Pradesh	West Bengal
Total replenishable ground water resources (Mha m yr^{-1})	0.85276	1.86550	8.38210	2.30923
Provision for domestic industrial and other uses (Mha m yr^{-1})	0.12792	0.18652	1.25743	0.34642
Available ground water resources for irrigation (Mha m yr^{-1})	0.72484	1.67898	7.12467	1.96281
Net draft (Mha m yr^{-1})	0.60798	1.57576	2.68354	0.47452
Balance ground water potential available for exploitation (Mha m yr^{-1})	0.11686	0.10322	4.44113	1.48829
Utilisable irrigation potential for development (Mha)	1.46170	2.91715	16.79896	3.31794

periods during the cropping season owing to breaks in monsoon rains or early withdrawal of the monsoon. The within-season drought periods at times extend to more than one month and often occur during the latter part of the rain season.

In some high productivity areas not only is there evidence of changing water tables, but also of ground water pollution. The possible factors responsible for this decrease may be relentless pressure of increasing human and cattle population, destruction of forest, faulty irrigation systems leading to water-logging and salinity, and improper agricultural practices on marginal lands.

Hydrological changes will be an important component of future global change pressures on South Asia. Changes in the magnitude, frequency, and duration of hydrological events influence directly the overall availability of water resources and there is little doubt that the present situation in the region will be exacerbated in future (Lal 1994). The major river systems in South Asia, including those in the Indo-Gangetic Plain, are largely unregulated and vulnerable to hydro-meteorological changes associated with greenhouse warming. A recent analysis of global water resources (Vorosmarty et al. 2000), identifies South Asia as a highly stressed region. Clearly, in the face of a changing global environment, especially climate and its regional manifestations, the issue of sustainable agricultural production in the Indo-Gangetic Plain is a fundamental food security issue for the South Asia region.

3.8 Conclusions

Intensive resource use by the rapidly growing population of South Asia, especially in the energy and food sectors, contributes significant fraction of emissions in the atmosphere. Regional atmospheric composition changes, transport and deposition of emission products and other global changes have substantial implications for the region as well as the globe. The following conclusions can be drawn from the forgoing discussion:

- Monsoon rainfall in South Asia has been relatively stable over the last past millennium. However, notable failures have characterized the record, the most catastrophic of which was from 1790–1796. Palaeoclimatic evidence indicates that the monsoon circulation was more intense during the Holocene warm period around 6 000–9 000 B.P.
- Observations of mean annual surface air temperature obtained from instrumental records indicate a significant present warming of 0.4 °C per century. Unlike in other parts of the world, this warming is mainly due to an increase in maximum temperatures and not nocturnal minima.
- Although more than 20% of the global population lives in South Asia, greenhouse gas emissions at present are low (3%) as a proportion of global emissions. However, projections indicate substantial increases (six-fold by 2020) in such emissions from the region. Enhanced emissions will significantly impact the biogeochemistry and the regional radiative balance. Emission products from the region can be transported over substantial distances (also from other regions to this region) due to meteorological factors affecting climatic conditions. Changes in regional radiative balance and hydrology and may have extra-regional to global consequences.
- Apart from greenhouse gases, aerosols are a major component of the emission products. A substantial amount of aerosol production arises from anthropogenic activities like fossil fuel and biomass burning. While acid deposition resulting from long-range transport and deposition of such aerosols is not a major problem at present because of neutralizing effect of dust in the atmosphere and the alkaline soils in the region, the potential exists for acidification of soils in the region in future. Consequent implications for regional food and human health security follow.
- Biogeochemical cycling in the Arabian Sea region is significant. The process of denitrification has the potential to alter the combined nitrogen inventory, biological productivity and atmospheric CO_2 content on a wider oceanic scale. Oxygen deficiency in the Arabian Sea affects the production and consumption of other climatically important gases such as N_2O and CH_4.
- The suboxic Arabian Sea is a region with delicate biogeochemical balance and is a region that is a sensitive indicator of global change effects.
- The Arabian Sea is a net emitter of CO_2, while the Bay of Bengal is a net sink. In the Arabian Sea, the emission rate of CH_4 is several times higher than the global average. However, at present it is not large enough to affect the global CH_4.
- It appears that regional warming may enhance biological production in the Bay of Bengal, whereas it would be deleterious to that in the Arabian Sea.
- Changing atmospheric greenhouse gases and aerosols will definitely have consequences for agricultural yields. Impacts on food production will also be considerable. Key agricultural production areas, such as the Indo-Gangetic Plain, which is the bread basket of South Asia, are likely to be adversely affected, especially if changes in the hydrological regime and acidification of soil continues.
- Projected increases in temperature under global warming, taken independently, will reduce grain yields on the order of 10%. On the other hand, grain yields will respond positively to an increase in atmospheric CO_2 owing to the CO_2 fertilization effect.
- Aerosol-induced sunlight reduction may already reduce agricultural yields; any increase in the aerosol concentration of the South Asian haze layer will exacerbate the situation.
- Anticipated changes in already-intense UV-B levels will inhibit growth in many crop species.
- Decadal and longer-term changes may manifest as changes in frequencies of extreme events to which the region is especially vulnerable.
- Wheat and rice in the region's bread basket, the Indo-Gangetic Plain, are stagnating, most likely because of changes that may include soil structure, chemistry, surface salinity, and a decline in underground water tables. The vulnerability of the region to the multiple stresses of global change, including potential changes in the variability of the monsoon, is a matter of profound concern.

The effect of natural climatic change in the region, the consequences of changing atmospheric composition, anthropogenic forcing due to land cover and land use changes, and other ways in which global change impacts on the region, are complex and multi-dimensional factors in a system that needs to be considered as a whole. To gain a better understanding of regional manifestations of such changes and to develop appropriate strategies for mitigation, adaptation to global change and reduce vulnerabilities in food security, human health and to improve socio-economic responses to global change, it is necessary to take a more holistic view by the way of attempting regional-scale integrated assessments (Shukla 2000). Such an effort for South Asia is in early stage, but deserves concerted study and consideration by the scientific as well as the policy communities.

References

Abrol IP (1996) India's agriculture scenario. In: Abrol YP, Gadgil S, Pant GB (eds) Climate variability and agriculture. New Delhi, Narosa Publishing House, 19–25

Abrol IP (1998) Indo-Gangetic plains: changing land use and food security issues. In: Tiwari MK, Velayutham M (eds) Proceedings of the workshop on Indo-Gangetic Plains Region Land-use Land-cover Change. New Delhi. SASCOM Scientific Report N°10, 22–48

Abrol YP, Gadgil S (eds) (1999) Rice – In a variable climate. New Delhi, APC Publishing House

Abrol YP, Ingram (1996) Effects of higher day and night temperature on growth and development of major crops. In: Neue D, Sowsroek WG (eds) Global change and agricultural production. New York, John Wiley, 123–140

Abrol YP, Sharma A (1996) Climate variability and Indian agriculture. Science and Culture 62:196–206

Abrol YP, Wattal PN, Gnanam A, Govindjee, Ort DR, Teramura AH (eds) (1991) Impact of global climate changes on photosynthesis and plant productivity. New Delhi, IBH Publishing House

Abrol YP, Gadgil S, Pant GB (eds) (1996) Climate variability and agriculture. New Delhi, Narosa Publishing House

Aggarwal PK (1991) Simulated growth, development and yield of wheat in warmer areas. In: Saunders DA (ed) Wheats for nontraditional warm areas. Mexico, CIMMYT, 429–446

Aggarwal PK, Kalra N (eds) (1994) Simulating the effect of climatic factors, genotype and management on productivity of wheat in India. New Delhi, IARI Publication

Aggarwal PK, Sinha SK (1993) Effect of probable increase in carbon dioxide and temperature on productivity of wheat in India. Journal of Agricultural Meteorology 48:811–814

Ahmed SM, Patil D, Rao PS, Nath BN, Rao BR, Rajagopalan G (2000) Glacial-interglacial changes in the surface water characteristics of the Andaman Sea: Evidence from stable isotopic ratios of planktonic foraminifera. Proceedings of the Indian Academy of Sciences (Earth and Planetary Sciences) 109:153–156

ALGAS (1998) India, Pakistan, Bangladesh. Manila, Asian Development Bank, October 1998

Altabet M, Francois R, Murray DW, Prell WL (1995) Climate-related variations in denitrification in the Arabian Sea from sediment $^{15}N/^{14}N$ ratios. Nature 373:506–509

Anderson D (1999) Extremes in the Indian Ocean. Nature 401: 337–339

BBS (Bangladesh Bureau of Statistics) (1975–1995) Yearbook of the Agricultural Statistics of Bangladesh, Dhaka

Beig G, Mitra AP (1997a) Atmospheric and ionospheric response to trace gas perturbations through the ice age to the next century: Part I. Chemical composition and thermal structure. Journal of Atmospheric and Terrestrial Physics 59:1245–1259

Beig G, Mitra AP (1997b) Atmospheric and ionospheric response to trace gas perturbations through the ice age to the next century: Part II. Ionization. Journal of Atmospheric and Terrestrial Physics 59:1261–1275

Bengtsson L, Botzet M, Esch M (1996) Will greenhouse gas induced warming over the next 50 years lead to higher frequency and greater intensity of hurricanes? Tellus 48(A):53–57

Bhalme HN, Jadhav SK (1984) The southern oscillation and its relation to the monsoon rainfall. Journal of Climatology 4:509–520

Bhalme HN, Rahalkar SS, Sikdar AB (1987) Tropical quasi-biennial oscillation of the 10 mb wind and Indian monsoon rainfall – Implications for forecasting. Journal of Climatology 7:345–353

Bhaskar Rao (1997) Tropical cyclone simulation with Emanuel's convection scheme. Mausam 48:113–122

Bhaskaran B, Mitchell JFB (1998) Simulated changes in the south East Asian monsoon precipitation resulting from the anthropogenic emissions. International Journal of Climatology 18: 1455–1462

Bhaskaran B, Mitchell J, Iavery JFB, Lal M (1995) Climate response of the Indian subcontinent to doubled CO_2 concentrations. International Journal of Climatology 15:873–892

Bhattacharyya T, Pal DK, Velayutham M (2000) Soil carbon stock in the Indo-Gangetic Plains (IGP) and its role in decision support system for land resource management. Paper presented in International Conference on managing natural resources for sustainable agricultural production in the 21st century, 14–18 February 2000, New Delhi, India

Bhatti ZI (1998) Acid rain monitoring and atmospheric modelling: the case of Pakistan. In: Ileperuma OA (ed) Proceedings of Work-shop on Acid Rain Monitoring and Atmospheric Modelling, Sri Lanka, 63–67

Biswas SK, Kaliquzzaman M, Tarafdar SA, Islam A (1999) Nature and extent of airborne particulate matter pollution in urban and rural areas of Bangladesh during 1993–1998. Country report presented during the Regional-Cum-Interregional Workshop on Aerosol Data Synthesis and Integration, National Physical Laboratory, New Delhi

Brandes JA, Devol AH, Jayakumar DA, Yoshinari T, Naqvi SWA (1998) Isotopic composition of nitrate in the central Arabian Sea and eastern North Pacific: A tracer for mixing and nitrogen cycles. Limnology & Oceanography 43:1680–1689

Caratini C, Fontugne M, Pascal JP, Tissot C, Bentaleb I (1991) A major change at ca 3500 years B.P. in the vegetation of the western Ghats in north Kanara, Karnataka. Current Science 61:669–672

Carmichael GR, Streets DG, Van Anardenne J, Arndt RL (1998) Anthropogenic fossil fuel sulfur and nitrogen emissions in Asia: results from RAINS ASIA. In: Parashar DC, Sharma C, Mitra AP (eds) Proceedings of APN Aerosol Scoping Workshop – SASCOM Scientific Report No 11. New Delhi, Centre on Global Change, National Physical Laboratory, 162–176

CBS (Central Bureau of Statistics) (1994) National sample census of agriculture, Nepal, 1991/92. Kathmandu, HMG/NPC

CBS (Central Bureau of Statistics) (1996) Statistical year book of Nepal. Kathmandu, IIMG/NPC

Chemeides WL, Yu H, Liu SC, Bergin M, Zhou X, Mearns L, Wang G, Kiang CS, Saylor RD, Luo C, Steiner A, Giorgi F (1999) Case study of the effects of atmospheric aerosols and regional haze on agriculture: An opportunity to enhance crop yields in China through emission controls. Proceedings of the National Academy of Sciences of the United States of America 96 (24):13626–12633

Chowdhury A, Dandekar MM, Raut PS (1989) Variability in drought indices in India – A statistical approach. Mausam 37:471–482

Chowdhury AM, Haque MA, Quadir DA (1997) Consequences of global warming and sea level rise in Bangladesh. Marine Geodesy 20:13–31

Cullen JL (1981) Microfossil evidence for changing salinity patterns in the Bay of Bengal over the last 20000 years. Palaeogeography, Palaeoclimatology and Palaeoecology 35:315–356

Das SN, Thakur RS, Mitra AP (1999) Acid rain studies: Indian scenario. Global Change Series Number 16. New Delhi, Centre on Global Change, National Physical Laboratory

Dong B, Valdes PJ, Hall NMJ (1996) The changes of monsoonal climates due to earth's orbital perturbations and ice age boundary conditions. Palaeoclimates 1:203–240

Douville H, Royer JF, Polcher J, Cox P, Gedney N, Stephenson DB, Valdes PJ (2000) Impact of CO_2 doubling on the Asian summer monsoon: Robust versus model-dependent responses. Journal of the Meteorological Society of Japan 78:421–439

Duplessy JC (1982) Glacial to interglacial contrast in the North Indian Ocean. Nature 295:494–498

Gadgil S (1995) Climate change and agriculture – An Indian perspective. Current Science 69:649–659

Gadgil S (1996) Climate change and agriculture – An Indian experience. In: Abrol YP, Gadgil S, Pant GB (eds) Climate variability and agriculture. New Delhi, Narosa Publishing House

Gadgil S, Abrol YP, Seshagri Rao PR (1999a) On growth and fluctuation of Indian food grain production. Current Science 76: 548–556

Gadgil S, Sheshagiri Rao PR, Sridhar S (1999b) Modelling impact of climate variability on rainfed groundnut. Current Science 76:557–569

Gangadhar Rao D, Sinha SK (1994) Impact of climate change on simulated wheat production in India. In: Implications of climate change for international agriculture: crop modelling study. USEPA 230-B-94-003. USEPA, Washington DC p1–1

George MD, Kumar MD, Naqvi SWA, Banerjee S, Narvekar PV, Sousa SN de, Jayakumar DA (1994) A study of the carbon dioxide system in the northern Indian Ocean during premonsoon. Marine Chemistry 47:243–254

Golitsyn GS, Semenov AI, Shefov NN, Fishkova LM, Lysenko EV, Perov SP (1996) Long term temperature trends in middle and upper atmosphere. Geophysical Research Letters 23: 1741– 1744

Goyet C, Coatanoan C, Eischeid G, Amaoka T, Okuda K, Healy R, Tsunogai S (1999) Spatial variation of total CO_2 and total alkalinity in the northern Indian Ocean: A novel approach for the quantification of anthropogenic CO_2 in seawater. Journal of Marine Research 57:135–163

Gupta PK, Mitra AP (1999) Greenhouse gas emissions in India: ADB-methane Asia campaign. Global Change Scientific Report No 19. New Delhi, Centre for Global Change, National Physical Laboratory

Gupta SK, Minhas PS, Sondhi SK, Tyagi NK, Yadava JSP (2000) Water resource management. In: Natural Resource Management for Agricultural Production in India. Special Publication in International Conference on Managing Natural Resources for Sustainable Agricultural Production in the 21st century. 14–18 February 2000, New Delhi, India, 139–244

Handler P (1986) Stratospheric aerosols and the Indian monsoon. Journal of Geophysical Research 91:14475–14490

Hassel D, Jones R (1999) Simulating climatic change of the southern Asian monsoon using a nested regional climate model (HadRM2). Hadley Centre Technical Note No. 8

Hingane LS, Rupa Kumar K, Ramana Murthy BhV (1985) Long-term trends of surface air temperature in India. Journal of Climatology 5:521–528

Ittekkot V, Arain (1986) Nature of particulate organic matter in the river Indus, Pakistan. Geochimica et Cosmochimica Acta 50:1643–1653

Ittekkot V, Safiullah S, Mycke B, Seifert R (1985) Seasonal variability and geochemical significance of organic matter in the River Ganges. Nature 317:800–802

Ittekkot V, Nair RR, Honjo S, Ramaswamy V, Bartsch M, Manganini S, Desai BN (1991) Enhanced particle fluxes in Bay of Bengal induced by injection of fresh water. Nature 351:385–387

Jayakumar DA (1999a) Biogeochemical cycling of methane and nitrous oxide in the Northern Indian Ocean. Goa, Goa University

Jayaraman A (1999b) Results on direct radiative forcing of aerosols obtained over the tropical Indian Ocean. Current Science 76:924–930

Jayakumar DA, Naqvi SWA, Narvekar PV, George MD (2000) Methane in coastal and offshore waters of the Arabian Sea. Marine Chemistry 74:1–13

Karisiddaiah SM, Veerayya M (1996) Potential distribution of subsurface methane in the sediments of the eastern Arabian Sea and its possible implications. Journal of Geophysical Research 101:25887–25895

Karl TR, Jones PD, Knight RW, Kukla G, Plummer N, Razuvayev V, Gallo KP, Lindesay J, Charlton RJ, Peterson TC (1993) Asymmetric trends of daily maximum and minimum temperature. Bulletin of the American Meteorological Society 74:1007–1023

Kitoh A, Yukimoto S, Noda A, Motoi T (1997) Simulated changes in the Asian summer monsoon at times of increased atmospheric CO_2. Journal of the Meteorological Society of Japan 75: 1019–1031

Kripalani RH, Inamdar SR, Sontakke NA (1996) Rainfall variability over Bangladesh and Nepal: Comparison and connection with features over India. International Journal of Climatology 16:689–703

Krishna Kumar K, Soman MK, Rupa Kumar K (1995) Seasonal forecasting of Indian summer monsoon rainfall. Weather 50: 449–467

Krishna Kumar K, Rajagopalan B, Cane MA (1999) On the weakening relationship between the Indian monsoon and ENSO. Science 284:2156–2159

Kumar MD, Naqvi MD, George MD, Jayakumar DA (1996) A sink for atmospheric carbon dioxide in the northern Bay of Bengal. Journal of Geophysical Research 101:18121–18125

Kumar MD, Sarma VVSS, Ramaiah N, Gauns M, Sousa SN de (1998) Biogeochemical significance of transparent exopolymer particles in the Indian Ocean. Geophysical Research Letters 25:81–84

Lal M (1994) Water resources of the South Asian region in a warm atmosphere. Advances in Atmospheric Sciences 11:239–246

Lal M, Cubasch U, Voss R, Waszkewitz J (1995) Effect of transient increase in greenhouse gases and sulphate aerosols on monsoon climate. Current Science 69:752–763

Lal M, Srinivasan G, Cubasch U (1996) Implications of greenhouse gases and aerosols on the diurnal temperature cycle of the Indian subcontinent. Current Science 71:746–752

Lal M, Whetton PH, Pittock AB, Chakraborty B (1998) The greenhouse gas induced climate change over the Indian subcontinent as projected by general circulation model. Terrestrial, Atmospheric and Oceanic Sciences 9:673–690

Lal M, Singh KK, Srinivasan G, Rathore LS, Naidu D (1999) Growth and yield response of soybean in Madhya Pradesh, India to climate variability and change. Agriculture and Forest Meteorology 93:53–70

Lelieveld J, Ramanathan R, Crutzen PJ (1999) The global effects of Asian Haze. IEEE Spectrum 36:5–54

Lelieveld J, Crutzen P, Ramanathan V, Andreae MO, Brenninkmeijer CAM, Campos T, Cass GR, Dickerson RR, Fischer H, Gouw JA de, Hansel A, Jefferson A, Kley D., Laat ATJ de, Lal S, Lawrence MG, Lobert JM, Mayol-Bracero O, Mitra AP, Novakov T, Oltmans SJ, Prather KA, Reiner T, Rodhe H, Scheeren HA, Sikka D, Williams J (2001) The Indian Ocean experiment: widespread air pollution from South and Southeast Asia. Science 291: 1031–1036

Lighthill J, Holland G, Gray WM, Landsea C, Craig G, Evans J, Kurihara Y, Guard C (1994) Proceedings of the Third WMO/ICSU International Workshop on Tropical Cyclones (IWTC-III), TMRP Report No. 49, WMO/TD No. 624

Mandal GS (1992) Tropical cyclones and their warning system in the North Indian Ocean. WMO TCP Report No.28, Annex II-1

Mandal TK, Kley D, Smith HGJ, Srivastava SK, Peshin SK, Mitra AP (1999) Vertical distribution of ozone over the Indian Ocean (15oN–15oS) during First Field Phase INDOEX-1998. Current Science 76:938–943

Mani A (1981) The climate of the Himalaya. In: Lal JS, Moddie AD (eds) The Himalaya: Aspects of changes. New Delhi, Oxford University Press, 3–15

Meehl GA, Washington WM (1993) South Asian summer monsoon variability in a model with doubled atmospheric carbon dioxide concentrations. Science 260:1101–1104

Mitra AP (2000) Issues and perspectives of the South Asian region. Global Change Report No. 18, New Delhi, National Physical Laboratory

Mooley DA (1981) Increase in the frequency of the severe cyclonic storms of the Bay after 1964 – Possible causes. Mausam 32: 35–40

Mooley DA, Mohile CM (1984) Cyclonic storms of the Arabian Sea, 1877–1980. Mausam 35:127–134

Mooley DA, Parthasarathy B (1984) Fluctuations of all-India summer monsoon rainfall during 1871–1978. Climatic Change 6: 287–301

Mooley DA, Shukla J (1989) Main features of the westward moving low pressure systems which form over the Indian region during the summer monsoon season and their relation to the monsoon rainfall. Mausam 40:137–152

Mukherjee BK, Indira K, Reddy RS, Ramana Murty BhV (1985) Quasi-biennial oscillation in stratospheric wind and Indian monsoon. Monthly Weather Review 118:1421–1430

Nair RR, Ittekkot V, Manganini SI, Ramaswamy V, Haake B, Degens ET, Desai BN, Honjo S (1989) Increased particle flux to the deep ocean related to monsoons. Nature 338:749–751

Naqvi SWA (1987) Some aspects of the oxygen-deficient conditions and denitrification in the Arabian Sea. Journal of Marine Research 49:1049–1072

Naqvi SWA, Noronha RJ (1991) Nitrous oxide in the Arabian Sea. Deep Sea Research 38:871–890

Naqvi SWA, Charles CD, Fairbanks RG (1994a) Carbon and oxygen isotopic records of benthic foraminifera from the Northeast Indian Ocean: implications on glacial-interglacial atmospheric CO_2 changes. Earth and Planetary Science Letters 121: 99–110

Naqvi SWA, Jayakumar DA, Nair M, George MD, Kumar MD (1994b) Nitrous oxide in the western Bay of Bengal. Marine Chemistry 47:269–278

Naqvi SWA, Yoshinari T, Jayakumar DA, Altabet MA, Narvekar PV, Devol AH, Brandes JA, Codispoti LA (1998) Budgetary and biogeochemical implications of N_2O isotope signatures in the Arabian Sea. Nature 394:462–464

NASA (2001) *http://www.earthobservatory.nasa.gov/*

Nigam R (1993) Foraminifera and changing pattern of monsoon rainfall. Current Science 64:935–937

Nigam R, Hashmi NH, Menezes ET, Wagh AB (1992) Fluctuating sea levels off Bombay (India) between 14 500 and 10 000 years before present. Current Science 62:309–311

Overpeck J, Anderson D, Trumore S, Prell W (1996) The southwest Indian monsoon over the last 18 000 years. Climate Dynamics 12:213–225

Owens NJP, Law CS, Mantoura RFC, Burkill PH, Llewellyn CA (1991) Methane flux to the atmosphere from the Arabian Sea. Nature 354:293–296

Pal M, Sharma A, Sengupta UK, Abrol YP (1997) Exclusion of UVB radiation from normal solar spectrum on growth of soybean and maize. Agriculture Ecosystem Environment 61:29–36

Pant GB, Maliekal JA (1987) Holocene climatic changes over northwest India: An appraisal. Climatic Change 10:183–194

Pant GB, Parthasarathy B (1981) Some aspects of an association between the southern oscillation and Indian summer monsoon. Archiv fur Meteorologie, Geophysik und Bioklimatologie, Series. B, 29:245–251

Pant GB, Rupa Kumar K, Parthasarathy B, Borgaonkar HP (1988) Long term variability of the Indian summer monsoon and related parameters. Advances in Atmospheric Sciences 5:469–481

Pant GB, Rupa Kumar K (1997) Climates of South Asia. Chichester, John Wiley

Parthasarathy B (1984) Interannual and long-term variability of Indian summer monsoon rainfall. Proceedings of the Indian Academy of Sciences (Earth and Planetary Sciences) 93: 371–385

Parthasarathy B, Sontakke NA, Munot AA, Kothawale DR (1987) Droughts/floods in the summer monsoon season over different meteorological subdivisions of India for the period 1871–1984. Journal of Climatology 7:57–70

Parthasarathy B, Rupa Kumar K, Munot AA (1991) Evidence of secular variations in Indian summer monsoon rainfall-circulation relationships. Journal of Climate 4:927–938

Patra PK, Lal S, Venkataramani S (1998) Seasonal variability in distribution and fluxes of methane in the Arabian Sea. Journal of Geophysical Research 103:1167–1176

Phadtare NR (2000) Sharp decrease in summer monsoon strength 4 000–3 500 cal yr B.P. in the Central Higher Himalaya of India based on pollen evidence from Alpine peat. Quaternary Research 53:122–129

Prasad S, Kusumgar S, Gupta SK (1997) A mid to late Holocene record of palaeoclimatic changes from Nal Sarovar: A palaeodesert margin lake in western India. Journal of Quaternary Science 12:153–159

Prell WL, Kutzbach JE (1987) Monsoon variability over the past 1 500 000 years. Journal of Geophysical Research 92:8411–8425

Qamar-uz-Zaman Ch, Farooqi AB, Rasul G (1998) Tendency of Pakistan's current climate and climate change scenarios. Climate Change Impact Assessment and Adaptation Strategies in Pakistan (UNEP), Pakistan Meteorological Department, Islamabad

Ramana Roa BV, Katyal JC, Bhalra PC (1994) Final Report on Indo-USAID-Sus-Project, CRIDA, Hyderabad, India

Ramaswamy C (1968) Monsoons over the Indus valley during the Harappan period. Nature 217:628–629

Rao BVR, Rao MR (1996) Weather effects on pests. In: Abrol YP, Gadgil S, Pant GB (eds) Climate Variability and Agriculture. New Delhi, Narosa Publishing House, 281–296

Ravindranath NH, Hall DO (1994) Indian forest conservation and tropical deforestation. Ambio 23:521–523

Rupa Kumar K, Ashrit RG (2001) Regional aspects of global climate change simulations: Validation and assessment of climate response of the Indian Monsoon region to transient increase of greenhouse gases and sulphate aerosols. Mausam, Millennium Special Issue, 52:299–244

Rupa Kumar K, Hingane LS, Ramana Murthy BhV (1987) Variation of tropospheric temperatures over India during 1944–1985. Journal of Climate and Applied Meteorology 26:304–314

Rupa Kumar K, Pant GB, Parthasarathy B, Sontakke NA (1992) Spatial and subseasonal patterns of the long-term trends of Indian summer monsoon rainfall. International Journal of Climatology 12:257–268

Rupa Kumar K, Krishna Kumar K, Pant GB (1994) Diurnal asymmetry of surface temperature trends over India. Geophysical Research Letters 21:677–680

Sabine CL, Key RM, Johnson KM, Millero FJ, Poisson A, Sarmiento JL, Wallace DWR, Winn CD (1999) Anthropogenic CO_2 inventory of the Indian Ocean. Global Biogeochemical Cycles 13:179–198

Saji NH, Goswami BN, Vinayachandran PN, Yamagata T (1999) A dipole mode in the tropical Indian Ocean. Nature 401:360–363

Sarin MM, Rengarajan R, Ramaswamy V (1996) ^{234}Th scavenging and particle export fluxes from the upper 100 m of the Arabian Sea. Current Science 71:888–893

Sarma VVSS (1999) Variability in forms and fluxes of carbon dioxide in the Arabian Sea. Goa, Goa University

Sarma VVSS, Kumar MD, George MD (1998) The central and eastern Arabian Sea as a perennial source to atmospheric carbon dioxide. Tellus 50B:179–184

Saseendran ASK, Singh KK, Rathore LS, Singh SV, Sinha SK (2000) Effects of climate change on rice production in the tropical humid climate of Kerala, India. Climate Change 44:495–514

Satheesh SK, Ramanathan V, Li-Jones, Xu, Lobert JM, Podgorny IA, Prospero JM, Holben BN, Loeb NG (1998) INDOEX first field phase results: An aerosol model for the tropical Indian Ocean during NE Monsoon. INDOEX Publication 24, C4 Publication 207, USA

Schafer P, Ittekkot V (1996) Mitteilungen aus dem Geologisch-Paläontologischen Institut der Universität Hamburg, SCOPE/UNEP Sonderband, Heft 78:67–93

Sen Gupta R, Rajagopal MD, Qasim SZ (1976) Relationships between dissolved oxygen and nutrients in the Northwestern Indian Ocean. Indian Journal of Marine Science 5:201–211

Shankar D, Shetye SR (1999) Are interdecadal sea level changes along the Indian coast influenced by variability of monsoon rainfall? Journal of Geophysical Research 104:26031–26042

Shukla PR (2000) Future energy trends and carbon mitigation strategies for India. In: Audinet P, Shukla PR, Grare F (eds) India's energy – Essays on sustainable development. Dehli, Manohar Publishers, 21–50

Sikka DR (1980) Some aspects of the large-scale fluctuations of summer monsoon rainfall over India in relation to fluctuations in the planetary and regional scale circulation parameters. Proceedings of the Indian Academy of Sciences (Earth and Planetary Sciences) 89:179–195

Singh G, Joshi RD, Chopra SK, Singh AB (1974) Late quaternary history of vegetation and climate of the Rajasthan desert, India. Philosophical Transactions of the Royal Society, London 267B, 467–501

Singh P, Booha KJ, Rao AY, Iruthayaraj MR, Sheikh AM, Hundal SS, Narang RS, Singh P (1994) Evaluation of groundnut model PNUTGRO for crop response to water availability, sowing dates and season. Field Crop Research 39:147–162

Sinha SK, Swaminathan MS (1991) Deforestation, climate change and sustainable nutrition security. Climate Change 16:33–45

Sirocko F, Sarnthein M, Erlenkeuser H, Lange H, Arnold M, Duplessy JC (1993) Century scale events in monsoonal climate over the past 24000 years. Nature 364:322–324

Somasundar K, Rajendran A, Kumar MD, Sen Gupta R (1990) Carbon and nitrogen budgets of the Arabian Sea. Marine Chemistry 30:363–377

Sontakke NA, Pant GB, Singh N (1993) Construction of all India summer monsoon rainfall series for the period 1844–1991. Journal of Climate 6:1807–1811

Srivastava HN, Dewan BN, Dikshit SK, Rao GSP, Singh SS, Rao KR (1992) Decadal trends in climate over India, Mausam 43:7–20

Stephenson DB, Chauvin F, Royer J (1998) Simulation of the Asian summer monsoon and its dependence on model horizontal resolution. Journal of the Meteorological Society of Japan 76:237–265

Stephenson DB, Douville H, Rupa Kumar K (2000) Searching for a fingerprint of global warming in the Asian summer monsoon. Mausam, Millennium Special Issue

Swain AM, Kutzbach JE, Hastenrath S (1983) Estimates of Holocene precipitation for Rajasthan, India, based on pollen and lake level data. Quaternary Research 19:1–17

Swathi PS, Sharada MK, Yajnik KS (2000) A coupled physical-biological-chemical model for the Indian Ocean. Proceedings of the Indian Academy of Sciences (Earth and Planetary Sciences) 109:503–537

TEDDY (TERI Energy Data Directory and Year Book) (1998–1999) New Delhi, Tata Energy Research Institute

Thamban M, Rao VP, Schneider RR, Grootes PM (2000) Glacial to Holocene fluctuations in hydrography and productivity along the southwestern continental margin of India. Palaeogeography, Palaeoclimatolology and Palaeoecology 165:113–127

Thompson LG, Yao T, Mosley-Thompson E, Davis ME, Handerson KA, Lin PN (2000) A high-resolution millennial record of the south Asian monsoon from Himalayan ice cores. Science 289: 1916–1919

UNEP (1998) Study on climate change impact assessment and adaptation strategies for Pakistan. Government of Pakistan. Ministry of Environment, Local Government and Rural Development, Islamabad

Uprety DC, Garg SC, Tiwari MK, Mitra AP (2000) Crop response to elevated CO_2: Technology and Research (Indian Study). Global Environmental Research 3:155–167

Van Campo E (1986) Monsoon fluctuations in two 20000-yr B.P. oxygen-isotope/pollen records off southwest India. Quaternary Research 26:376–388

Van Campo E, Duplessy JC, Rossignol-Atrick M (1982) Climatic conditions deduced from 150-kyr oxygen isotope-pollen record from the Arabian Sea. Nature 296:56–59

Velayutham M, Pal DK, Bhattacharyya T (1999) Organic carbon stock in soils of India. In: Lal R, Kimble JM, Eswaran H, Stewart BA (eds) Global climate change and tropical ecosystems. Boca Ratan, Lewis Publishers, 71–96

Velayutham M, Pal DK, Bhattacharyya T, Srivastava P (2000) Soils of the Indo-Gangetic plains, India – the historical perspective. Special IGP Publication (in press)

Vorosmarty CJ, Green P, Salisbury J, Lammers R (2000) Global water resources: Vulnerability from climate change and population growth. Science 289:284–288

Warrick RA, Ahmed QK (eds) (1996) The implications of climate and sea level change for Bangladesh. Dordrecht, Kluwer

Wasson RJ, Smith GI, Agrawal DP (1984) Late quaternary sediments, minerals and inferred geochemical history of Didwana Lake, Thar Desert, India. Palaeogeography, Palaeoclimatology and Palaeoecology 46:345–372

Webster PJ, Magaña VO, Palmer TN, Shukla J, Tomas RA, Yanai M, Yasunari T (1998) Monsoons: Processes, predictability, the prospects for prediction. Journal of Geophysical Research 103: 14451–14510

Webster PJ, Moore AM, Loschnigg JP, Leben RR (1999) Coupled ocean-atmosphere dynamics in the Indian Ocean during 1997–1998. Nature 401:356–360

Weerakkody U (1996) Vulnerability and adaptation assessments for Sri Lanka. In: Smith JB, Lenhart S, Huq S, Mata LJ, Nemesova I (eds) Vulnerability and adaptation to climate change – Interim results from the US Country Studies Program. Netherland, Kluwer Academic Publishers, 207–223

Wei HL, Fu CB (1998) Study of the sensitivity of a regional model in land cover change over northern China. Hydrological Processes 12:2249–2285

Zhao Z, Kellogg WW (1988) Sensitivity of soil moisture to doubling of carbon dioxide in climate model experiments. Part II The Asian monsoon region. Journal of Climate 1:367–378

Zhisheng A, Kutzbach JE, Prell WL, Porters SC (2001) Evolution of Asian monsoons and phased uplift of the Himalayas-Tibetan Plateau since Late Miocene Times. Nature 411:62–66

Chapter 4

Regional-Global Interactions in East Asia

Congbin Fu · Hideo Harasawa · Vladimir Kasyanov · Jeong-Woo Kim · Dennis Ojima · Zhibin Wan · Shidong Zhao[1]

4.1 Introduction

Temperate East Asia comprises a major portion of the Earth's largest continent. The region is bordered by the planet's highest mountains and largest ocean and its climate and ecosystems are uniquely dominated by the East Asian monsoon. Moreover, East Asia is the homeland of some of the world's oldest, most populous, most advanced and most rapidly evolving human civilisations. It is therefore not surprising that this region plays a major role in global processes, nor that human activities have powerfully influenced these processes on both regional and planetary scales. The overarching story of the region is of massive human-induced changes in every aspect of the environment and their consequences for natural resources, the global environment, and human welfare. The major conclusion is that the region's past, present, and future may be understood and predicted only in the context of a comprehensive integrated framework that includes the physical, chemical, biological and human components of the system. Given the central role of the monsoon in the region, such a framework may best be termed the General Monsoon System.

4.1.1 Characteristics of the Region

The East Asia region includes China, the Democratic Peoples Republic of Korea, Japan, Mongolia, the Republic of Korea, and the Asian part of Russia. About 25 million km^2 in area, it is located in the north-eastern part of the Eurasian continent, the world's largest continent, and includes the northwestern borders of the Pacific, the world's largest ocean. The region includes the world's largest plateau, the Tibetan Plateau, with a mean elevation of more than 4 000 m, and the world's highest peak (Mt. Qomolanma (Everest), 8 848 m) (Fig. 4.1).

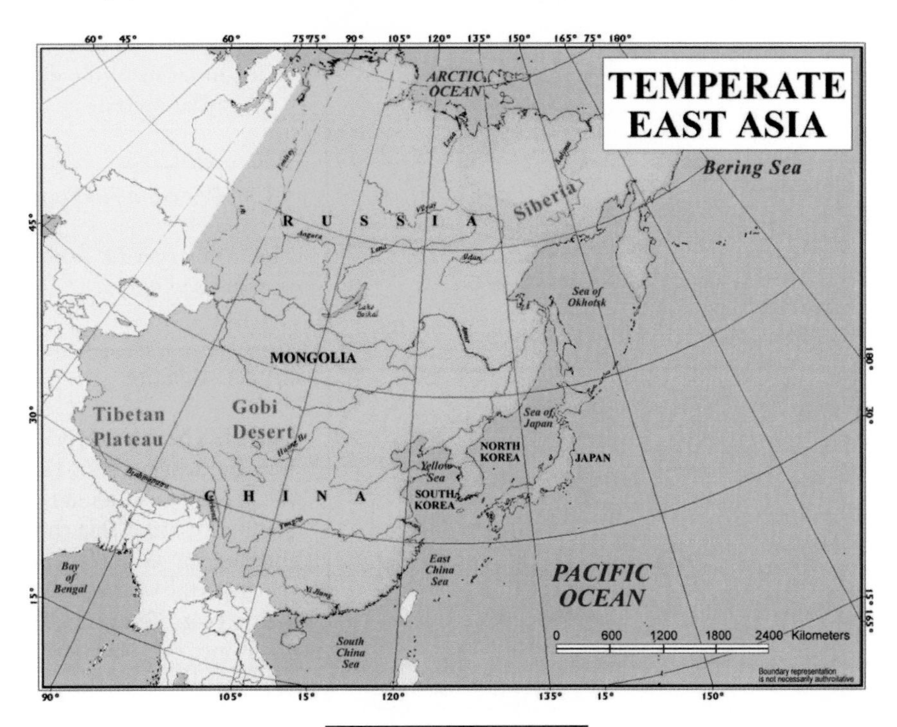

Fig. 4.1. The region

[1] *Contributing authors:* X. Li, A. Korotky, A. Makhinov, T. Orlova, V. Petrenko, V. Radchenko, TEACOM LUTEA SSC.

Mainly due to strong land-ocean thermal contrast and the dynamic and thermal effects of the Tibetan Plateau, East Asia has a well-developed monsoon climate system, which releases huge amounts of latent heat from monsoon rainfall and plays an important role in the global energy and water balances. The surface monsoon flow patterns in winter and summer over the region are distinctly different (Fig. 4.2). In winter the region is dominated by a dry-cloud continental air mass, while it is dominated by a warm and humid air mass in summer. The monsoon system is characterised not only by this strong seasonal change, but also by high inter-annual and inter-decadal variability that have profound impacts on economic development and human life in the region.

Diversity in climate, terrestrial ecosystems, peoples and culture is a major feature of the region. For example, East Asia encompasses examples of nearly all the major terrestrial ecosystems recognised on the planet, ranging from cold deserts to humid rainforests and from permafrost to rich coastal forest areas. The terrestrial ecosystems of East Asia comprise a large portion of the global biomass and constitute a huge evapotranspiration source in the global hydrological cycle. They also serve as a major carbon sink in the global biogeochemical cycle.

East Asia is one of the most populated areas of the planet, with a current human population of more than 1.5 billion. The large population and continuing demand for rapid economic growth result in ever-increasing industrialisation and intensive use of land and biotic resources. The human activities producing a rapid increase of anthropogenic emission of greenhouse gases and sulphate aerosols, deforestation, and desertification constitute the major driving force causing environmental changes in East Asia. Urbanisation is another significant feature of anthropogenic regional changes. Most of the world's cities with a population of more than 5 million are located in East Asia and other parts of Asia. Thus, the natural geographic features and socio-economic conditions in East Asia make it a priority region for global change studies.

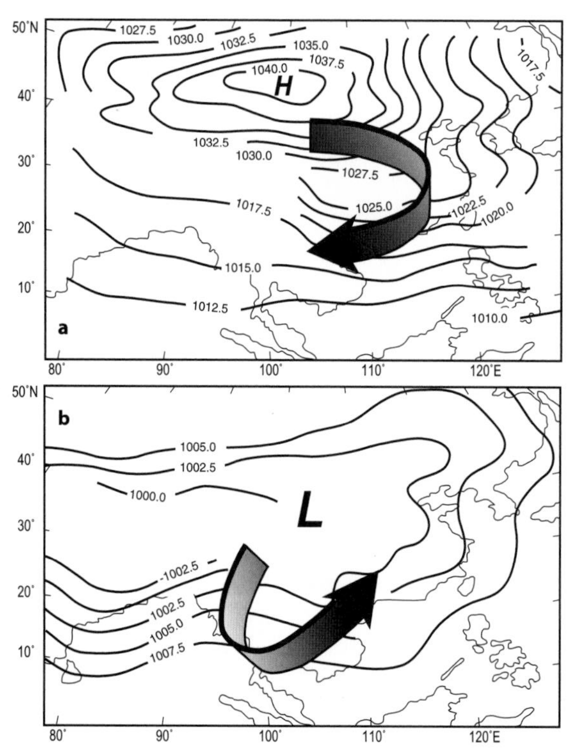

Fig. 4.2. Monsoon circulation patterns over East Asia: **a** winter and **b** summer

4.1.2 Central Questions

This synthesis attempts to answer the following questions in relation to regional aspects of global changes in East Asia:

- What environmental changes have taken place in the region?
- What are the driving forces for such changes, natural or anthropogenic?
- What will be the future changes on the decadal time scale?
- What will be the impacts of such changes on the region, including water resources, agriculture and human health, etc?
- How should human society adapt to such changes and impacts?
- What role does East Asia play in earth system dynamics, in terms of the global hydrological and biogeochemical cycles?

4.1.3 Principal Findings

The major specific regional findings of this study are summarised below.

4.1.3.1 *Climate*

There has been a nearly 1 °C warming over the past century in most parts of east Asia, including Japan, with the exception of an area of cooling in eastern half of southern China. This cooling is perhaps related to anthropogenic aerosols. The warming has occurred primarily in the winter season and has been greatest in the last 20 years. Temperatures in other seasons also show a slight increase. No significant trend of precipitation is apparent in regionally averaged temperatures, but such a trend is evident in the humidity index, which takes account of both precipitation and evaporation in relation to temperature changes.

Shortage of water resources is one of the most urgent problems for sustainable development of the region. The impacts of the El Niño-Southern Oscillation (ENSO) phenomenon on the region are significant, but different as an El Niño develops and dissipates. In the year before onset, the main rain belt is located to the north of the Yangtze River, with lesser rainfall areas to the southeast and in Northwest China. In the year of onset, most parts of the country have less rainfall. In the year after onset, most parts of the country receive more rainfall.

Cold-summer disasters in Northeast China and Japan are associated with the SST anomalies in the eastern equatorial Pacific, while there are statistical relationships between El Niño and tropical cyclone activity in the western Pacific. The East Asian monsoon is weak (strong) when ENSO is in its warm (cold) phase.

The region's climate has exhibited abrupt changes on time scales from decades to millennia. The abrupt change in the 1920s, characterised by a rapid global warming, is perhaps the strongest signal in the last century on the global scale. This change was characterised by increased tropical rainfall, a more active tropical monsoon, accompanied by a weaker subtropical monsoon and dry middle latitudes. The abrupt change in the 1960s was characterised by a weakening of the African and Indian monsoons, leading to the persistent drought in the Sahel region of North Africa. Yet another abrupt change occurred in the late 1970s, perhaps related to accelerating global warming. Thus, there are strong linkages between regional and global climate changes.

4.1.3.2 *Terrestrial Ecosystems and Land Cover*

East Asia includes a wide variety of natural terrestrial ecosystems – forests, grasslands, deserts, and wetlands, together with many cultivated ecosystems. Natural ecosystems in the south-eastern part of the region either have totally been replaced by man-made ecosystems or have been seriously impacted by human activities associated with the large population, increase and rapid development of urbanisation and industrialisation. Ecosystems in the northern and north-eastern part more or less retain their natural characteristics.

In China, due to increasing population, industrialisation, and urbanisation, demand for land resources has grown steadily. Farmland, pastures, and other managed ecosystems have encroached on natural vegetation on a very large scale, while land degradation has been very serious. After a rapid increase in the first half of the 20th century, farmland area is now actually decreasing, mainly due to spreading urbanisation. A continuing shortage of land resources has resulted, and environmental problems have affected economic and social development. Grassland covers over 40% of China's area, and is currently decreasing in area and degrading in quality, largely due to expansion of agriculture and overgrazing. Large areas of arid and semiarid range are becoming desert. Forests, which once covered a third of China, declined to 8% coverage by mid-20th century, but are now increasing in area and productivity due to improved national management programmes.

Over a quarter of China's area is desert or is undergoing desertification. The area of desertification is increasing due to wind and water erosion, freezing and melting processes, and salinisation. Desertification in many parts of East Asia is related to climatic variation, such as in parts of western China and Mongolia. Human activities such as overgrazing, fuelwood collection, deforestation, and inappropriate farming practices are exacerbating the harsh environmental conditions. In parallel, natural wetlands have been continuously decreasing for a long time because of agricultural development, urbanisation, and industrialisation.

Turning to Japan, human-settled areas more than tripled in area over the 20th century. Agricultural land also increased modestly, accompanied by large increases in productivity. Forest occupies almost two thirds of Japan's territory; its area is remaining stable, although its species composition is changing.

The ecosystems of East Asia have experienced massive transformations in historic times due to both natural processes and, increasingly, human activities.

4.1.3.3 *Marine and Coastal Systems*

The marginal seas of East Asia span a wide range of latitudes from the Bering Sea to the South China Sea. The shallow marginal seas play an important role as a buffer in land-ocean interaction in this monsoon region. Longitudinal transfers of energy and matter from land to sea and within the ocean are of great importance for regional climate and natural dynamics. The variable discharges of the great Asian rivers greatly influence the hydrology, hydrochemistry and coastal dynamics of the marginal seas.

Observed sea surface temperature increase in the last century is being followed by sea-level rise that is most pronounced in the heavily populated lowlands of the southern part of East Asia. The biological productivity of East Asian seas is now in a transitional state caused by natural factors in the north and mostly by anthropogenic factors in the south. Temperature increases, together with considerable inflow of nutrients from the land is leading to eutrophication of coastal waters with high incidence of algal blooms and possible decrease in marine biodiversity. On shorter time scales, ENSO produces significant fluctuations in sea level, with implications for biological systems. Indeed, the region's oceanic biota are strongly affected by both human and climatic factors.

4.1.4 Driving Forces for Environmental Change

Four major factors have converged to produce the considerable environmental changes in East Asia outlined above:

4.1.4.1 *Population Growth*

Population grew explosively during the 19th and 20th centuries, but has slowed markedly in recent years. China, for example, is expected to reach its peak population around 2030. At the same time, there is large-scale migration into urban areas, with massive impacts on croplands, fuel resources, and water supply.

4.1.4.2 *Economic Liberalization*

Centralised economic structures are giving way to free-market economies, promoting rapid economic growth and increasing globalisation. This is leading to more intensive land use, changes in consumption patterns, a transition from food production to cash crops and growth of urban areas. Access to modern technologies has led to increased productivity of croplands has been achieved by use of synthetic fertilisers, mechanisation, and increasing livestock. However, in some areas, such as Mongolia and eastern Russia, abusive farming practices and declining infrastructure has lead to increased degradation of agricultural ecosystems.

4.1.4.3 *Environmental Protection Policies*

Measures to protect the environment are coming into force. For example, the Protection of Natural Forest programme launched in 1998 is intended to protect all of the natural stands and make Chinese forestry sustainable in the future.

4.1.4.4 *Climate Change*

A general warming trend is predicted for the region due to anthropogenic greenhouse gas emissions. The warming is expected to be more pronounced in the arid, semi-arid and Siberian regions than in the coastal monsoon region. Precipitation is expected to increase marginally in winter throughout the region, and in summer in a non-uniform pattern. Summer increases are expected in Siberia, the Korean peninsula, the Japanese islands, and southwestern China. However, substantial declines in rainfall may occur in the northern, western, and southern parts of China.

Models suggest large shifts in the distribution and productivity of boreal forests, with increases in grassland and shrubland, and decrease in tundra. Coupled with increases in demand, serious shortages of boreal industrial timber may result. A major fraction of the existing mountain glacier mass will probably disappear, accompanied by increases in runoff. At the same time, permafrost is expected to disappear in northeast China, Mongolia and large parts of Tibet.

Uncertainties characterise the modelling of future runoff. Some scenarios have melting glaciers increasing runoff. Some models indicate possible future decreases in runoff in all major river basins of China. Should this occur, water shortages produced by population growth and economic development will be exacerbated. Climate change may be expected to have major impacts on agricultural production; however, quantitative projections are highly uncertain and cover a wide range of both positive and negative estimates. Sea-level rise will exacerbate existing problems of coastal erosion and saltwater intrusion in deltaic regions and coastal plains. Since major cities are located in coastal zones, and include large flood-prone areas, expansion of vulnerable areas due to sea-level rise will be a significant problem. In terms of human health, several-fold increases in heat-related deaths may be expected and the range of disease vectors such as mosquitoes may enlarge.

4.1.5 An Integrated Framework for Understanding Global Change in the Context of East Asia

It is clear that major changes have occurred in the atmosphere, terrestrial ecosystems, and marine and coastal ecosystems of East Asia in the past. These are linked with each other, and also with human society, that both serves as a forcing factor and experiences the consequent impacts of environmental change. Moreover, regional environmental changes affect global-scale aspects of the planetary climate system, producing another larger-scale system of linkages and feedbacks.

Dominated by the East Asian monsoon, the region is characterised by climate change on time scales ranging from seasonal jumps to inter-annual and inter-decadal variations and abrupt changes between climate regimes. These fluctuations strongly influence ecosystems in complex patterns. On the longest time scale, variations in the strength of the monsoon on time scales of several millennia have largely determined the historical patterns of vegetation over China.

The terrestrial ecosystems of the East Asian monsoon region constitute a major fraction of the Earth's biomass. They serve as a huge evapotranspiration source and also play a crucial role in the global hydrological cycle and biogeochemical cycles. Consequently, it should be expected that changes in regional ecosystems would have significant feedbacks on climate at regional as well as global scales through the fluxes of energy, water and trace gases.

Numerical simulations confirm these linkages. Using a regional climate model, it is found that removal of all vegetation over the East Asia monsoon region would weaken the summer monsoon and reduce rainfall significantly. Ecosystem models permit estimation of the potential vegetation that would be expected to accompany the current climate. Because of human land use, this is markedly different from actual current vegetation. These differences imply significant effects on surface energy and water balances, and therefore on the physical climate system. Comparison of contemporary climate with the simulated climate corresponding to estimated potential vegetation shows highly significant effects attributable to human land-use changes, e.g., warming and drying over east central and northwest China, and moisture reduction over most of east China. This suggests the possibility of improving environmental conditions by recovering the natural vegetation and controlling atmospheric properties through management of human activities. It is therefore necessary to couple the biological component with the physical monsoon climate in order to understand and predict changes in the Asian monsoon. Moreover, climate-ecosystem-land use interaction is an important component of global change in the East Asia monsoon system.

Thus, climate and ecosystems in the Asia monsoon region are coupled with each other in two-way interactions. Ecosystem variations are mainly driven by monsoon variations and changing vegetation cover due either to climate or human land use has significant feedback on monsoon climate.

Greenhouse gas emissions are increasing in the region at over twice the global rate. Moreover, human induced tropospheric aerosols, principally sulphate aerosol from coal combustion, have had complex measurable effects on regional climate patterns. Indeed the cooling due to anthropogenic aerosols over East Asia is of the same order of magnitude as the average global warming due to increased carbon dioxide.

In summary, anthropogenic modification of the East Asia monsoon system appears to function through two major processes:

- Changes of distribution, structure and function of terrestrial ecosystems introduced by land use change, which directly impact the physical processes and the biogeochemical processes of the land surface;
- Anthropogenic emission of greenhouse gases and aerosols due to the development of industry and agriculture, which impacts directly atmospheric chemistry and physics.

It is very likely that the anthropogenic modification of the monsoon system will continue to occur. Therefore, there is a need to introduce the human component to couple with the natural monsoon system, and when the anthropogenic effect is taken into account, two basic processes need to be introduced: chemical processes coupled with the biological and physical processes, and social-economic processes coupled with natural processes.

4.1.6 The Concept of a General Monsoon System

The East Asian monsoon may best be considered as a physical-biological-chemical-social coupled system, *the General Monsoon System*, to describe the integrated behaviour of environmental and human changes in East Asia (Fig. 4.3). This system includes the physical processes of the monsoon system (atmospheric dynamics,

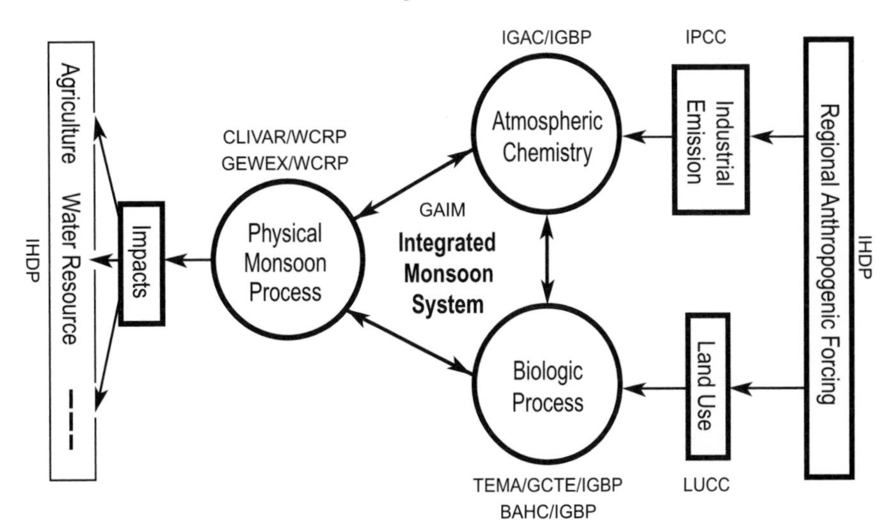

Fig. 4.3. Linkages within the General Monsoon System

air-sea and air-land physical interactions), biological processes such as ecosystem dynamics, and chemical processes in atmosphere, rivers and ocean. Coupling of physical, biological and chemical processes is a core concern. Anthropogenic forcing principally comprises industrial emissions and land use/land cover changes. Changes in the monsoon system in turn have significant impacts on agriculture, water resources and socio-economic development. Such a conceptual model provides a powerful co-ordinating and integrating research, analysis, and policy development tool for the region.

4.1.7 Conclusions

Clearly, human activities have had large and measurable effects on the environment of East Asia. Land use intensification, changes in consumption patterns, conversion of land from subsistence food production to cash crops and to urban use have greatly impacted environmental conditions through Temperate East Asia. Human land use in the region has greatly reduced the area of forest, and converted much of the forested area to managed plantations. Livestock numbers have greatly increased at the same time that grassland area has decreased and degraded; there is now a serious imbalance between the numbers of livestock and their sources of nutrition. Farmland area has remained relatively constant in recent years, with loss due to urbanisation being balanced by conversion of forest and grassland to agriculture; agricultural productivity has been markedly increased through irrigation, increased inputs of fertilisers, and genetic improvements. The ability of the region to maintain increases in crop production is questionable, and in many areas soil fertility and water availability have declined; the loss of nitrogen, phosphorus, and potassium through erosion in China exceeds the annual applied amount of chemical fertilisers. Over a quarter of China is threatened by desertification caused by over-grazing, deforestation, inappropriate farming practices, and other human interventions. Although population growth in the region has slowed, large-scale migration into urban areas continues, with tremendous impacts on croplands, fuel resources, and water supply.

On a larger scale, the current east Asia monsoon system has been modified to a considerable extent by human activities, both the change of vegetation cover from its natural conditions to the current intensively exploited urban and agricultural landscapes, and anthropogenic emission of greenhouse gases and aerosols due to the development of industry and agriculture. The weakening of the summer monsoon due to these processes is perhaps one of the main reasons for the development of aridification and desertification trends in many parts of East Asia.

The conceptual model of a *general monsoon system* that includes physical, chemical, biological, ecological, *and* socio-economic processes thus provides a useful theoretical framework for the regional aspects of global change in East Asia from the standpoint of earth system science. When finally developed, a Regional Integrated Environmental System Model should be a useful tool for further synthesis. In the meantime, preliminary assessments on the impact of such changes on the regional social and economic development were made based on scenarios of projected climate change over the next half-century. These studies demonstrate that scientific knowledge of climate and environmental changes is a necessary ingredient for wise policy making and essential for designing preventive, adaptive and remedial measures toward a sustainable world.

Changes in the environment of East Asia have been of very significant magnitude in comparison with those experienced in other regions of the world. These human-induced changes have had major impacts not only on regional conditions, but also on a vital component of the planet's general circulation system. Regional-scale processes may have consequences in the global earth system.

4.2 The Atmosphere, Climate Variability and Change

4.2.1 Climate Variability

The climate of East Asia is typical of that of a monsoon region, and is characterised by high variability on seasonal, inter-annual, inter-decadal and longer time scales. The monsoon regions of Asia and West Africa, and to a lesser extent Australia, have the world's highest climate variability on all time scales (Fu and Zheng 1998).

As an example, the mean absolute variability of annual rainfall in the Australian monsoon region in the period 1951–1990 reached more than 200 mm yr^{-1}, while in East Asia it was be as high as 300 mm yr^{-1} (Fig. 4.4). The high rate of precipitation variation in the monsoon regions shows that these regions make a substantial contribution to the total inter-annual variability of the global climate (Webster and Yang 1992). If differences between mean values of precipitation during each decade across the period 1910–1990 are considered, large inter-decadal changes are evident in the Asian, West African and Australian monsoon regions, where they may reach more than 1 000 mm 10 yr^{-1}.

4.2.2 ENSO Variability

ENSO is a well-known coupled atmosphere-ocean phenomenon centred in the Pacific and Indian Oceans, but with worldwide impacts on climate. It has been observed that there is a close linkage between the sea surface tem-

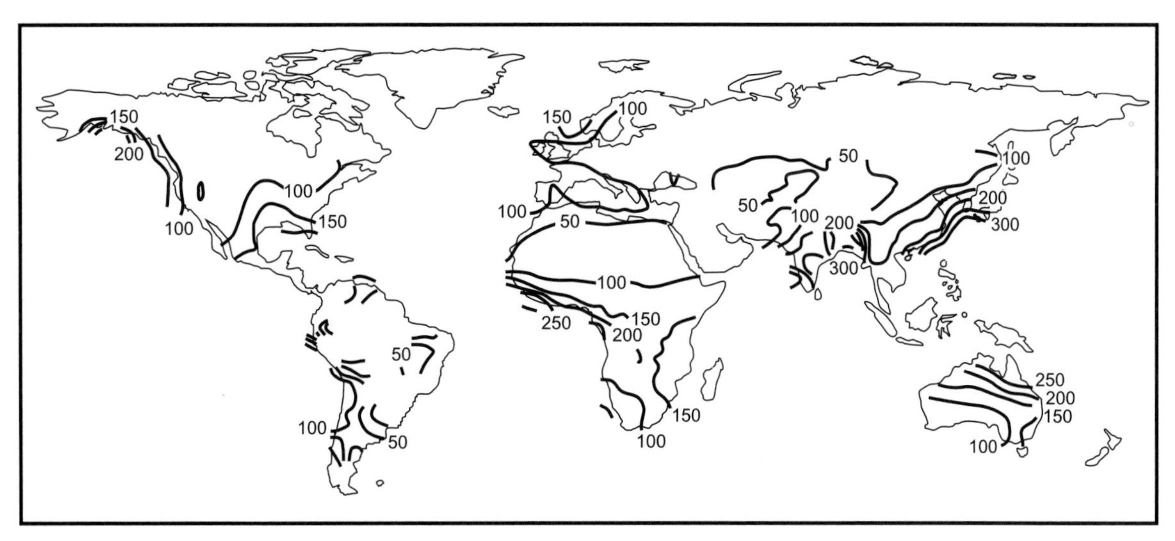

Fig. 4.4. Interannual global precipitation variability (mm yr^{-1}) (Fu and Zeng 1998)

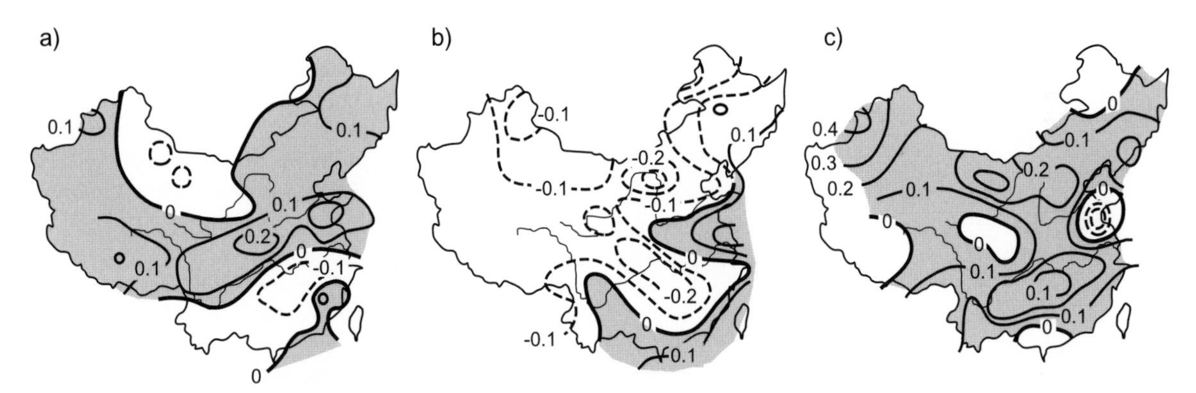

Fig. 4.5. Summer monsoon rainfall anomalies (%) over China **a** during the year before the onset of El Niño, **b** the year of occurrence and **c** the year after onset (Fu and Fan 1999)

perature anomalies in the eastern equatorial Pacific and the intensity and location of the western Pacific subtropical high, which is the action centre that mainly controls the location of summer rain belts in East Asia (Fu and Teng 1993). The relationship between the El Niño events and rainfall patterns in China is distinctive. Rainfall distribution patterns differ significantly during the stages of El Niño. In the year before onset (Fig. 4.5), the main rain belt is located to the north of the Yangtze River, while a lesser rainfall area is located to the south of the Yangtze. In the year of onset, most parts of the country have less rainfall. In the year after onset, most areas receive more rainfall, with a major rain belt in the Two-Lake Basin and another in northern China plain. The association of these spatial patterns with different stages of El Niño has permitted the development of a scheme to predict the situation of the western Pacific high and then the summer precipitation patterns of East China based on winter SST anomalies (Fu and Zeng 1988).

Cold-summer disasters in Northeast China and Japan are associated with higher than normal SSTs in the eastern equatorial Pacific (Xu et al. 1983; Fu and Teng 1993). Prior to 1990, nine out of twelve El Niño onset years were associated with cold summers in northeast China and all of the severe cold-summer disasters have occurred in El Niño onset years. During the 1990s, when the global land surface temperatures increased, this relationship seemed to weaken, suggesting that other large-scale factors might be playing a role.

Fewer typhoons appear to occur during El Niño years (Pan 1982), and the East Asian monsoon is weak (strong) when ENSO is in its warm (cold) phase (Fu 1993; Huang et al. 1994; Wang 1999). During an El Niño, the sea surface temperature is higher than normal in the Northwest Pacific and its surrounding marginal seas, reducing land-sea contrasts across the coastline of East Asia and correspondingly weakening the monsoon.

4.2.3 Palaeomonsoons

The Tibetan Plateau serves as a seasonally varying heat source for the entire Asian monsoon system. Indeed, its uplift over geological time has to a great extent controlled the development of the palaeomonsoonal circulation over China. Correspondingly, the Tibetan Plateau region is exceptionally sensitive to climate change. Recent cooling and warming periods appeared 5–8 years earlier over the Plateau than in eastern China (Tang 1989). On a longer time scale, the Holocene Altithermal began earlier on the Tibetan Plateau than in other regions of China, and during the last glacial maximum the expansion of glaciers on the Plateau occurred earlier than along the margins (Li and Shi 1995). Thus, variations in the regional climate of the Tibetan Plateau may be useful in predicting future trends of temperature in monsoonal China, and perhaps even in the entire Northern Hemisphere.

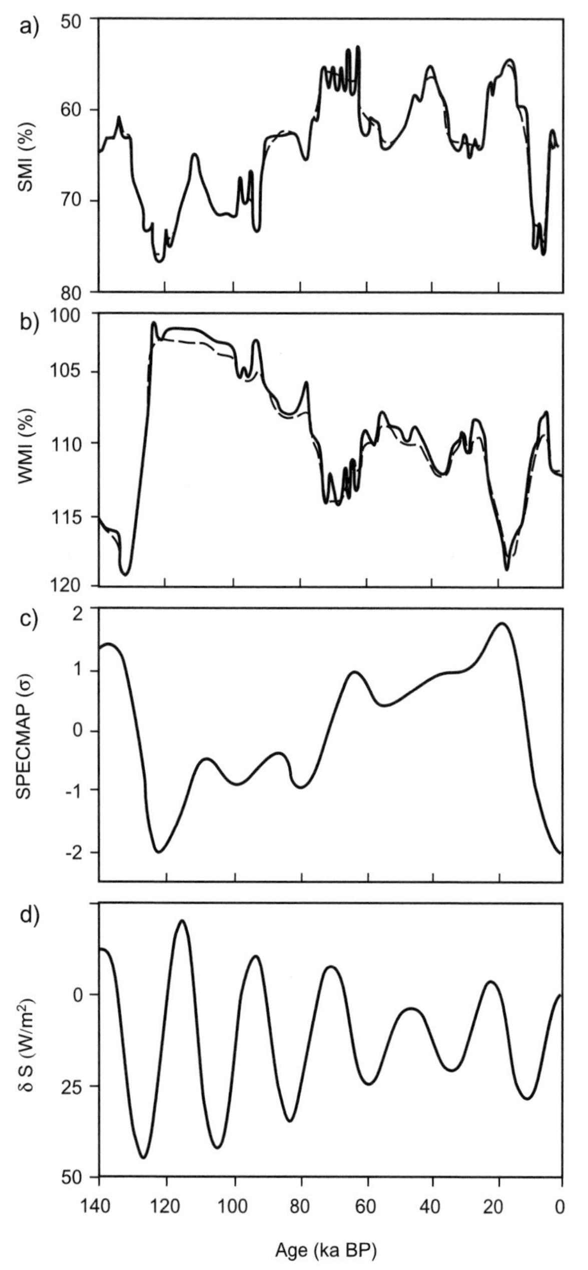

Fig. 4.6. Variation of **a** summer monsoon index (SMI), **b** winter monsoon index, **c** global ice volume and **d** summer solar insolation over the past 14 000 years in central China (Liu et al. 1995)

East Asia is rich in loess, lacustrine, and ice-core records that provide a unique natural archive of palaeoclimatic history. For example, studies of the loess-soil sequence suggest that Wenan loess deposition in central China was primarily forced by variations of global ice volume through changes in the winter monsoon, while soil formation was more closely coupled with insolation changes through the summer monsoon (Liu et al. 1996) (Fig. 4.6). Loess deposition studies also show that the climate system experienced rapid and high-amplitude changes in the last glacial period.

Extending back to as early as about 3 000 years before the present, Chinese literature contains a huge amount of valuable information of climate change. For example, drought and flood indices have been derived over a 500-year period, representing the long-term fluctuations of the summer monsoon over East China (Wang and Zhao 1981). These data have been extended back 2 000 years in combination with other proxy data. Figure 4.7 is the time series of reconstructed wet and dry indices for the last 1 100 years, representing long-term variation of the East Asian monsoon in the historical period (Liu et al. 1998).

4.2.4 Abrupt Climatic Changes

Highly variable climates may also exhibit abrupt changes from one apparently stable climatic regime to another (Fu and Wang 1992). Application of catastrophe theory and quantitative schemes for the definition, classification and detection of abrupt change of climate has permitted identification of the major abrupt climate change events in China. About 3 000 years ago (1300–1250 B.C.), central China experienced an abrupt change from wet to dry, and in about A.D. 300, northern China abruptly went extremely dry. In about A.D. 680 the climate abruptly reverted back to wet (Ye and Yan 1992) (Fig. 4.8).

On decadal scales, three major events occurred in the last century (Fu 1994b):

- The event in early 1900 was characterised by rapid cooling, dry conditions in the global tropics and in the African and South Asian monsoon region, and wet conditions in middle latitudes and the East Asia monsoon region. At the same time, a rapid transition occurred in the meridionality of the monsoon current in the summer monsoon over the Indian and Pacific Ocean (Fu and Fletcher 1988).
- The abrupt change in the 1920s was the strongest signal in the last century on a global scale. It was characterised by rapid global warming, increased tropical rainfall, and a more active tropical monsoon, to-

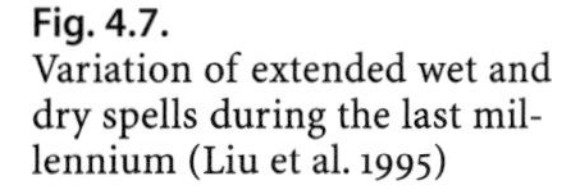

Fig. 4.7. Variation of extended wet and dry spells during the last millennium (Liu et al. 1995)

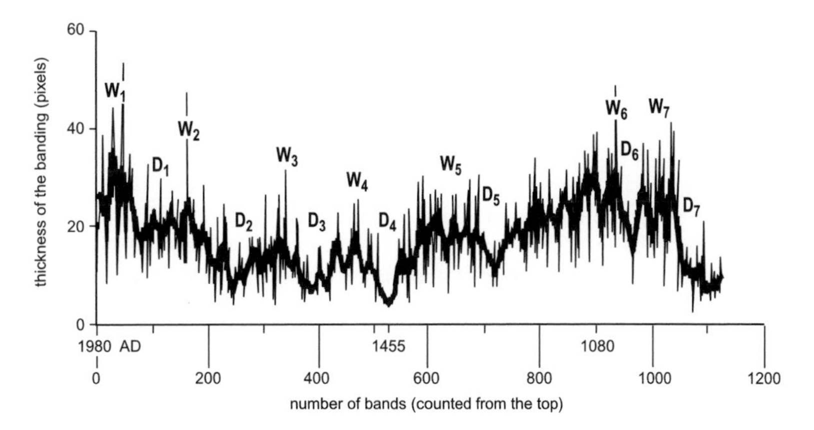

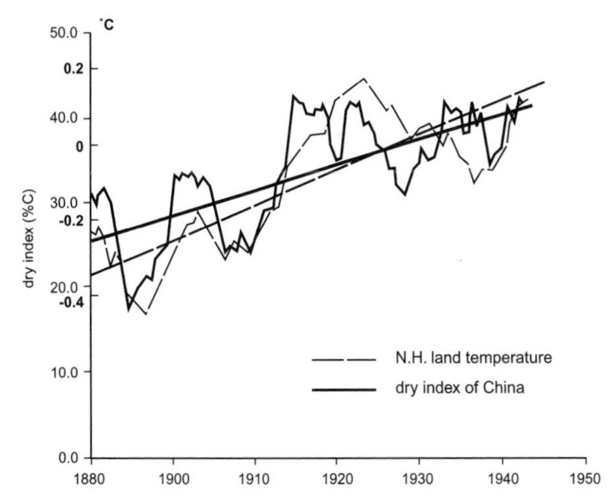

Fig. 4.8. Variation in the aridity index of China and the northern hemisphere land temperatures (Fu 1994b)

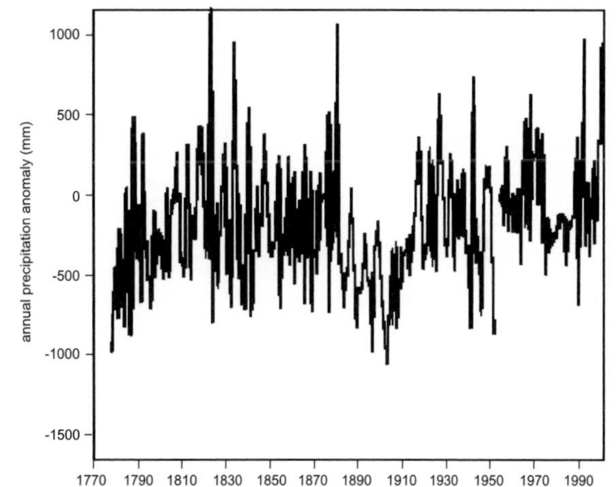

Fig. 4.9. Monthly precipitation variation for Seoul, 1771–1998 (Lee and Lim 2001, pers. comn.)

gether with a weaker subtropical monsoon and dryer middle latitudes. The Indian summer monsoon abruptly changed around 1922 from a weak period to a strong one. The long-term change of monsoon climate and the temperature over much of the Northern Hemisphere are closely linked. Warming over land was stronger and earlier than over the ocean, enhancing land-sea thermal contrast and therefore favouring development of the monsoon low and disturbances in the monsoon flow, possibly producing the abrupt enhancement of the summer monsoon over India (Fu 199a; Fu et al. 1999). Around 1920, the Chinese monsoon also exhibited an abrupt change, but in the opposite direction from that over India.

- The third event around the 1960s was characterised by a weakening of the African and Indian monsoons. The former led to the persistent drought in the Sahel region of North Africa. A multi-variable analysis of this abrupt change (including temperature, precipitation, sea surface temperature, 500 hPa height, and sea level pressure fields) has revealed a planetary structure in its spatial pattern. In the 1960s, there was an arid zone from northern Africa extending northeastward to Japan sandwiched between two relative wet zones. Such a spatial pattern appeared also in the other fields and can be used to explain many features of the regional diversity of climate change superimposed on the global scale mean patterns (Yan et al. 1990, 1991). The abrupt change in later 1970s, perhaps related to recent rapid global scale warming, has received more attention (Fu et al. 1998). The results show possible linkages between global warming and strong El Niño events.

Similar changes occurred in Korea. Figure 4.9 shows a time series of the monthly precipitation as obtained from the rain gauge data in Seoul over the period of 1771–1998 (Kim and Ha 1987; Lee and Lim 1998) A drought is shown to have persisted for over 20 years around 1910 in Seoul.

Abrupt changes on the decadal scale in Asia have several important common features: close linkage with global temperature changes, indicating the thermally-driven nature of abrupt change on this time scale; sensitivity of the monsoon region to abrupt change; and the planetary scale structure of the abrupt change (Fu 1994b).

4.2.5 Past and Future Trends

4.2.5.1 *Temperature and Precipitation*

As for the globe as a whole, the most prominent climatic change in East Asia in the past hundred years has been significant warming: the average annual temperature has increased by more than 1 °C (Fig. 4.10). Most evident since the 1970s, this increase primarily reflects warming of the winter season, although other seasons also show slight warming. Large decadal variability in precipitation seems to have masked a smaller positive trend (Karl 1998).

Over smaller areas, there has been a 2–4 °C temperature increase over the past 100 years in eastern and north-eastern parts of the region, while there has been a 1–2 °C temperature decrease in the eastern half of southern China, except for the coastal area (Karl 1998) (Fig. 4.11). This is a notable feature, since it occurred in an overall warming period. In Japan, annual average temperature has increased during the last century by about 1 °C and even more in large urban areas.

Warming trends in the last two decades have been greater than in any comparable periods in the last 100 years. This period also included two record-breaking warm phases of El Niño and a record-breaking La Niña cold phase. A significant drying trend has occurred in the inland areas of East Asia (Ma and Fu 2000) (Fig. 4.12).

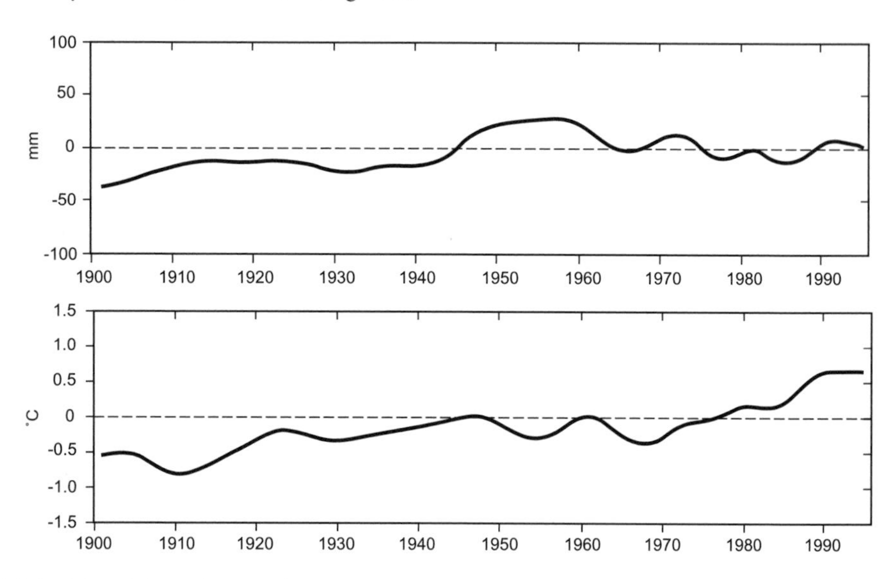

Fig. 4.10. East Asia regional trends in annual **a** precipitation and **b** temperature during the twentieth century twentieth century (Karl 1998)

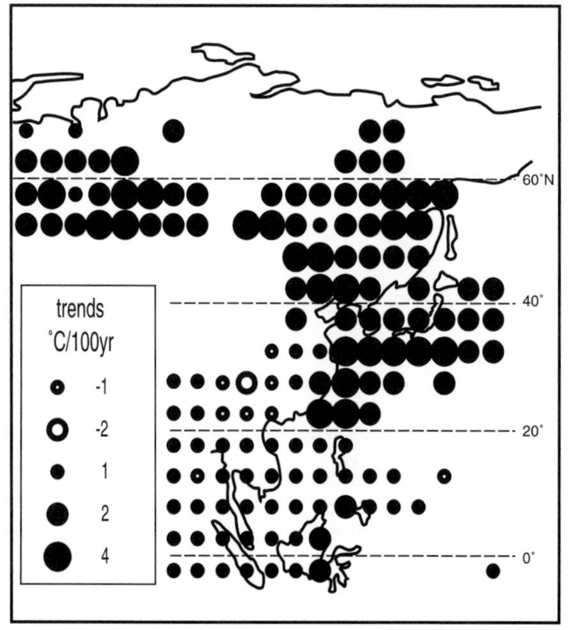

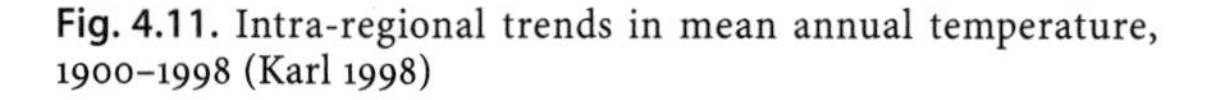

Fig. 4.11. Intra-regional trends in mean annual temperature, 1900–1998 (Karl 1998)

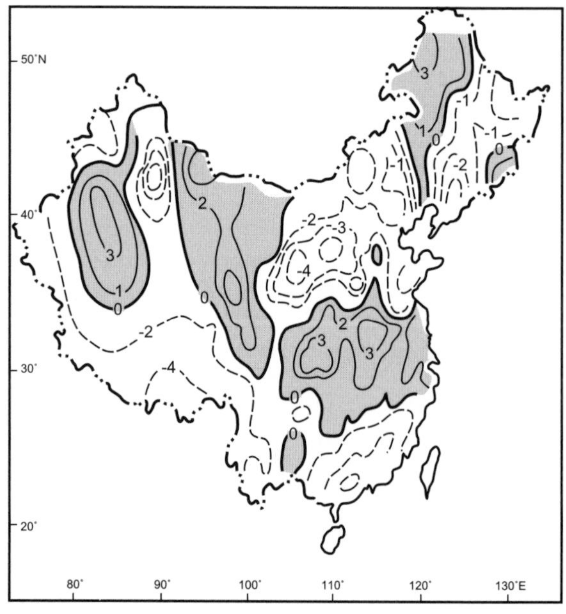

Fig. 4.12. Distribution of the trend of surface humidity index over China, 1951–1997 (positive (*shaded*) denotes becoming more moist; negative indicates drying) (Ma and Fu 2000)

4.2.5.2 *Future Trends*

Coupled atmosphere and ocean Global Circulation Models (GCMs) are the best currently available tools for development of scenarios of future climate. There is general agreement among current models that warming of 2–3 °C will occur over the region with CO_2 doubling. This is likely to be more pronounced in the arid and semi-arid and Siberian regions than in the coastal monsoon region. Recent simulations, which include the offsetting effects of sulfate aerosols, project a rise in annual mean temperature of about 0.8 °C over the eastern part of the region, about 1 °C over most parts of eastern China, and close to 2 °C in the Siberian region by the middle of the next century. However, the uncertainties associated with projections including sulfate aerosols are considerable.

In equilibrium and transient-response numerical experiments with GCMs, precipitation throughout the region is projected to increase marginally (<0.5 mm d^{-1}) at the time of CO_2 doubling during winter (DJF). In the summer (JJA), the spatial pattern of projected changes in precipitation is not uniform over the region. Model projections suggest that precipitation will increase slightly (0.5–1.0 mm d^{-1}) in the northern part of the region (Siberia), and by more than 1 mm d^{-1} over the Korean peninsula, the Japanese islands, and the southwestern part of China. In contrast, precipitation may decline in the northern, western, and southern parts of China. The projected decline in rainfall over most of China is substantial in numerical experiments that include the effects of sulphate aerosols (IPCC 1998).

4.2.5.3 *Atmospheric Chemistry*

The rapid increase of anthropogenic emission of greenhouse gases (CO_2, CH_4, NO_x, CFCs), SO_2 and sulphate aerosols is particularly significant for the region. For instance, although the per capita emission of CO_2 in Asia is the second lowest in the world with 0.4 Gt C yr^{-1}, a third of North America, the yearly increase rate in Asia is the highest (4.9%), more than twice the global average (2.2%) (Fig. 4.13). Asia is also the major producer of anthropogenic CH_4, since it has the largest area of rice paddy fields. Recent studies have concluded that human-induced tropospheric aerosols have had significant effects on climate change. At present, some projections of climate change with coupled atmosphere-ocean models include the radiative effects of aerosols. Those that have been run include a very simplified representation of aerosol effects. However, the distribution of human-induced tropospheric aerosols varies regionally, and the impacts of sulphate aerosols on regional climate change appears particularly significant in East Asia. The transport of aerosols and trace gases out of East Asia is often extensive (Fig. 4.14). It is estimated that the troposphere aerosols resulting from combustion of fossil fuels, biomass burning and other anthropogenic sources have led to a negative direct forcing as high as –1.7 W m^{-2} in East Asia, compared to the global average of about –0.5 W m^{-2}. On global average, the negative forcing of anthropogenic aerosols over East Asia is of the same order as direct positive radiative forcing by CO_2 (1.5 W m^{-2}) (Qian and Fu 1996). A simple analysis by a two-dimensional model can partially explain the cooling area observed in Southwest China over the last 100 years (1900–1998) (see Fig. 4.11).

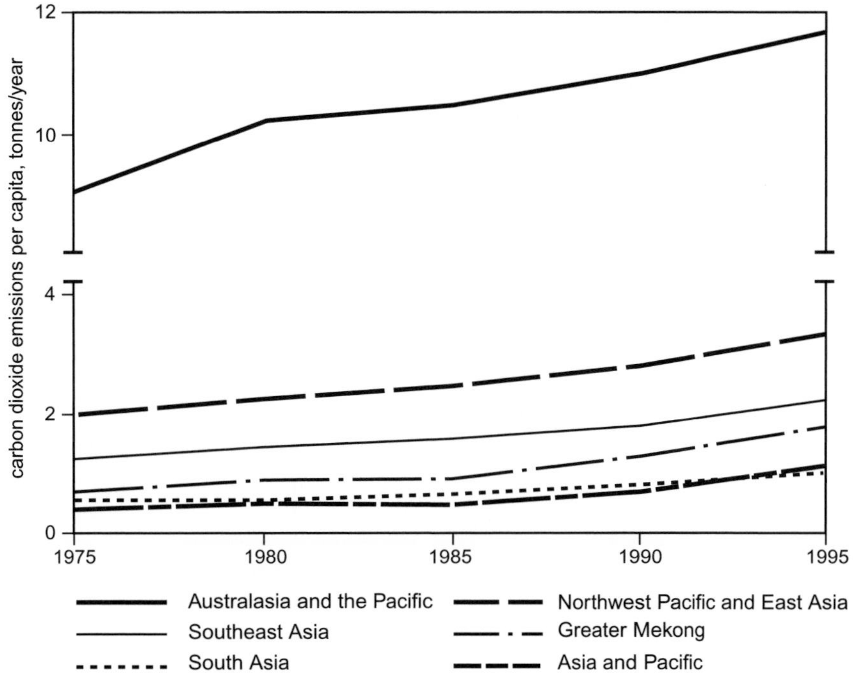

Fig. 4.13.
Per capita carbon dioxide emissions in Asia and the Pacific (UNEP 2000)

To understand further the role of sulphate aerosols in regional climate change, a climate-aerosol interaction module has been developed (Qian et al. 1999) (Fig. 4.15). This model includes interactions between regional climate and sulphate aerosol transfer and deposition through the atmospheric radiation transfer scheme and atmospheric dynamic parameters.

By interactive coupling of regional climate and sulphate aerosol models, the impact of increasing sulphate aerosol on the regional climate in East Asia has been simulated, including both direct and indirect aerosol effects. Findings include the suggestions that:

- aerosol distribution and cycling processes show substantial regional spatial and temporal variability;
- both direct and indirect aerosol forcing have regional effects on surface climate;
- aerosol-induced feedback processes can affect the aerosol burdens at the subregional scale.

Fig. 4.14. Transport of aerosols and trace gases over and out of East Asia (NASA 2001)

The most interesting result of this simulation is that the simulated cooling patterns are very close to those observed over southwest China (Fig. 4.16). It would appear that the increase of anthropogenic emission of sulphate aerosol on the regional climate in East Asia should not be neglected as an agent of change. The increase of sulphate aerosol is mainly caused by the continuous increase of coal combustion in the region related to industrial development. The effect of sulphate aerosols is not only partially to offset greenhouse warming, but also to produce more complex effects on regional climate, such as changes in precipitation.

4.2.6 Human Influences and Impacts

Anthropogenic modification of the monsoon system, due mainly to industrial emissions, land use/land cover changes and urbanisation, may occur through two major processes:

1. changes of distribution, structure and function of terrestrial ecosystems introduced by land use change, which directly impact the physical processes and the biogeochemical processes of the land surface;
2. anthropogenic emission of greenhouse gases and aerosols due to the development of industry and agriculture, which impacts directly atmospheric chemistry and physics.

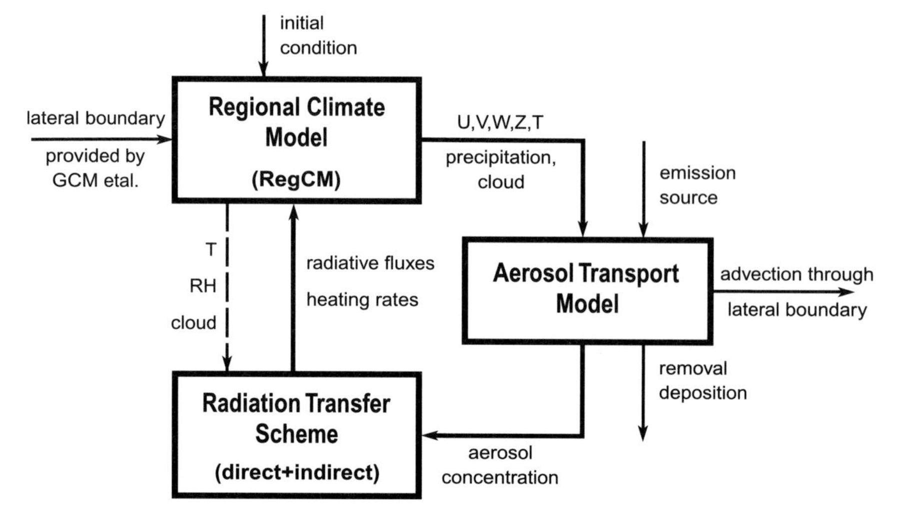

Fig. 4.15.
Regional climate and sulphate aerosol interactions (Qian et al. 1999)

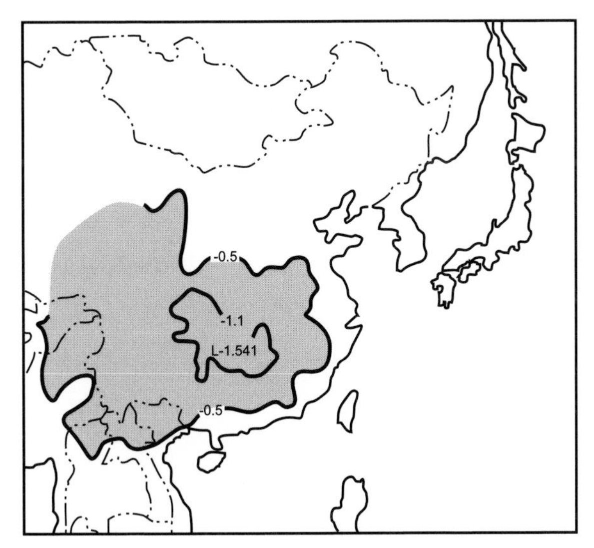

Fig. 4.16. Simulated sulphate-aerosol-induced temperature changes over East Asia (cooling indicated by *shading*) (Qian et al. 1999)

It is very likely that the anthropogenic modification of the monsoon system will continue to occur. A pressing need exists to introduce the human component into coupled regional models of the natural monsoon system. When this is done, two basic processes need to be recognised: chemical processes coupled with the biological and physical process and social-economic processes coupled with natural processes.

Changes in the climate system, whether natural or human-induced have a wide range of impacts relevant to human welfare. For example, an increase in the frequency or severity of heat waves/episodes would cause an increase in (predominantly cardio-respiratory) mortality and illness (IPCC 1996). If summertime daily peak temperatures rise due to global warming, rising death rates are predicted (Fig. 4.17). Because this relationship is seen particularly among the elderly, a thorough response to global warming-induced severe heat and heat waves and their health effects must be considered. Apart from the direct impacts of heat waves, there is also the problem of a potential increase in the habitats of disease vectors. For example, malaria-carrying mosquitoes able to live in relatively low temperatures have been discovered.

Climate change also impacts other elements of the earth system, producing changes that are of importance to human welfare.

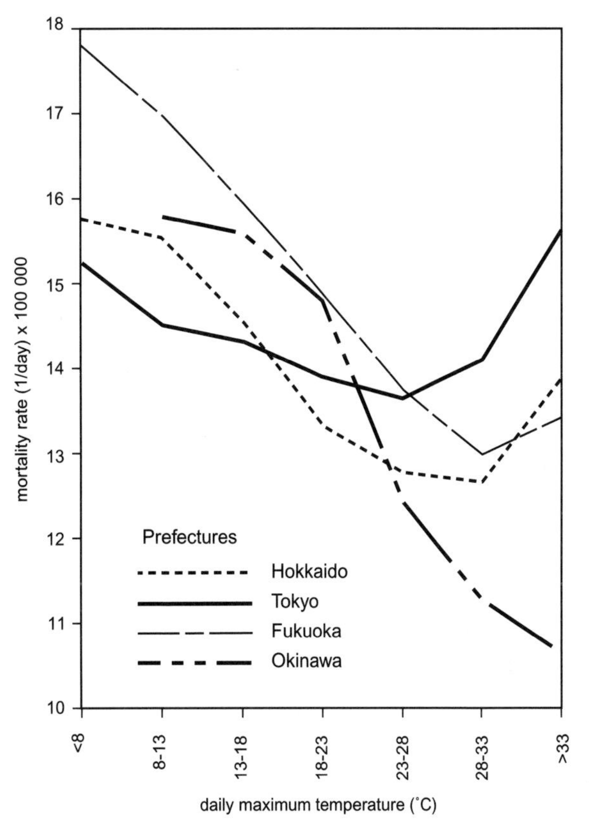

Fig. 4.17. Relationship between daily maximum temperature and mortality rate for males aged 65 years or older in the Japanese prefectures of Hokkaido, Tokyo, Fukuoka and Okinawa (Honda et al. 1995)

4.3 Land-Ocean Interactions

The coastline of the region is about 73 000 km in length, and the related ocean and sea surface exceeds 10 000 km^2, with depths exceeding 9 000 m. This gigantic water mass plays a crucial role in all regional processes, including global climate change and natural ambient systems. Stretching from the Polar Circle to the Tropic of Cancer, these waters include a great variety of marine and coastal ecosystems (Table 4.1). There are six major seas in the region: Bering, Okhotsk, Japan, Yellow, East China, and South China. Each of these has unique oceanographic, hydrographic, biological, and anthropogenic regimes, all subject to changing environmental conditions.

4.3.1 Oceanic Consequence of El Niño

One of the most dominant oceanic fluctuation phenomena affecting all of the basins of Temperate East Asia is ENSO. Among its oceanic consequences is a drop in sea level in the Western Pacific, including South and East China seas, because of the eastward migration of warm surface waters. This decreases the intensity of the dominant Kuroshio Current, resulting in negative temperature anomalies in the SSTs of the Sea of Japan (Savelyev 1998). The correlation between cycles of solar activity, ENSO events, and cycles of biological productivity in Asian marginal seas has been emphasized by Shuntov et al. (1996, 1997). ENSO also has a significant effect on the ice cover in the Bering Sea (Niebauer 1998).

Table 4.1. The geography of the seas of East Asia (Zenkevich 1963; Oceanographic Encyclopaedia 1974)

	Bering Sea	Sea of Okhotsk	Sea of Japan/ East Sea	Yellow Sea	East China Sea	South China Sea (northern portion)
Area (km^2)	2.3×10^6	1.6×10^6	1.0×10^6	4.2×10^6	8×10^5	3.5×10^6
Mean depth (m)	1640	859	1536	44	349	1024
Coastline (km)	5000	460	7600			
Climate	North: artic/sub-artic South: temperate marine	Temperate monsoon	North: boreal South: tropical		Monsoon	Tropical monsoon
Summer sea surface temperatures	10 °C	18 °C	North: 15 °C South: 26 °C	>28 °C	30 °C in Kuroshio area; 20–24 °C at coast	28 °C
Biologic regime	Abundant phyto- and zooplankton, significant benthic biomass in the North	High phytoplankton biomass, numerous commercial fisheries for both pelagic and benthic species	High productivity in coastal waters	Highly productive	Productive and species rich along eastern coast	Productive, species rich, numerous coral reefs
Major rivers	Anadyr			Huanghe, Yangtze, Heihe	Yangtze	Pearl
Coastal population density (persons km^{-2})	<1	1–10	North: 1–10 South: 50–100	>100	>100	>100
Dominant current system/circulation pattern	Cyclonic		North: Primorskoye South: Tsusima		Kuroshio	Oceanographic features driven by monsoon winds
Other characteristics		Highly stratified water column		Large riverine outflow increases turbidity and decreases salinity during summer		

4.3.2 Coastal Processes and Sea-Level Changes

Marine transgressions and regressions significantly influenced all natural processes in the TEA region during the Pleistocene and Holocene. The zone between 38–42° N, with the biggest shifts in land areas and maximum temperature contrasts, has been especially sensitive to sea level changes. In the late Pleistocene, the climate in this zone changed from oceanic (warm) to continental (cold) phases (Korotky et al. 1997). During the climate optimum, the sea level rose 20–22 m in two thousand years. At that time coastal dynamics were characterised by attenuation of erosion in valleys and development of sediment accumulation in the near-mouth parts of rivers submerged by the sea and in inner parts of bays and bights (Korotky 1994; Korotky et al. 1997). During the Holocene climate optimum, the sea level was up to 2 m above the present level at Amursky Bay, Peter the Great Bay, and the Sea of Japan. The coastline was indented by shallow bights with rich biota and broad-leaved trees grew in coastal forests (Rakov 1995). In the last 300 000 years, the Bohai Sea transgressed to and above its present level 5–7 times; around 8 000 years ago, during the Hyang Hua transgression, many lowlands were flooded, among them the region of Tianjin (Zhao and Chin 1985) (Fig. 4.18). On the eastern Chinese coast many towns were again flooded 2 500 years ago, although there was only a small amplitude transgression (Kaplin and Selivanov 1999).

Sea-level rise results from thermal expansion of ocean water and the melting of mountain glaciers as well as increased evaporation, precipitation, and water discharge from the land. Evaporation, precipitation, cloudiness and river discharge influence and are affected by changing land and sea temperatures. An increase in surface land temperature of 1.5 °C, which corresponds to the minimum estimated value for doubled carbon dioxide concentrations, might raise sea level 20–25 cm by 2050. Given the complexity of connections between temperature increase and sea-level rise, it is assumed that a potential temperature increase of 3.5 ±2.0 °C during the next 100 years will result in a sea-level rise of 35–55 cm. According to IPCC (1996) estimates, the sea level will rise 50–60 cm by 2100 if current conditions persist. These figures are 4–5 times greater than changes in the 19th century, when sea-level rise was only 1.5 mm per year (Klige 1982; Wigley and Raper 1987, 1992; Kostina 1997). Previously observed natural eustatic sea-level rise, 1–2 mm yr^{-1} in the Pacific, can be predicted to extend into the 21st century, independent of the greenhouse gas effect (Nunn 1998). In addition to long-term effects from greenhouse gas emissions, increased human activity also directly influences sea-level rise in the short term. The

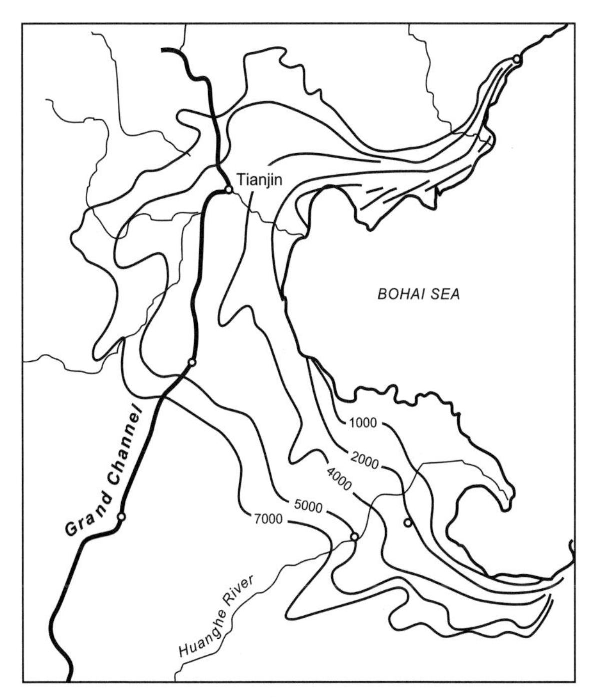

Fig. 4.18. Isolines of marine transgressions around the Bohai Sea during the last 8 000 years (Chen et al. 1995)

combination of ground water withdrawal, surface water diversion, and land-use changes have caused much of the observed sea-level rise. In East Asia, deforestation-through the release of water normally stored either in plant material or in the high soil moisture maintained by plant cover-is also contributing to current sea-level rise (Sahagian et al. 1994).

Throughout the 20th century, sea level has been rising with potentially dangerous consequences for large areas of low-lying coastal regions in Temperate East Asia (Aibulatov and Artyukhin 1993). Around 20% of sandy and pebble coasts are now advancing at a rate exceeding 1 m yr^{-1} (Bird 1990). One such coast is the Khasan sandy beach to the north of the Tumen River at the Sea of Japan (Petrenko 1998). If the Tumen River Economic Development Plan is implemented, the impact of increased human activity in the coastal zone will contribute increasingly to coastal erosion and other significant changes (Petrenko 1998; Kasyanov 1998). In eastern Sakhalin, the Sea of Okhotsk, one of the most sensitive areas to sea-level rise, has a low-lying lagoon coast. Ingression of sea water into river valleys and shifting of coastal borders will affect development of the oil industry there (Mikishin 1998). Sea-level rise also strongly affects Chinese coastal areas and can induce disaster hazards as a result of increased frequency of storms and floods, destruction of drainage systems, erosion and retreat of the coast, and ingression of marine salt waters in freshwater reservoirs (Chen 1997). The deleterious effects of sea-level rise will impact retreating salt marshes and mangroves. These are areas of major ecological importance for commercially valuable fish species, and act as barriers to floodwaters and as filters for water pollutants, such as heavy metals (Leatherman 1989; Ellison and Stoddart 1991; Edgerton 1991).

Despite difficulties and uncertainties associated with modelling future climate change, most models broadly concur that climate warming and concomitant changes in annual amount and distribution of precipitation may lead to drying of mid-continental regions and increased rainfall in coastal areas of the southern part of the Russian Far East. The influence of the cold-water Primorye Current is likely to become weaker, whereas the influence of warm water transported to the Sea of Japan from the Tsusuma Strait will probably become stronger. Sea-level rise may exceed 1.5 m for a temperature increase of 2–3 °C above present values. Even such a relatively insignificant transgression will result in inundation of river mouths with consequent development of bogs, lakes and lagoons. Sedimentation will increase and an increase in oak forests and reduction of the coniferous taiga may occur. Where rapid transgressions occur, boglands will increase and sediment accumulation zones will shift upstream from the mouths of rivers, up to 10–15 km in the case of the Partizanskaya River and more than 100 km in the Tumen and Amur Rivers.

4.3.3 River Discharge

The unique chain of marginal seas significantly affects the material transport and energy fluxes between land areas of East Asia and the Pacific Ocean. During glacial cycles, the decrease of sea area and changing SSTs in the marginal seas had a profound impact on regional climate. As the Western Pacific marginal seas trap terrigenous material supplied by East Asia, coastal water sedimentation rates can be one to two orders of magnitude higher than in the deep ocean (Wang 1999).

Four of the largest rivers of east Asia, the Yangtze, Amur, Huanghe and Xijiang, each with a huge catchment area, produce major effects in the coastal zone and on marine ecosystems as a result of the discharge of water, dissolved matter and alluvium. All these parameters are affected by global change. Large rivers respond more sensitively to climatic transformations in comparison with average-sized and smaller river systems as small impacts are integrated over their extended basin areas.

In the Amur River basin (1 855 000 km^2), century-scale changes of water dynamics have occurred against the background of a general increase of temperature and precipitation. Anthropogenic effects on the hydrology of Amur River outflow are, for the present, insignificant in comparison with the other large rivers of the region. With increasing precipitation, an increase in outflow variability is likely. High discharge levels may result in

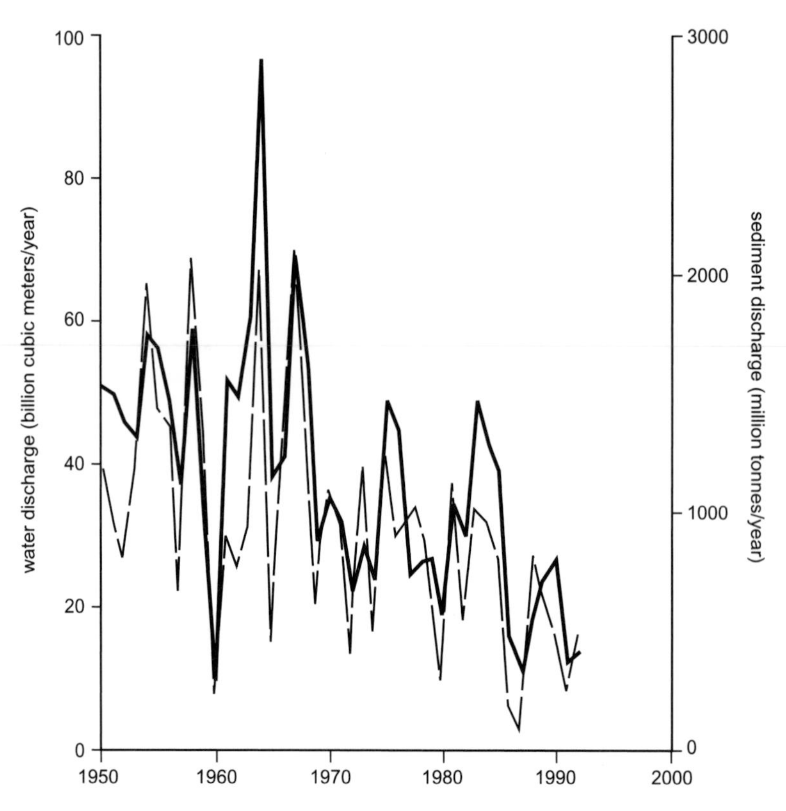

Fig. 4.19.
Variations in the Yellow River discharge of water (*solid line*) and sediment (*broken line*) at Lijin, 1950–1992 (Hu et al. 1998)

riverbed deformation, as happened during high outflow periods in the last century. Currently, splitting of riverbeds and lateral erosion is in progress.

Increased outflow variability and activation of riverbed processes in the Amur basin results in higher loads of suspended and dissolved matter within the coastal zone. This in turn results in a mosaic of highly vulnerable bottom sediments and undersea natural assemblages. An increase of dissolved materials, especially nitrates and phosphates, has caused eutrophication of Amur water and significant increases in biogenic discharges into the coastal zone, affecting the productivity of marine ecosystems.

The Huanghe and Yangtse Rivers are both undergoing the same changes as the Amur, but these are magnified by human activities to a much greater extent. The Huanghe (Yellow River) contributes 8% of total global sediment discharge from land to ocean. Over the last 40 years, water and sediment discharge has decreased by half. This represents both natural fluctuations (drought) and human activity in the river basin. Three thousand years ago, discharge of sediments was only about 10–20% of present levels because of the lack of soil erosion at the Loess Plateau. Since 1965, construction of dams and reservoirs has acted to decrease water and sediment discharge into the sea (Fig. 4.19). In addition, about 50% of all water flow is used now for irrigation, industry, household consumption etc. As a result of regulation of the Huanghe and construction of reservoirs, the coastline of the region has changed significantly. The role of climate change and anthropogenic activity in reduction of sediment discharge is almost equal. After accounting for water consumption, human activity becomes the dominant factor for reduction of overall river discharge in recent decades (Yang and Hu 1995).

In the last 130 years, the Yangtze River has shown a decrease of water and sediment discharge due to construction of irrigation systems and reservoirs (Shen 1995). As a result, accumulation of sediments on the marine part of Yangtze Delta has decreased and the delta will in the future be subject to erosion at a rate of some tens of meters per year (Chen and Zong 1998).

The long history of Chinese agriculture is mirrored in the delta dynamics of the Xijiang (Pearl) River. Six thousand years of excessive agricultural production and deforestation has increased soil erosion in valleys and solid matter discharge to the sea. This has led to growth of new delta areas and the shifting of Guangzhou City away from the sea (Chen and Li 1995) (Fig. 4.20). The increasing population density in different regions of East Asia has accelerated soil erosion and sedimentation rates to produce a significant anthropogenic influence on the marginal seas.

4.3.4 Impact on Water Resources

Modelling suggests that the areas most vulnerable to climate change will be in the northern part of China. Table 4.2 shows the impacts of climate change on the

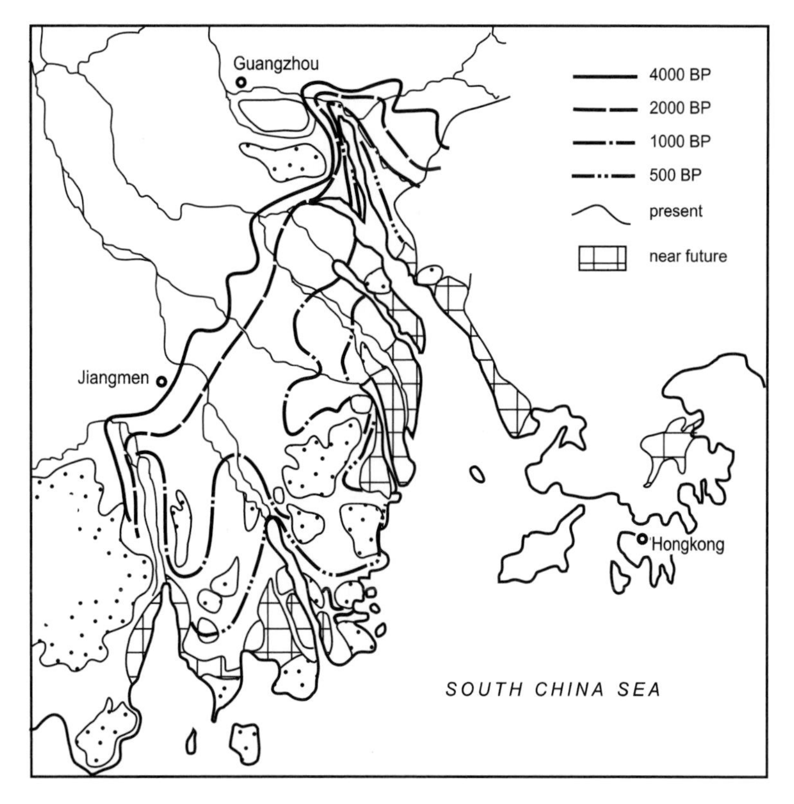

Fig. 4.20. Sedimentation rates for the Holocene (oxygen isotope stage 1) in the Western Pacific and its marginal seas. Horizontal axis denotes water depth (Wang 1999)

Table 4.2. Percentage change of annual runoff in seven river basins in China as a consequence of possible future climate change (Liu 1997)

	General Circulation Model			
	LLNL[1]	UKMO-H3	OSU-B1	GISS-G1
Dongjiang	0.4	-3.1	8.1	-4.9
Hangjiang	-7.7	-2.6	4.4	-0.7
Huaihe	-14.7	-6.3	7.8	-3.9
Huanghe	-7.2	-4.6	-2.6	4.0
Haihe	-16.0	7.2	1.0	-7.3
Liaohe	-14.0	17.4	-2.3	5.9
Songhuajiang	3.4	12.1	-7.5	3.8

[1] L. Gates, personal communication.

annual runoff of seven river basins in China, located in different climate zones from south to north. Because of the strong influence of the monsoon climate, all four GCMs project changes in runoff due to changes in precipitation in spring, summer and autumn (Liu 1997). The greatest uncertainty relates to the lack of credible projections of the effect of climate change on the Asian monsoon and the ENSO phenomenon, both of which strongly influence river runoff. In moderately and extremely dry years, the projected potential water deficiency caused by climate change, although less than that caused by population growth and economic development, may seriously exacerbate the existing water shortage.

4.3.5 Cycles of Carbon and Nutrients

The ocean serves as both source and sink of greenhouse gases, but is also an inertial system damping the greenhouse effect. The ocean contains 65 times more CO_2 than the atmosphere and small changes in the carbon cycle in the ocean may induce considerable atmospheric consequences (Bas 1987). About half of all anthropogenic CO_2 is absorbed by the ocean (Budyko 1991). This absorption is facilitated by the physical processes of solution, subduction and mixing. The gas is thus removed from contact with atmosphere. Ocean carbon exists in both inorganic and organic forms. Inorganic CO_2 is transferred by gravitational settling of marine organisms from surface oceanic waters to the deep sea and marine sediments. The carbon cycle, to great extent, is determined by oceanic circulation.

Biological processes also pump CO_2 into the ocean through algal photosynthesis. The resulting organic carbon becomes food for other organisms, including man, and only a small part of it is buried in sediments as $CaCO_3$. Most of the carbon is oxidized again in the deep sea. If the ocean were lifeless, the carbon content of the atmosphere would be much higher (Kondratyev and Grassl 1993). It is important to keep in mind that although marine biota comprises only 0.5% of terrestrial biomass, 30–35% of all atmospheric carbon passes through marine plants. It is noteworthy that the biological part of the marine carbon cycle does not stay at equilibrium (Pahlow and Riebesell 2000). Anthropogenic pollution, including oil pollution, suppresses phytoplankton activity and, as a result, reduces ocean capacity to sequester carbon (Degterev et al. 1992). Annual absorption of carbon by the ocean is about 0.5×10^{15} g C yr^{-1}, which combined with that absorbed by temperate forests, does not equal the amount of C emitted into atmosphere. The missing sink is actually several smaller sinks (Schindler 1999).

Asian marginal seas of the Pacific can absorb around 1 Gt of excess carbon, including 0.19 Gt for Bering Sea, 0.31 Gt for Sea of Japan, 0.07 Gt each for Yellow Sea and East China Sea, and 0.43 Gt for South China Sea. It is noteworthy that in high latitudes the ocean absorbs carbon, whereas in low latitudes carbon is released from the ocean. The high-latitude Bering and Okhotsk Seas serve as a conveyer exporting excess CO_2 into northern Pacific intermediate waters. As a result, the Bering Sea is unsaturated with respect to carbon. There is a similar dynamic in the Sea of Okhotsk, the East China Sea, and South China Sea. In the Sea of Japan, the excess carbon does not come out of the ocean, but remains at depth. Phosphorus, silicon and nitrogen cycles echo the pattern of the carbon cycle. Anthropogenically derived nitrogen is not stored on the shelf, rather, nutrients from East Asian seas, particularly the Bering Sea and Sea of Okhotsk, are entrained into the Kuroshio Current (Chen and Tsunogai 1998).

Independent of latitudinal processes, ocean carbon dynamics in the region are influenced by El Niño activity. During El Niño events CO_2 is released, whereas during La Niña, the ocean can absorb additional carbon (Kondratyev 1993).

4.3.6 Marine Biota Changes

Global change in the ocean, the atmosphere and on land is strongly modulated by the feedback between marine organisms, nutrients and greenhouse gases. The marked coherence observed between the distributions of physical, chemical and biological patterns suggests that the processes involved in this feedback are linked with pelagic community structure (Krause and Angel 1994).

4.3.6.1 *Primary and Secondary Production*

Primary production in the open ocean recycles biogenic elements from the deep-sea and sequesters organic carbon from surface photic layers to deep waters and sediments. The marginal seas of East Asia are highly productive with regional maxima of primary production at the Yellow and East China Seas and Kamchatka and Sakhalin shelves (Fig. 4.21). The hypereutrophic and eutrophic waters of the region, particularly in the Bering Okhotsk Seas, indicate high primary productivity potential.

There is little variability in annual productivity from phytoplankton, bacteria, protozoa and mesoplankton in the Pacific as a whole. This is the result of opposing seasonal tendencies in different regions of the Ocean. In temperate waters, productivity is high in spring and summer and low in winter. By contrast, the tropical-subtropical waters are subject to oligotrophic cyclonic gyres that limit production in the summer (Vinogradov et al. 1996). This spatial production stability is presumed to be important for the global cycling of carbon and nutrients through the Pacific (Shushkina et al. 1995).

4.3.6.2 *Red Tide Blooms*

The composition and seasonal dynamics of harmful algae and harmful algal bloom events can best be characterised in the coastal waters of Primorye and Kamchatka rather than in other areas on the Russian Far East coast. Long-term observations suggest that there has been a slight increase in the frequency, intensity and geographical distribution of harmful algal blooms during the last two decades in Russian coastal waters. Routine phytoplankton monitoring in Peter the Great Bay has revealed a peak of activity in harmful algae between the end of the 1980s and the beginning of the 1990s (Fig. 4.22). The appearance and massive blooms of raphidophytes and dinoflagellates of the genera *Gymnodinium* and *Prorocentrum* (uncommon for this area) were observed in this period. The decrease in the intensity of harmful algal blooms in the mid and late 1990s does not necessarily suggest that any positive change in the coastal ecosystem has occurred. Every year during the summer-autumn period recurrent blooms of nontoxic species, at densitites of greater than dozens of millions cells per liter, are observed in the hypereutrophic coastal waters of Amursky Bay and Avachinskaya Inlet.

Most of the harmful species and their bloom events occur in coastal waters subject to significant anthropogenic influence. It is believed that a complex of natural and anthropogenic factors favorable for outbreaks of harmful algae exists in those areas. Contributing factors include high levels of mineral and dissolved organic substances, as well as vertical stability of the water layers, associated with the substantial freshening and warming of the surface waters during the summer period. A through review is provided by Cloern (1996).

4.3.6.3 *Fish Production*

The population dynamics of pelagic fishes are mostly determined by natural factors operating at varying periodicities, from several years to several tens of years. The peak catch in Russian waters, and the Far Eastern Seas as a whole, occurred in the second part of the 1980s,

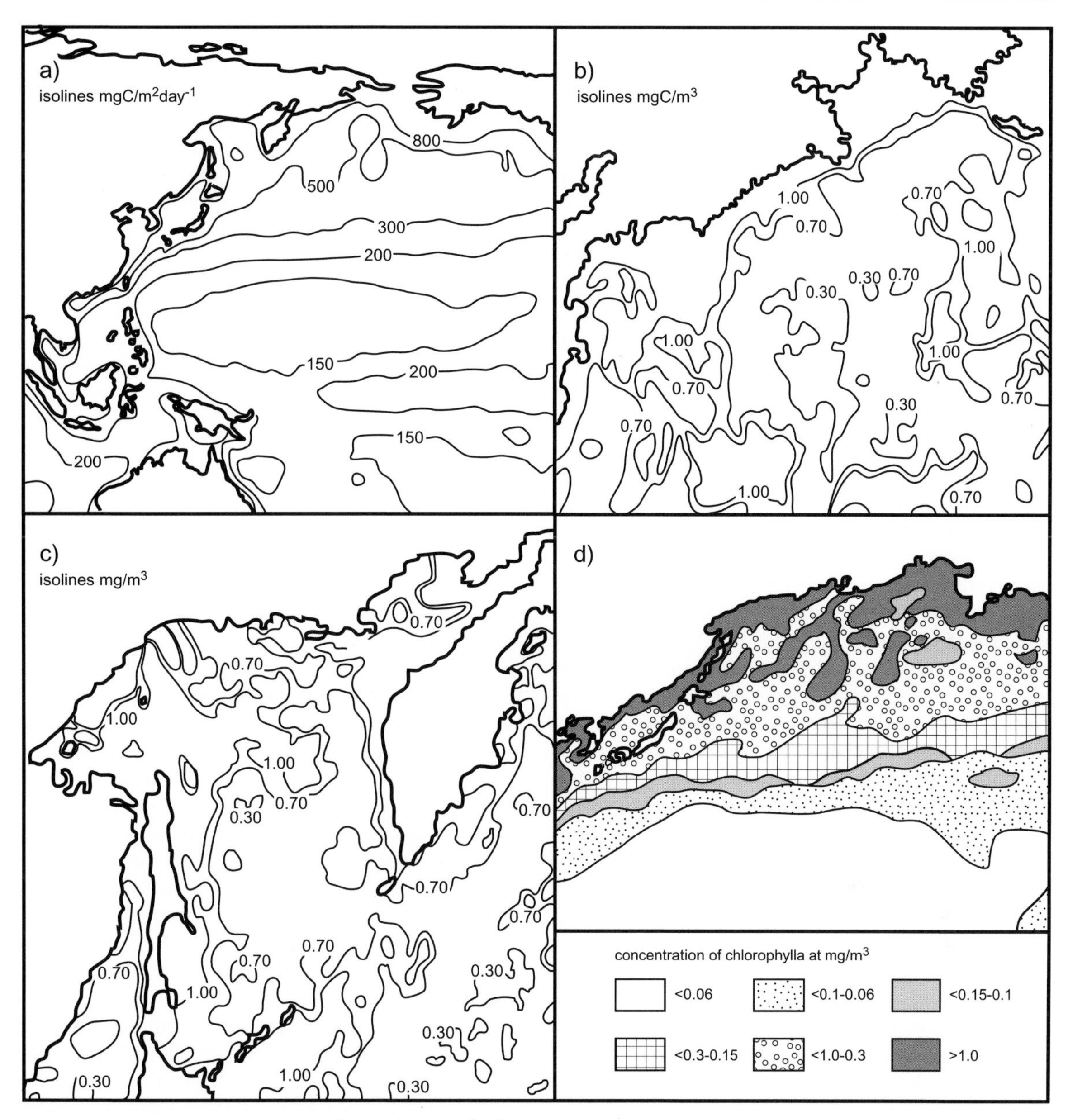

Fig. 4.21. **a** Average annual primary production (mg C m^{-2} d^{-1}) of the Pacific evaluated on the basis of both primary production field measurements and satellite-derived surface chlorophyll concentrations, 1978–1986 (Vinogradov et al. 1996); **b** July–August surface chlorophyll distribution (mg m^{-3}) in the western Bering Sea (Nezlin et al. 1997); **c** July–August surface chlorophyll distribution (mg m^{-3}) in the Sea of Okhotsk (Nezlin et al. 1997); **d** autumn chlorophyll concentrations (mg m^{-3}) (Vinogradov et al. 1996)

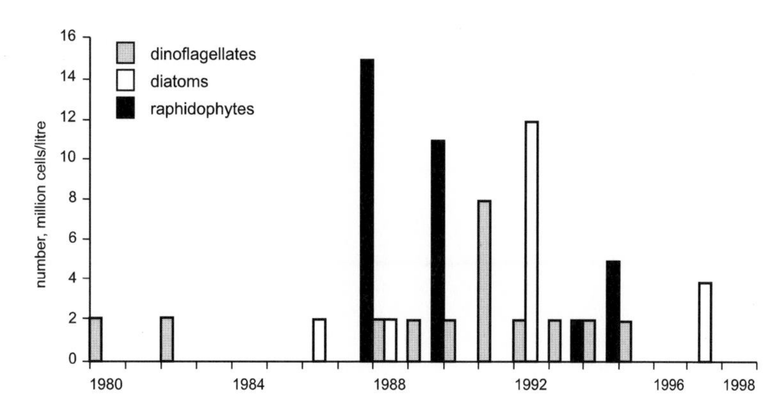

Fig. 4.22.
Harmful algal blooms in Peter the Great Bay, Sea of Japan, 1980–1998 (Orlova et al. 2000)

largely due to high catches of sardine and Alaskan pollack (1.4–1.5 t km^{-2}). Populations of both species are continuing to decline. This has been attributed to ecosystem changes, in the case of sardines, but overfishing is the likely culprit for pollack (FAO 1997). Compared to natural processes in pelagic ecosystems of the open seas, the anthropogenic influence, other than fishing, is rather small for these types of fast growing species. For species with slow population growth rates, such as marine mammals, crabs, gastropods and halibut, anthropogenic impacts are more pronounced (Shuntov et al. 1997).

The marine trophodynamics of the Far Eastern Seas now occur under the influence of climatic factors. There is a decrease in abundance of common epipelagic fishes (walleye pollack in the north and sardine in the south) that influences all lower tropic levels. Reduction in these species decreases trophic pressure on planktonic communities and increases the abundance of nonpredatory zooplankton. This, in turn, leads to increases in biomass and productivity of predatory zooplankton with subsequent decrease in abundance and productivity of non-predatory zooplankton (Dulepova 1997). The overall biological productivity of the system may be diminished at this stage. Increases in the biomass of alternative pelagic species-such as herring in the north and anchovy and squid in the south-will proceed, again followed by decreases in biomass and productivity of predatory zooplankton, with subsequent increases in nonpredatory zooplankton. In result, the overall fish productivity will increase, but at a lower trophic level and with a changed community (Shuntov et al. 1997).

In the early 1990s the biomass of pelagic fishes, walleye pollack and sardine, dropped by 15 Mt in Russian Pacific waters. Up to the mid 1990s this decrease in ichthyomass continued and reached 25–30 Mt. At present, virtually all of the major commercially valuable pelagic fish stocks, including walleye pollack, sardine, Pacific cod, yellow croaker, and Japanese jack mackeral, are either fully exploited or depleted (FAO 1997). There have been some increases in catch of salmon, crustaceans, and squid in the region. Fishing effort in the region has continued to rise. In some coastal seas, there has been a shift in catch from large high-value fish to lower-value smaller fishes, from demersal and pelagic predator fishes to pelagic plankton-feeding fishes, from mature to immature fish (FAO 1997). This has the potential to destabilise the community as evidenced from data that show large inter-annual changes of basic parameters including abundance of species and groups of zooplankton (Shuntov et al. 1996).

Total fish catch in the region, including all the East Asian marginal seas and adjoining open Pacific waters, is the largest in the world, and natural or anthropogenic variations in catch significantly influence the living standard of the human population of the region (Table 4.3). This extremely large population, while having a highly variable geographic distribution along the coastal zone, is highly dependent on fish production as a dietary staple (Table 4.4). There is also a proportionally large socio-economic involvement in the fishing industry along the coastal zone.

Table 4.3. Fish catch in billions of tons in the East Asian Marginal Seas and adjoining open Pacific (FAO 1997)

Year	Catch (billion t)
1988	24.5
1989	24.0
1990	23.0
1991	21.2
1992	20.6
1993	20.6
1994	20.4
1995	21.8
1996	23.5
1997	24.6

4.3.6.4 *Benthic Communities*

Sea-level rise and temperature increase, as well as pollution, eutrophication and other results of human activity, all affect marine benthic communities. Colonization of new substrata by marine organisms during sea-level rise not only changes the structure and functioning of previously existing coastal marine communities, but also has consequences of global importance. As a result of colonization, former coastal lowland forests, meadows, swamps and marshes with pronounced capability for carbon accumulation and methane production, will be replaced by marine ecosystems that produce, accumulate, and export calcium carbonate. This succession must be accounted for in ocean-atmosphere greenhouse gas dynamics. On the other hand, the new marine biota may participate in the provision and stabilization or erosion of sediment material at the coastal zone thus impacting subsequent shoreline morphodynamics. Coral reefs, coralline algae, molluscs, echinoderms and sea grass communities are of special importance in this respect (Kasyanov 1998).

Table 4.4. Per capita annual fish production for the East Asia region (kg $person^{-1} yr^{-1}$) (FAO 1997)

	1996	1997
DPRK	46.0	19.5
Republic of Korea	50.7	51.6
China	19.1	22.0
China-Hong Kong		56.8
China-Taiwan	38.2	35.8
Russia	14.6	16.1
Japan	69.9	70.7

Sea-level rise and higher temperatures in the 21st century may be similar to the conditions of the thermal optimum in the early Holocene. At that period there were rather insignificant changes, for instance, in malakofauna of the Sea of Japan, and some subtropical species shifted their northern limits further to the north (Zhirmunsky et al. 1997). The species composition of the sublittoral zone in the Sea of Japan is expected to change during the 21st century. Warm-water species will become more abundant and their distribution will be shifted 700–1 000 km north, similar to that observed four to five thousand years ago (Taira and Lutaenko 1993). The most pronounced effect of temperature rise will be seen in the tropics. Temperature increases of 2–3 °C (observed during the 1998 El Niño event) result in coral bleaching and death (Lasserre 1992; Wuethrich 2000). In addition, if interactions between species are sensitive to temperature, then small shifts could have rapid effects and cascade through communities (Wuethrich 2000).

Temperature changes may be counteracted by anthropogenically induced changes. Temperature increases, up to 4–5 °C, may have less impact for marine biota than environmental pollution (Stock 1992). Considerable alterations in bottom and pelagic communities of some coastal areas of Peter the Great Bay can be explained only by anthropogenic factors, such as chronic pollution and eutrophication. These alterations include significant changes in species composition, disturbance of seasonal population dynamics of phyto- and zooplankton, and increasing phytoplankton biomass and density (Belan 1997). The effect of human activity has been studied in the littoral zone of Shikotan Island, in the South Kuril Islands, which is subject to pollution from a fish factory. Prolonged anthropogenic pollution and eutrophication has led to decrease in species diversity despite an increase in biomass. A characteristic feature of communities under pollution stress is the increasing role of detritophagous animals, especially with direct development of sestonophagous animals with pelagic larva (Kussakin and Tsurpalo 1999). Under conditions of significant pollution, up to 99% of all biomass is produced by tiny polychaete *Capitella capitata*, its congeneres, and *Tharyx pacifica*. These are opportunistic species without pelagic larva that are abundant throughout all world's oceans in extreme environments caused by variety of damaging factors (Kasyanov et al. 2001). Loss of biodiversity due to careless human activity is extremely important in the ocean, where taxonomically high-rank biodiversity is much more pronounced, in comparison to terrestrial conditions.

The characteristics and productivity of the coast zone and marine ecosystems are profoundly affected by human activities, both proximate and remote. These effects are then acted upon by natural variability in the system. Which is the trigger for change, natural variability or anthropogenic forcing, varies with locality and time.

4.3.7 Impact of Sea-Level Rise on the Coastal Zone

4.3.7.1 *China*

Along most of the continental coast, relative sea level is an important factor for coastal environments. Deltaic coasts in China face severe problems from sea-level rise as a result of tectonically and anthropogenically induced land subsidence. In the next 50 years, the expected worldwide sea-level rise due to climate change will not be a major factor in relative sea-level rise for the Old Huanghe and Changjiang Deltas in China, although it may be for the delta of the Zhujiang. Sea-level rise due to global warming will, however, exacerbate problems in all three deltas, along with saltwater intrusion problems in deltaic regions and coastal plains. China, however, has a long history of defence against sea encroachment.

4.3.7.2 *Japan*

Tokyo, Osaka, and Nagoya are all located in the coastal zone. Together, they account for more than 50% of Japan's industrial production. In these metropolitan areas, about 860 km^2 of coastal land, an area supporting 2 million people and with physical assets worth $450 billion, are already below mean high-water level. With a 1-m rise in mean sea level, the area below mean high-water level would expand by a factor of 2.7, embracing 4.1 million people and assets worth more than $900 billion. The same sea-level rise would expand the flood-prone area from 6 270 km^2 to 8 900 km^2. The cost of adjusting existing protection measures has been estimated at about $80 billion.

One of the potential threats of increasing sea level is exacerbated beach erosion. Sandy beaches occupy 20–25% of the total length of the Japanese coast. About 120 km^2 of these beaches have been eroded over the past 70 years. An additional 118 km^2 of beaches – 57% of the remaining sandy beaches in Japan today – would disappear with a 30-cm sea-level rise. This percentage would increase to 82% and 90% if the sea level rose by 65 cm or 100 cm, respectively.

4.4 Terrestrial Ecosystems in East Asia

The East Asian region has long been recognised for its wealth of natural and human resources. Natural terrestrial ecosystems in East Asia include a variety of forest ecosystems (e.g., boreal forest, cold-temperate coniferous forest, temperate mixed forest, warm-temperate deciduous forest, subtropical evergreen forest, tropical seasonal rain forest and tropical rain forest), temperate grasslands and steppe ecosystems, continental deserts, and wetlands, both coastal and inland. The general distribution of ecosystems is shown in Fig. 4.23. Many cultivated and intensively managed ecosystems also are present. The need to produce food, maintain good air quality, and clean supply of water for the growing number of people in the region has inevitably produced changes in land use and land cover in the region. Increasing utilisation of the land and water resources will further tax the limits of the ecological systems and increase the risk of environmental degradation.

Economic changes related to relaxation of centralized economic structures and a move toward more liberal free-markets throughout the region have promoted rapid growth in East Asian countries. Land use intensification, changes in consumption patterns, conversion of land from subsistence food production to cash crops, have impacted environmental conditions throughout the region. Rapidly accelerating urbanisation is having an increasing impact on a wide scale. Transcending all these factors for change is the inexorable effect of population growth. The integrated effect of all these anthropogenic drivers of change is the modification of regional vegetation cover and land-use, which in turn affect the state of ecosystems. In the region, five land-cover types play a major role in characterizing land uses: forests, grasslands, croplands, desertified lands, and urban areas (Fig. 4.23).

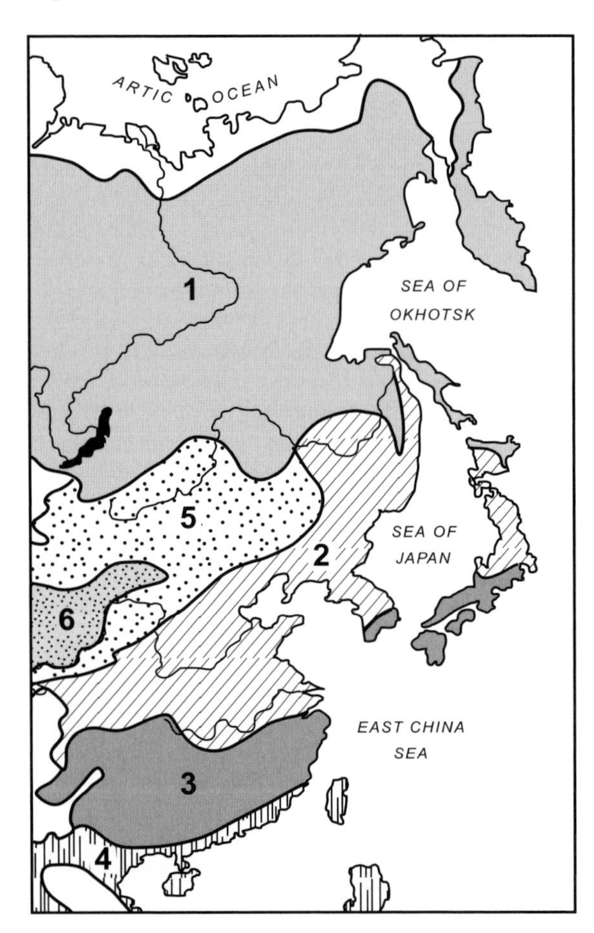

Fig. 4.23. East Asian terrestrial ecosystems (*1* denotes boreal forest, *2* temperate forest, *3* subtropical evergreen forest, *4* tropical rain forest, *5* temperate grassland and *6* desert) (UNESCO 1997)

4.4.1 Natural Land Cover

4.4.1.1 *Forest Ecosystems*

The boreal forest consists of *Picea abovata*, *Abies sibirica*, *Pinus sibirica* and *Larix dahurica*, distributed mainly in Siberia and the Russian Far East. The cold-temperate coniferous forest, a part of the boreal forest located in the south and in those inland areas with drier climate, is dominated by *Larix gmelini* and *Pinus sylvestris* var. *mongolica*. This forest is mainly found in the extreme north of northeast China and the adjacent parts of Mongolia and the Russian Far East. The temperate mixed forest consisted of *Pinus koraiensis* and several deciduous species like *Fraxinus mandshurica*, *Juglans mandchurica*, *Tilia amurensis*, *Ulmus japonica* and *Acer* spp. Rich in biodiversity, the forest is distributed over northeast China, the Russian Far East, northern Korea and northern Japan. Because of the high economic value of timber, the impacts of logging and other related human activities on forested areas are serious.

The warm-temperate deciduous forest over northern China between 32° and 42° N and 103° and 124° E consisted mainly of the genera *Populus*, *Salix*, *Ulmus*, *Quercus*, *Betula* and *Tilia*. This area has a long history of human occupation and high population levels; over the last several thousand years the natural ecosystems in this area have been seriously damaged and replaced by farmland and other man-made ecosystems. The subtropical evergreen forest distributed in central China, North Korea, South Korea, and Japan, between latitude 22° and 32° N consists mainly of the species in the families *Fargaceae*, *Lauraceae* and *Theaceae* and bamboo in the *Gramineae*.

Less than 25% of the land area remains forested. Of this, much is now managed and has been planted with plantation species. Little natural forest is left in East Asia. The value of wood and the shrinking supply of wood products from elsewhere will create increasing pressure to harvest more trees and shorten rotations. The impact of these changes on wildlife, water resources, and soil protection will be significant. Despite the importance of the remaining forest area to ecological dynamics and human welfare, too little is known about the factors affecting forest cover in the region or the extent of the forest cover area.

It is estimated that about one third of China was covered by dense forest ten thousand years ago. Until very recently forest cover has continuously declined, mainly because of increasing population numbers and agricultural development. This has had serious environmental consequences, including degradation of forests, loss of biodiversity, accelerated soil erosion, and catastrophic flooding. Over the past 50 years, the human population in forested areas increased five-fold. From the 1950s to 1970s, the natural forests of northeast and southwest

Table 4.5. Forest Changes in China (Fourth National Forest Resources Inventory 1993)

Years	Standing volume (10^9 m^3)	Forest area (10^3 Mha)	Forest coverage (%)
1973–1976	9532	121 860	12.7
1977–1981	10 261	115 280	12.0
1984–1988	10 572	124 650	12.98
1989–1993	11 785	128 530	13.92

Table 4.6. Land Use in Japan at around 1900, 1950 and 1985 (Himiyama 1998)

	Circa 1900		Circa 1950		Circa 1985	
	× 10 km^2	%	× 10 km^2	%	× 10 km^2	%
Forest	24 348	65.44	24 885	66.88	24 818	66.58
Broad leaved forest	9856	26.49	9484	25.49	5314	14.25
Coniferous forest	4410	11.85	4377	11.76	3854	10.34
Mixed forest	9773	26.27	10 757	28.91	15 138	40.61
Bamboo forest	309	0.083	268	0.72	512	1.37
Other	0	0.00	8	0.02	0	0.00
Total	37 207	100.00	37 207	100.00	37 275	100.00

China were heavily logged. At the same time many plantations were established. Both the area of forest cover and the standing volume have been increasing since the 1980s (Table 4.5). About 14% of China's total land area is currently forested (Fourth National Forest Resources Inventory 1993). The Natural Forest Conservation Program, launched in 1998 is intended to protect all of the natural stands and to make Chinese forestry sustainable by expansion of natural forests and increasing productivity of forest plantations (Zhao 2000).

Forest occupied about two-thirds of Japan's territory in 1985 (a 2% increase over 1900). This is remarkable, considering the rapid industrialisation and urbanisation in Japan during the 20th century. There was a sharp decrease of forest in Hokkaido, where major agricultural development took place, and in the area of urban development in the Pacific belt and elsewhere. However, the decrease was offset by the planting in western Japan during the 1950s of coniferous trees, such as Japan cedar or Japanese cypress. Table 4.6 shows the decrease of broad-leaved forest and increase of mixed forest.

4.4.1.2 *Grasslands*

Extensive temperate grasslands cover large portions of Mongolia, Inner Mongolia, and adjacent areas in China and Russia Far East. These grasslands comprise a major portion of the largest grassland in the world, extending from Eastern Europe to Eastern Asia. Temperate desert is found in the central part of this region, including southwest Mongolia and northwest China. Wetlands occur in various parts of this region and are important centres for wildlife; extensive wetlands are located in Siberia.

Approximately 4 million km^2 or 41.7% of China's land area is covered with grasslands and arid lands, including deserts. Most of these grasslands, 3.1 million km^2, are located in the continental semi-arid portions of northern China. In addition to these temperate grasslands, steppe and desert ecosystems, approximately 20% or 0.67 million km^2, are southern tropical or sub-tropical mountain grasslands. Large salt marshes or grasslands are also found along the coastal areas and surrounding freshwater bodies (approximately 0.2 million km^2).

In the semiarid regions of the East Asia, where the climate is highly variable, nomadic pastoralism has been the dominant agronomic activity for many centuries. Pastoral livestock systems are highly adapted to climatic variability in many ways. In general, there is a direct relationship between climate variability and the spatial scale of pastoral exploitation. Extensive nomadic systems are found in the climatically most variable regions; less extensive, more intensive modes of livestock management occur in less variable grazing lands. Climate change in drylands can thus be expected to have important implications for the dynamics and viability of pastoral people, their exploitation patterns and through these exploitation patterns, for land cover and land cover change.

Recently, changes in cultural, political and economic factors have caused changes in how pastoral systems operate within the region. Currently, various systems are operating in China, Mongolia, and Russia. These encompass a range of grazing patterns (i.e., frequency, intensity of grazing and the types of animals). Throughout the region, a tendency toward a more sedentary livestock system has been evident from the 1950s to 1990s. Larger-sized cattle and sheep breeds have also been introduced. However these are less suitable to certain climate regimes (e.g., drought conditions of the Gobi desert and severe winter storms in the Inner Mongolian and Mongolian steppe region) than the original stocks used by indigenous pastoralists. These changes in pastoral management have altered the nomadic patterns of the region.

The trend of livestock numbers for most countries in East Asia reflects a growth in meat production. Cattle numbers have increased markedly during the past 40 years in all countries except South Korea (Fig. 4.24). Although livestock trends have slowed or declined during the past decade, Mongolia is an exception. In that country, an increase in cattle production has continued with a rapid move towards a free-market system. The production of sheep in East Asia has varied depending on the availability of pasturage and agricultural preferences. China's production continues to increase, although recent land use policies are attempting to reduce herd sizes in Inner Mongolia to animal densities of the late 1980s. In South Korea and Japan, numbers have declined. In Mongolia and North Korea, numbers have held steady during the past 40 years. In contrast, goat production in China and Mongolia has increased markedly in response to market demand for cashmere. In the other countries, the goat production has not changed as dramatically, because of the lack of grasslands for increased production.

The changes in livestock production are indicative of cultural changes in food production and market forces, and affect the use of agricultural lands for fodder or pasturage for the livestock. As population numbers increase and economic development continues, agricultural production systems will shift. In the livestock sector, vulnerability of animals to climatic events such as heat stress, drought, and winter storms may be exacerbated if stocking densities are increased. The 2000 winter livestock kill in Mongolia demonstrates the vulnerability of the extensive rangeland system in the Mongolian Plateau.

In addition, recent political and economic changes (i.e., in the past 50 years) in land use management have resulted in extensive land degradation. These changes have resulted from intensive stocking rates in localised areas, change in the breeds of animals used, and intensification of agricultural practices in marginal areas. More recent changes in the social-economic setting have forced new changes in land use management due to the relaxation of governmental controls and the implementation of a more free-enterprise system.

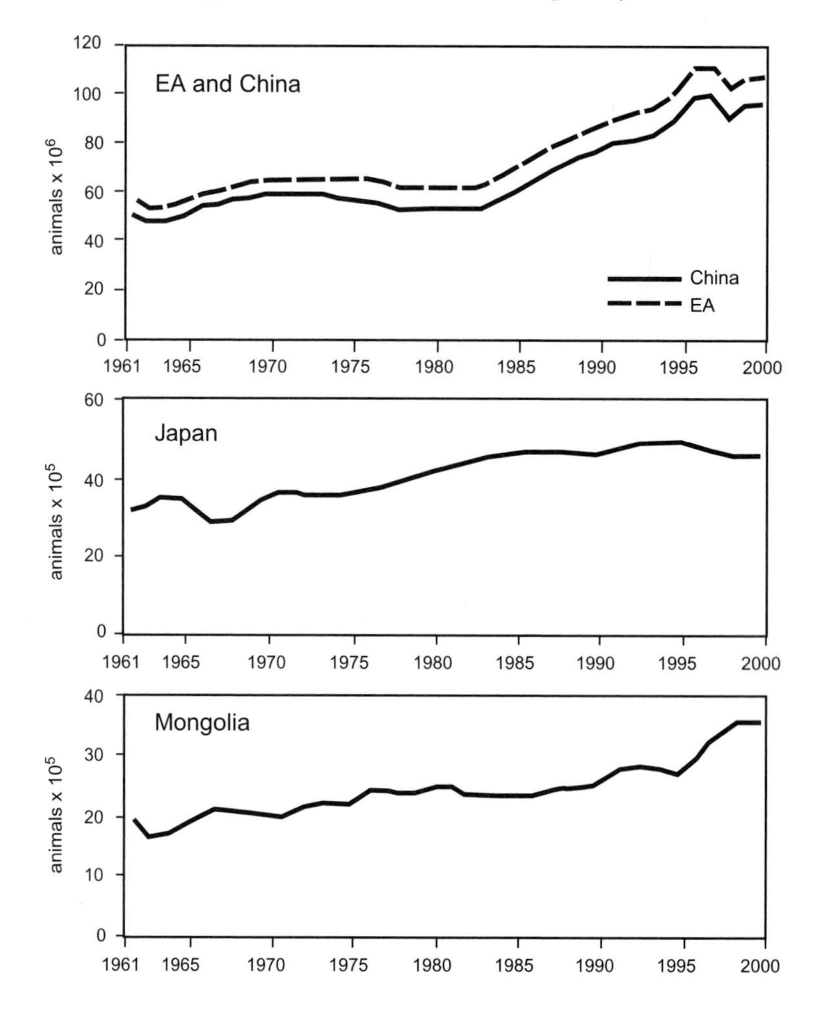

Fig. 4.24. Number of cattle in East Asia, China, Japan and Mongolia (FAO 2000)

There are two main trends evident in grassland: degradation of quality, and decrease in area. The former is the main problem with China's grassland. About a fourth, or 0.86 million km^2, of the grassland has degenerated. Compared with the beginning of the 1960s, productivity of grasslands has decreased by 30 to 50%. Degradation of these grasslands has resulted from wind and water erosion and overgrazing. By the end of 1989, 0.11 million km^2 of these areas had already become desert, and an additional 4.3 thousand km^2 of range are undergoing desertification each year. Moving sand dunes have appeared across a wide region.

Much of the northeastern plateau has suffered salinisation and become alkaline. Meadows that once provided good pasturage have been overstocked, resulting in a reduction in plant cover. These degraded grasslands provide good habitats for mice and insects that gnaw the vegetation and thus accelerate the downward spiral. An estimated 800 to 1 000 km^2 of China's grassland have been thus destroyed. The degradation of grassland also caused the recurrent and extensive dust storms, locally referred to as black storms, which have occurred every April to May since 1993. It was reported that 1.1 million km^2 of land were affected by the storms in 1993. Similar trends are observed in Mongolia.

4.4.2 Cropland Systems

Owing to increasing population, industrialisation, and urbanisation, the demand for land resources has been growing. During the past century, natural ecosystems such as forest, grassland and wetlands have been encroached upon by farmland and other man-made ecosystems on a large scale (Fig. 4.25). During the same period, land degradation has also been serious, resulting in a shortage of land resources, and leading to environmental problems such as desertification, deforestation and soil and water erosion affecting the sustainable economic and social development of the region.

In terms of land use, East Asia is experiencing mixed trends regarding expansion of croplands (Fig. 4.26). In China, although cropland peaked in 1956, over 0.2 million km^2 have been converted to croplands from other uses since the early 1980s. In Mongolia and North Korea, expansion of croplands occurred from the 1960s into the late 1980s. With privatisation in the 1990s, cropland expansion in Mongolia has reversed itself to a certain degree. In Japan and South Korea, a reverse trend has been observed for the past 30 years, with continued loss of arable lands. However, this loss has coincided with an ever-increasing intensification of land use practices on those lands still cropped, thereby maintaining overall crop productivity.

During the last century, farmland area in China increased in the first 60 years and decreased in the last 40 years (Table 4.7). The farmland area was 0.55 million km^2 in 1895, 0.96 million km^2 in 1915, and reached its peak of 1.1 million km^2 in 1956. During the past 40 years, arable lands continued to decline until the 1980s. In the 1980s, a slight expansion occurred, but since the 1990s the extent of arable land has remained relatively constant.

In China, between 1956 and 1995, up to 0.17 million km^2 of farmland were lost (Fig. 4.27), mainly through industrialisation and urbanisation. During this period, the number of cities increased from 161 in 1978 to 622 in 1995; and urban area increased from 6 500 km^2 to 17 400 km^2. The conversion of agricultural land to

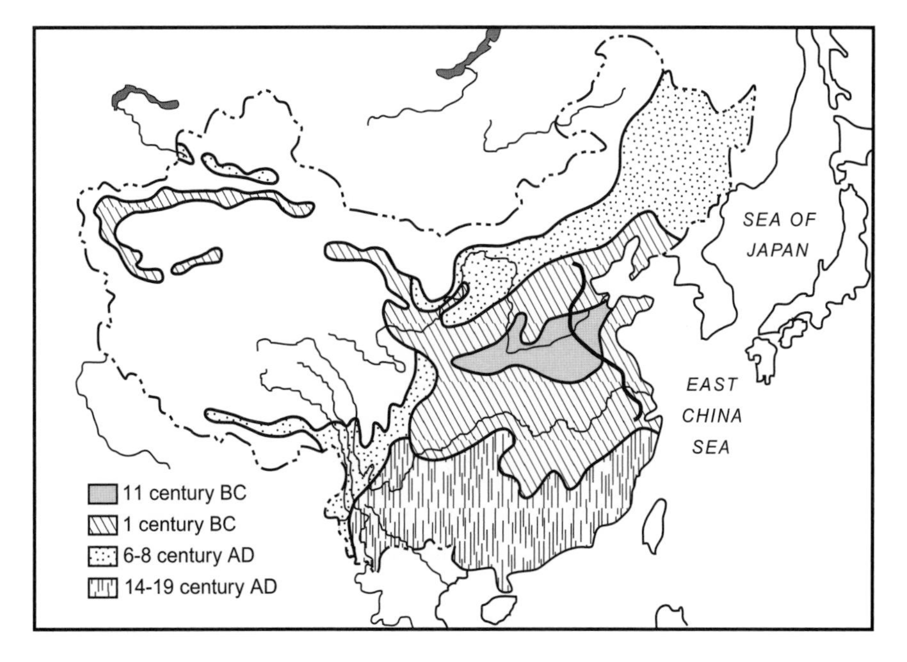

Fig. 4.25. Expansion of cropland in China from the 11th century B.C. (Deng 1981)

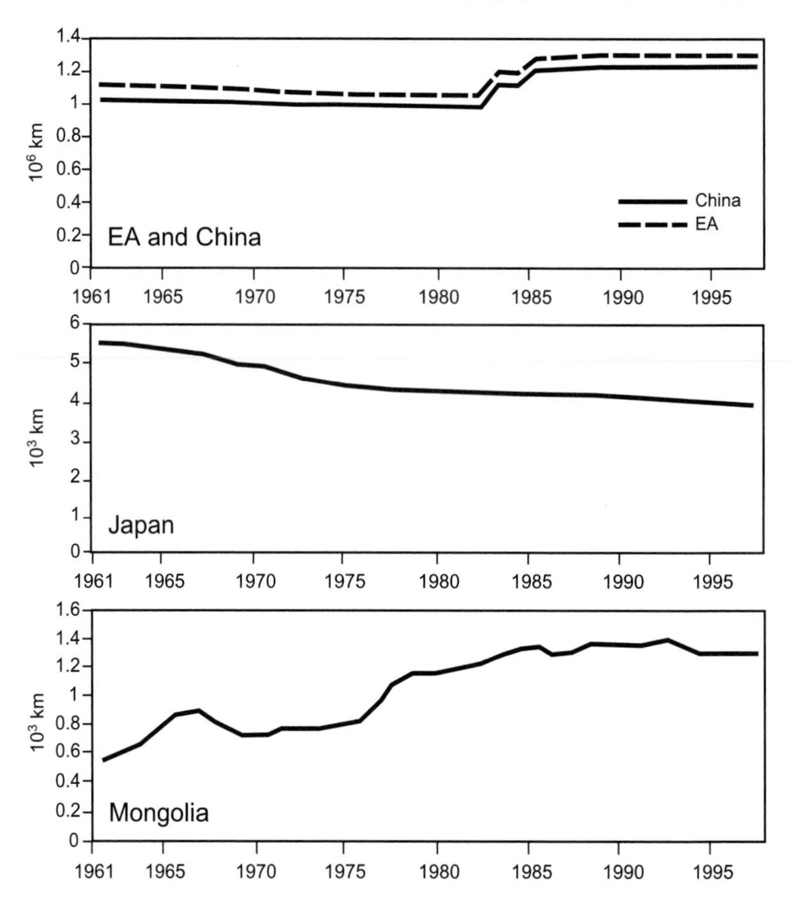

Fig. 4.26. Arable land in East Asia, China, Mongolia and Japan (FAO 2000)

Table 4.7. Composition of the land in China (Wu Chuanjun and Guo Huancheng 1994)

Category	Area (10^6 km^2)	Percentage of coverage (%)
Farmland	0.96	10.0
Garden	0.03	0.3
Forest land	1.33	13.9
Forest	1.15	12.0
Grassland	3.19	33.3
Used grassland	2.25	23.4
Water	0.29	3.0
Tideland	0.02	0.2
Inland-water	0.27	2.8
City, industry and communication	0.67	7.0
Unused land	3.16	32.9
Land suitable for agriculture	0.33	3.5
Land suitable for forest, bush and grassland	0.78	8.0
Marsh	0.07	0.8
Deserted land	0.17	1.8
Desert	0.60	6.3
Gobi	0.56	5.8
Cold desert	0.15	1.6
Barren rocky mountain	0.43	4.5
Permanent snow and glacier	0.05	0.6

forest or pasture use was also an important mechanism of farmland loss.

The percentage of Japanese agricultural land increased from 14.4% in 1850 to 16.7% in 1900, but has been stable since then at about 17%. The rate of increase was the highest in Hokkaido (+16.7%), and Aomori, Iwate, Akita and Fukushima Prefectures, all in the Northeast district, followed by Saga and Tottori in western Japan. These prefectures are all located in the periphery of the country. In contrast, Tokyo (–71%), Osaka, Kanagawa and other prefectures with large urban centres experienced high rate of decrease in agricultural land.

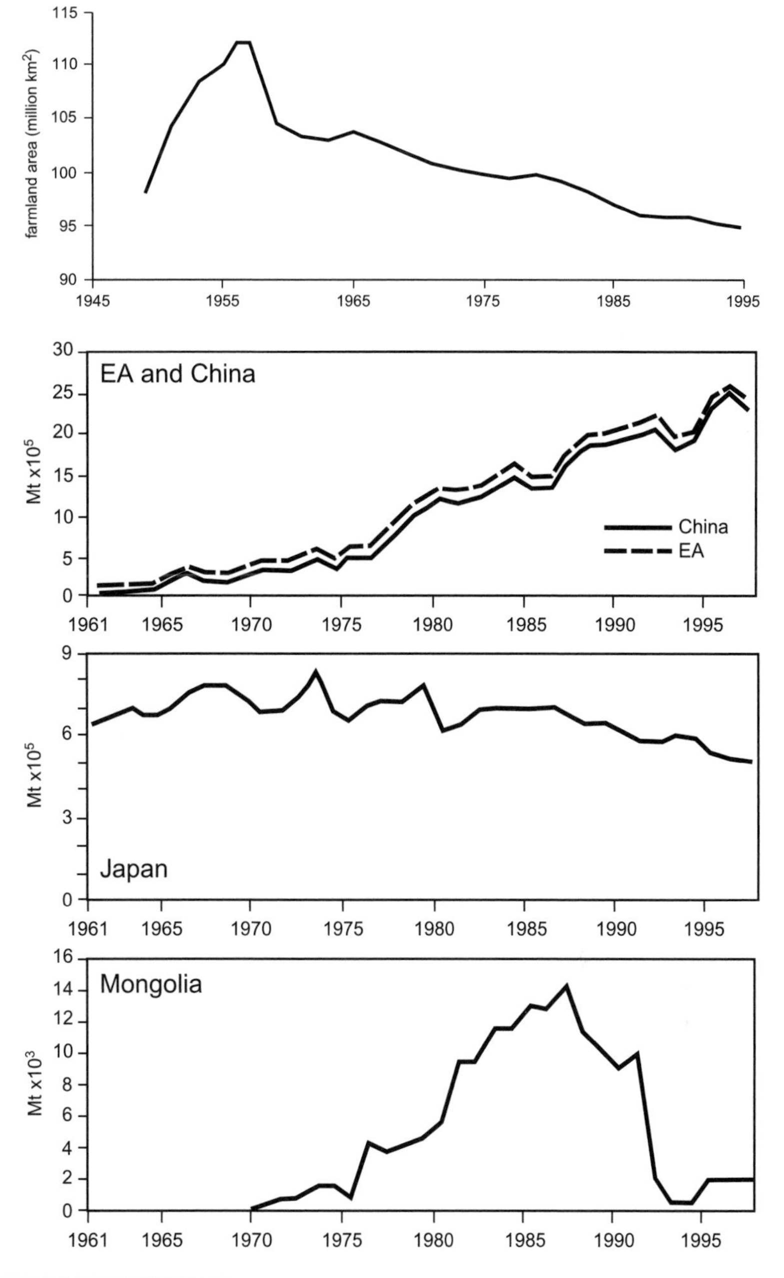

Fig. 4.27. Change in farmland area in China, 1949–1995 (China Statistics Bureau 1997)

Fig. 4.28. Use of nitrogenous fertilizers in East Asia, China, Japan and Mongolia (FAO 2000)

4.4.2.1 *Agriculture Output and Practices*

The agricultural output of East Asia, well known for its double and triple cropping systems, is one of the highest in the world. High soil fertility, good climatic conditions, and an industrious human resource base have all contributed to the rich agricultural production.

Agricultural practices in East Asia have undergone radical changes. During the past several decades the use of fertiliser has been markedly increased (Fig. 4.28); high yielding varieties have been widely introduced; new irrigation systems have been developed; and other changes in agricultural practices have been made. The result has been a significant increase in crop yields over the past few decades. However, the ability of the region to maintain increases in crop production is questionable and in many areas soil fertility and water availability are declining.

The use of nitrogenous fertiliser in the region has nearly quadrupled during the past 30 years. This increase has largely been driven by increased fertiliser use in China; declines have taken place in Japan, Mongolia, and North Korea. In Japan, a continued reduction in nitrogenous fertiliser use has been observed during the past 25 years, in part due to reduction of arable land area and more efficient application of fertiliser. In Mongolia and North Korea a different set of events produced a sharp decline in fertiliser use. In the 1990s, with the loss of trade relationships with the former USSR and the lack of hard currency, agricultural imports of fertiliser were greatly curtailed in both countries. A gradual increase in fertiliser use is now anticipated.

4.4.2.2 *Soil Nutrient and Carbon Loss in Farmland and Grasslands*

About 2 000 km^2 of farmland in China have been lost through soil erosion each year in the last decade. Thirty-five percent of farmland is threatened by soil erosion at present, mainly in the loess plateau and southwestern China. The loss of topsoil by soil erosion is as much as 3.3 billion t per year, containing 44 million t of nitrogen, phosphorus, and potassium. This loss exceeds the annual applied amount of chemical fertilisers in the country. Soil fertility has declined in many parts. Because of land degradation, the average soil organic matter content is only 1–2% generally. Nine percent of farmland has lower than 0.6% soil organic matter; 57% of farmland lacks phosphorus, 20% of farmland lacks potassium and 10% of farmland lacks both phosphorus and potassium. From 1950 to 1995, the farmland threatened by desertification doubled to 0.039 million km^2. This process continues. Saline-alkali land amounts to 0.050 million km^2, taking 4% of the total farmland area. Approximately 0.04 million km^2 of farmland has been polluted by industry, of which 0.13, 0.32 and 0.26 million km^2 are polluted by cadmium, hydrargyrum and fluorine. Farmland polluted by pesticide amounts to 0.13 million km^2 and 0.027 million km^2 of farmland is being threatened by acid rain.

Soil carbon in the semi-arid regions of temperate East Asia is a rich natural resource that has developed over thousands of years under natural grassland and steppe ecosystems. The Mongolian Plateau steppe ecosystems have soils with soil organic matter ranging between 1.5–5.5%, despite the sandy content of typical steppe soils. The current estimate of soil carbon pools found in the top 20 cm of Mongolian Plateau soil is approximately 4 Gt, or about 3% of the total grassland and dryland soil carbon of the world (Ojima et al. 1993). Soil carbon measurements taken in Inner Mongolia, China, and Mongolia reflect differences in climate and land-use gradients. For areas with similar grazing histories, the soil organic matter declines with precipitation. However, comparisons of the steppe soil organic matter of sites with similar rainfall, but with differing intensity of grazing pressures, show lower soil organic matter with increased grazing pressure. Changes in plant species composition was also indicated by the shift in δ^{13}C isotopic values of the organic matter. The drier sites usually have a greater density of the warm season grasses, *Cleistogenes squarosa*, and the surface samples taken indicate a less negative δ^{13}C value (i.e., δ^{13}C values ranging from –20 to –18). This change in soil organic matter δ^{13}C values was also observed in heavily grazed sites where *C. squarosa* increased in relative dominance. Overgrazing in this region threatens the maintenance of soil carbon due to reductions in carbon inputs, the removal of protective vegetative cover, and the physical disturbance of the soils. Desertification of these lands is already severely impacting the pastoral land use of the region.

4.4.3 Urban Expansion

Population growth in many countries in Asia has slowed considerably during the past decade owing to aggressive policy intervention, especially in China. The population trends indicate that the population of China will reach its peak around 2030 and level off at around 1.5 billion people. In other countries similar trends are observed, with the exception of Mongolia, where population growth continues to be high compared to global averages. However, the low population level provides a buffer against this increased population growth (Fig. 4.29). Nevertheless, a common feature among these countries is the large-scale migration of human population into the urban areas. Even in Mongolia, the urban population accounts for over 60% of the total population. All of the countries except China already have the majority of their population located in urban centres. This trend is expected to continue in all of the East Asian countries. In spite of these trends in urbanisation, over half of the population remains rural in China.

Future growth of Asian urban centres in the coming decades is staggering. It is projected that over 15 of the world's largest urban centres will exist in Asia in the next few decades. These will account for over 60% of the cities with populations greater than 10 million people. Urban growth is greatly modifying peri-urban and adjacent rural areas. The impacts of these changes on croplands, fuel resources, and water supply will be significant. Urban expansion, much of it industrially-based, will have serious implications for urban and regional air quality. Similarly, the growth of emission of greenhouse gases and sulphate aerosols will greatly impact regional and global climate change.

4.4.4 Impacts of Climate Change

4.4.4.1 *Forests, Grassland and Tundra*

For boreal forests in the Asian part of Russia, the UKMO-H3 and GFDL-A2 models suggest large shifts in distribution (e.g., area reductions of up to 50%) and productivity (Dixon et al. 1996). In the boreal region, grasslands and shrublands may expand significantly, whereas the

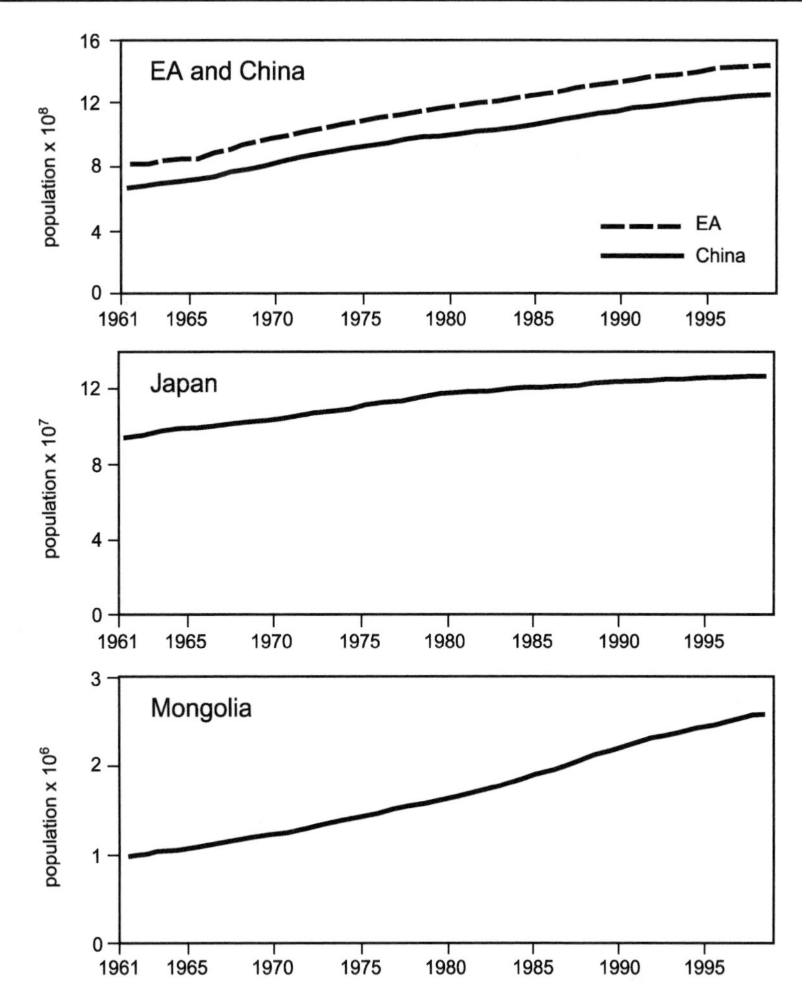

Fig. 4.29. Population trends for East Asia, Mongolia, Japan, and China (FAO 2000)

tundra zone may decrease by up to 50%. Climatic warming would increase the release of methane from deep peat deposits – particularly from tundra soils, since they would thaw and become wetter. The release of CO_2 is expected to increase, though not by more than 25% of its present level (IPCC 1996).

Projections for 2010 indicate that the need for industrial roundwood could increase by 38% (southern East Asia) to 96% (eastern East Asia). These requirements may result in serious shortages of boreal industrial roundwood, placing further stress on boreal forests (IPCC 1996).

4.4.4.2 *Cryosphere*

It is likely that by 2050 up to a quarter of the existing mountain glacier mass will disappear. For areas with very large glaciers, the extra runoff may persist for a century or more. By 2050, the volume of runoff from glaciers in central Asia is projected to increase threefold. Eventually, however, glacial runoff will taper off or even cease. Projected future glacier runoff is about 68 $km^3 yr^{-1}$. Permafrost in northeast China is expected to disappear if the temperature increases by 2 °C or more. Over the Qing-Zang (Tibetan plateau), estimates of the impact of climate change on permafrost will be a raising of its elevation limit to 4 600 m as a result of a warming of 3 °C (IPCC 1996).

4.4.4.3 *Agriculture*

Estimated impacts on agricultural yields using different GCMs vary widely (Table 4.8). Large negative impacts on rice production as a result of future change are of concern given expected population increases. In China, as a whole, yield changes by 2050 are projected to range from –78% to +15% for rice, from –21% to +55% for wheat, and from –19% to +5% for maize. An increase in productivity may occur if the positive effects of CO_2 on crop growth are considered, but the magnitude of the fertilisation effect remains uncertain.

Climate change is projected to cause a general northward shift of crop zones and is likely to have favourable impacts on agriculture in northern areas of Siberia. Grain production in southwestern Siberia is projected to fall by about 20% as a result of a more arid climate (IPCC 1996).

Table 4.8.
Agricultural yield changes by 2050

Country/region	Yield impact (%)	Direct CO_2 effect	Authors
Rice			
China	−6	Yes	Tao (IPCC, 1996)
China	−11 to −7	Yes	Zhang (IPCC, 1996)
China	−78 to −6[a]	No	Jin et al (IPCC, 1996)
China[b]	−37 to +15	No	Jin et al (IPCC, 1996)
China	−18 to −4	Yes	Matthews et al. (IPCC, 1996)
China	−21 to 0	No	Lin (1996)
Japan	+10	No	Sugitara (IPCC, 1996)
Japan	−11 to +12	No	Seino (IPCC, 1996)
Japan	−45 to +30	Yes	Horie (IPCC, 1996)
Japan	−28 to +10	Yes	Matthews et al. (IPCC, 1996)
Korea, Rep	−37 to +16	Yes	Yun (1996)
Korea, Rep	−40	Yes	Oh (1995)
Taiwan	+2 to +28	Yes	Matthews et al. (IPCC, 1996)
Wheat			
China	−8	Yes	Tao (IPCC, 1996)
China	−21 to +55	No	Lin (1996)
Japan	−41 to +8	Yes	Seino (IPCC, 1996)
Mongolia	−67 to −19	No	Lin (1996)
Russia[c]	−19 to +41	Yes	Menzhulin and Koval (IPCC, 1996)
Maize			
China	−4 to +1	Yes	Tao (IPCC, 1996)
China	−19 to +5	No	Lin (1996)
Japan	−31 to +51	Yes	Seino (IPCC, 1996)
Pasture			
Mongolia	−40 to +25	No	Bolortsetseg et al. (1996)

[a] The large negative percentage is a result of the modeled result at a single site in southwest China.
[b] For irrigated rice.
[c] Including European Russia.

4.4.4.4 *Desertification*

The causes of desertification in the region are mainly related to climatic variation and human activities. Desertification has accelerated through land-use activities associated with population growth; pressure from economic development; lack of awareness of the importance of the protection of ecosystems; overgrazing; excessive collection of fuelwood; deforestation and destruction of vegetation caused by reclamation on the steppe, desert steppe and range land; inappropriate farming practices on slopes; and the degradation of vegetative coverage.

The seriousness of desertification can be illustrated by looking at China. The area of desertification is approximately 2.6 million km^2, covering over 27% of its total territory. It is estimated that the annual increase in desertification is 2 460 km^2 Most of the areas affected by desertification are located in the arid, semi-arid and dry sub-humid areas, covering 471 counties in 18 provinces and autonomous regions in the west part of the Northeast China, the northern part of the North China and most regions of the Northwest China.

The area of desertification caused by wind erosion is 1.6 million km^2, accounting for 61.3% of the total land area affected by desertification, mainly distributed in the arid and the semi-arid areas. The desertified area caused by water erosion is about 0.2 million km^2, accounting for about 7.8% of total desertified area. The total area of desertification caused by freezing and melting processes in the cold plateau is 0.36 million km^2, accounting for 13.8% of the total desertified area in China. This kind of desertification is mainly found in the high-altitude regions. The area of desertification caused by salinisation is 0.23 million km^2, accounting for 8.9% of the total desertified in China. Soil salinisation is concentrated in oases, at the pediment of the northern foothills, along riverbanks and in delta regions.

The absence of a rational and coherent system for managing grazing lands and grazing livestock contributes to desertification problems. There is no integrated grassland management system; nor are there mechanisms to ensure an effective balance between grasslands and grazing livestock, and among animal husbandry, agriculture and forestry in China.

The total scope of the land under the threat of desertification is much wider. Degradation affects 80% of the total areas of the arid, semi-arid and dry sub-humid areas of China. The ratio of rangeland degradation is as high as 56.6%, and 1.05 million km^2 of rangeland and steppe areas have been degraded. The ratio of arable land degradation exceeds 40%, and there are 0.07 million km^2 of farmland threatened by deserti-fication.

Due to the rapid increase of the population since 1949, an estimated 67 million km^2 of high-quality grassland have been converted into agricultural land. There is now a serious imbalance in China between the numbers of livestock and their sources of nutrition. The number of animals in China has grown rapidly in recent years, while little has been done to increase the productivity of the grassland. In fact, the available grazing area has been reduced by the expansion of agriculture and the impacts of the related human activities, whereas the productivity of the remaining grassland has declined due to overgrazing and degradation. The result is a downward spiral with more animals feeding on less range, which in turn causes more serious degradation of grassland, which is therefore able to support fewer and fewer animals. This process has reduced the carrying capacity of natural grasslands in Inner Mongolia by 29 million sheep units over the last 40 years.

4.4.5 Potential Vulnerability to Climate Change

A north China case study will be considered. North China (including Beijing, Tianjin, the four provinces of Hebei, Henan, Shandong and Shanxi, part of Anhui, as well as part of inner Mongolia) is a major economic region. It also is a topographic and climatological entity. Because this region already is at risk from normal climate variability, it is likely to be vulnerable to long-term secular shifts. The impacts of climate change on this area may be discussed in terms of four factors:

- *Water resources*: Water resources in north China are vulnerable to climate change because of already low levels of available per capita water supplies, water projects that already are highly developed, large changes in river runoff related mainly to variability in flood season, and rapid economic development. Water resources also are sensitive to climate change because of the critical dependence of floods on the Asian monsoon and the ENSO phenomenon.
- *Agriculture*: The region appears to be especially sensitive to climate change because of potential increases in the soil moisture deficit. Warming and increased evapotranspiration, along with possible declines in precipitation, would make it difficult to maintain the current crop pattern in areas along the Great Wall and would limit the present practice of cultivating two crops in succession in the Huang-Hai Plains. Although climate warming may cause northward shifts of subtropical crop areas, frequent waterlogging in the south and spring droughts in the north may inhibit the growth of subtropical crops.
- *Forests*: Demand for agriculture as a result of population increases and changes in the characteristics of arable lands due to climate change will likely result in large reductions of forest area.
- *Coastal zones*: Climate change will exacerbate the already serious problem of relative sea-level rise because of tectonic subsidence and heavy groundwater with-

Table 4.9. Vulnerability to climate change, Northern China (Lin et al. 1994)

Sector	Scenarios	Method	Most vulnerable region	Summary of results	Cross-sector impact
Water Resources (W)	LLNL[a] UKMO-H3 OSU-B1 GISS-G1	Climatic, hydrological, and socioeconomic indices	Hai-Luan River Basin, followed by the Huaihe River Basin	Runoff change of −16 to +17%	Decreased supply (agric. prod. and reduction of food)
Agriculture (A)	GFDL-A3 UKMO-H3 MPI-K1	CERES and other crop models, moisture deficit and socio-oeconomic indices	Hebei, Shanxi, inner Mongolia, and along the Great Wall	Yield change (%) of wheat (−6 to +42), maize (−9 to +5), rice (−21 to −7), cotton (+21 to +53)	Increased risk for food and increased demand for water
Forests (F)	LLNL[a] UKMO-H3 OSU-B1 GISS-G1 GFDL-A3 MPI-K1	Aridity and fuelwood supply indices	All areas	Productivity increase of +1 to 10%; area change of −57 to +12% (varying with species)	Increased risk from (A) and effect on (W)
Coastal Zone (Z)	Sea-level rise of 30–65 cm	IPCC 7-step method	Jing-Jin-Tang and Yellow River Delta	Likely and viable strategy of dikes and seawalls	Increase risk to (A) and (W)

[a] L. Gates, pers. comm.

drawal. Defence against sea encroachment would be the only viable response because of the high concentration of population and economic activities. Contamination of groundwater by seawater intrusion would further worsen the water resource shortage problem.

Integrated vulnerability to climate change for northern China, is summarised in Table 4.9.

4.5 Land Use and Cover Changes, Human Interactions and Climate

Past and future changes in the atmosphere, terrestrial ecosystems, marine and coastal ecosystems are closely linked in an environmental system. These changes are also associated with the human system (Schellnbuber 1999), which serves as a forcing factor of the changes on one hand and which experiences the consequent impacts of environmental changes on human society on the other. Among the most powerful anthropogenic forces are changes in land use and land cover, which in turn respond sensitively over time to environmental changes.

East Asia is characterised by a high rate of climate change on all time scales: seasonal jumps, high inter-annual and inter-decadal variability and abrupt changes between climatic regimes (Fu and Zeng 1998). Changes of vegetation cover on seasonal, inter-annual, and inter-decadal scales are mainly driven by variations of the monsoon climate at those scales (Fu and Wen 1999). This is seen in the seasonal variation of the first eigenvectors of vegetation index, precipitation and air temperature (Fig. 4.30). The correlation coefficient between the vegetation index (NDVI) and temperature reaches 0.954 with no time lag whereas that between the vegetation index and precipitation reaches 0.788 with a lag of 6 weeks.

On the inter-annual scale, the relationship between vegetation and climate varies from region to region. In the humid region, where moisture can nearly always meet the needs of vegetation, temperature plays a more important role in affecting vegetation than precipita-

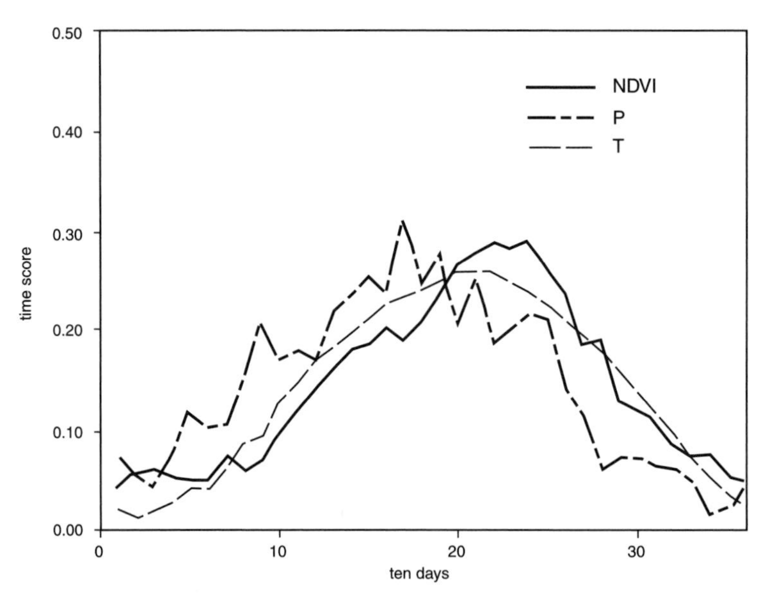

Fig. 4.30. First eigenvectors of vegetation index (NDVI), temperature and precipitation over China (Fu and Wen 1999)

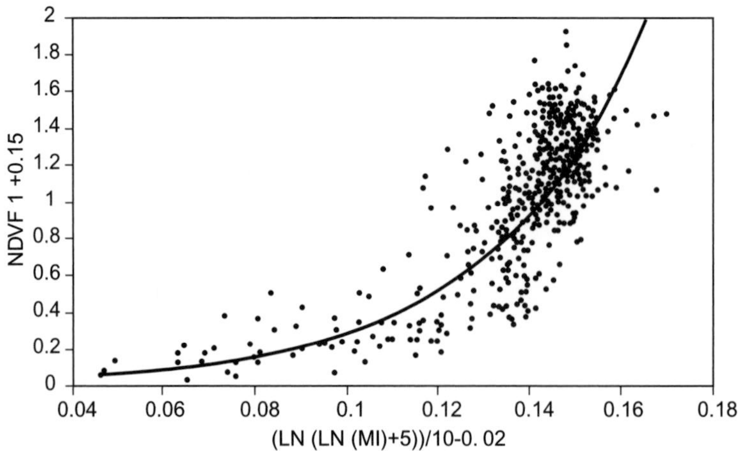

Fig. 4.31. Variation of the first eigenvector of NDVI with the moisture index in China (Zhao and Fu 2001)

tion, whereas in the semi-arid region located in the northern edge of the summer monsoon, precipitation dominates. Figure 4.31 shows the association between vegetation and moisture indexes. Strong summer monsoons usually bring more moisture to northern China, improving growing conditions, whereas weak monsoons result in drier conditions. In southern China, however, no significant correlation of this sort holds. In general, however, the association is sufficiently good to be useful in modelling. A simple ecosystem model shows good agreement between reconstructed NDVI and observed vegetation cover (Fig. 4.32 and 4.33) (Zhao and Fu 2000).

The monsoon has driven ecosystems in East Asia over the entire Holocene and probably for much longer (An

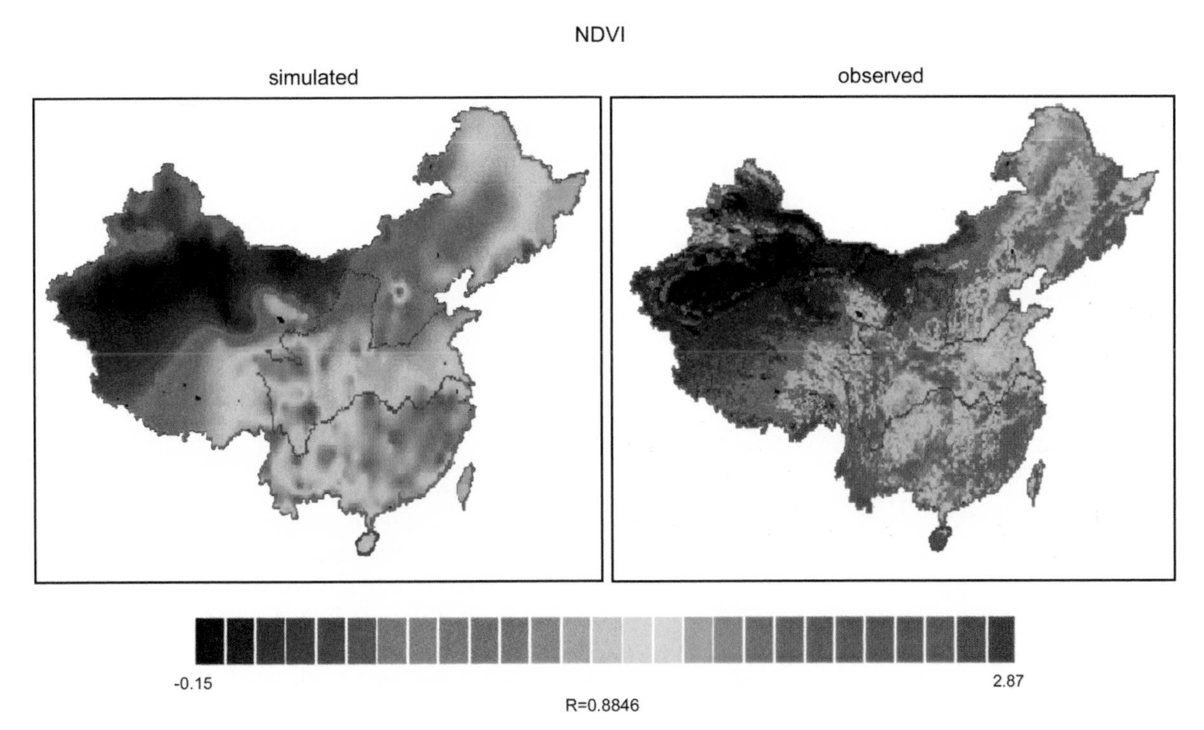

Fig. 4.32. Simulated and observed vegetation indices over China (Zhao and Fu 2001)

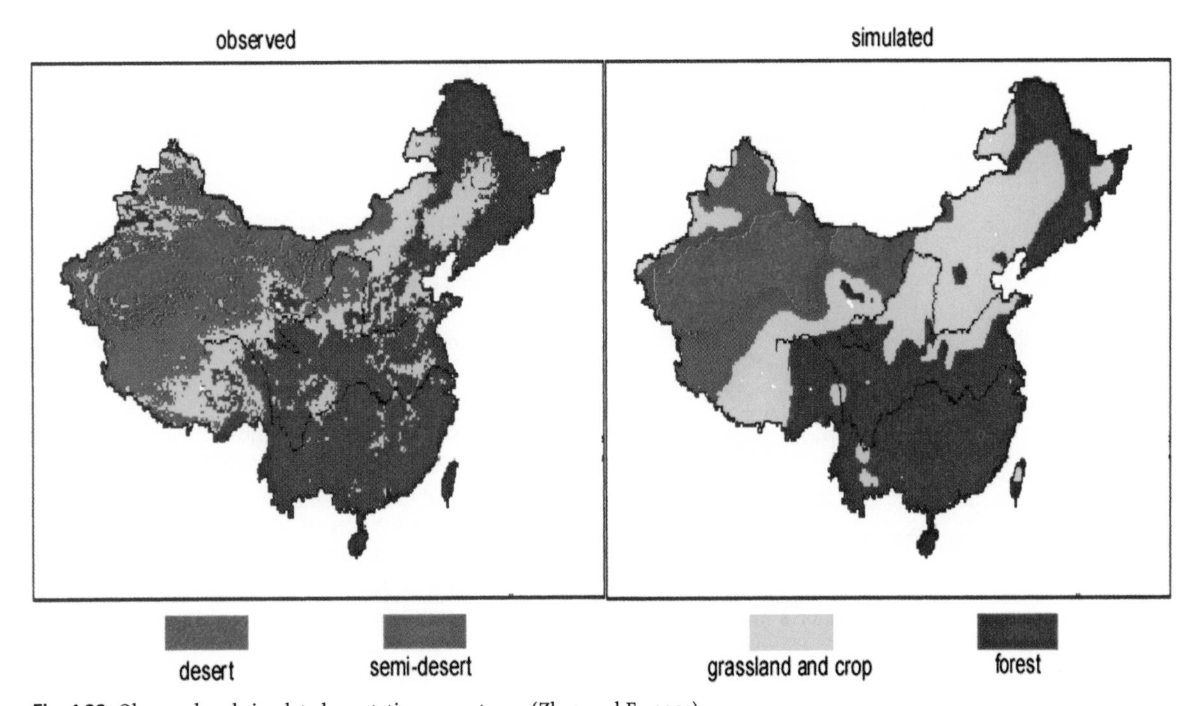

Fig. 4.33. Observed and simulated vegetation cover types (Zhao and Fu 2001)

et al. 1990). The active monsoon period of 9 000–5 000 B.P. was a period of wet and warm climate, while the weak monsoon period of 18 000–15 000 B.P. was a period of dry and cold climate. As precipitation and temperature fluctuated, patterns of vegetation over China shifted north and south accordingly (Fig. 4.34). Terrestrial ecosystems in the Asian monsoon region as a whole constitute more than 50% of the total biomass of the globe. They serve as a huge evapotranspiration source and play a crucial role in the global hydrological and biogeochemical cycles. Changes in these ecosystems have major feedbacks on climate, on both regional and global scales, through changes in fluxes of energy, water and trace gases.

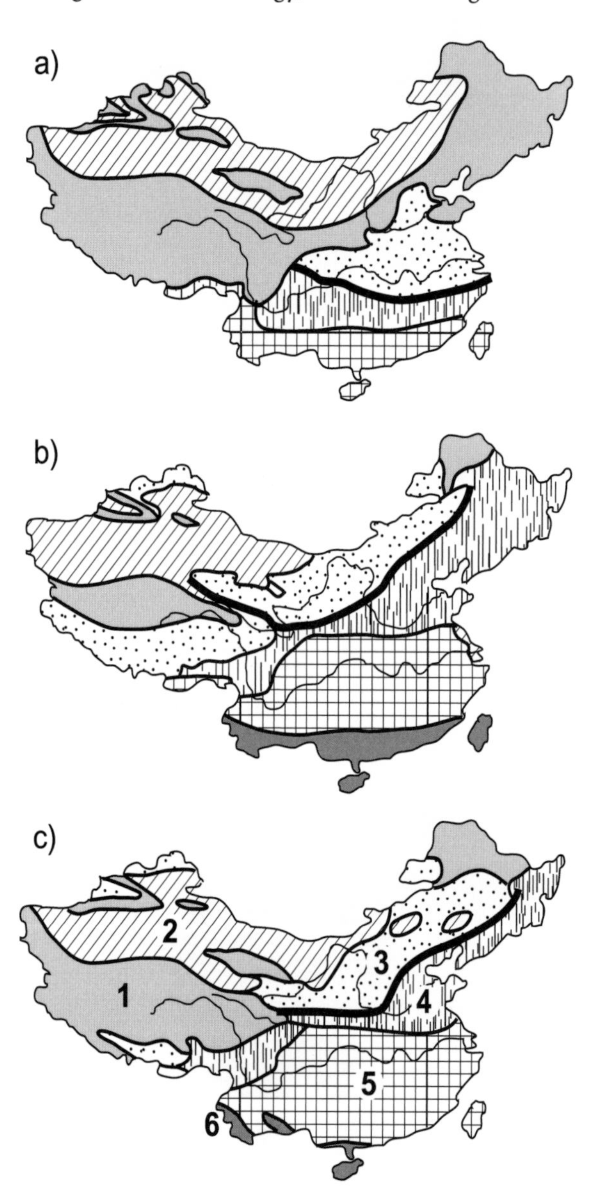

Fig. 4.34. Reconstructed vegetation cover for: **a** strong monsoonal conditions at 9 000 B.P., **b** weak monsoonal conditions at 15 000 B.P. and **c** present-day conditions (*1* denotes frozen earth, *2* desert, *3* grassland, *4* temperate steppe, *5* subtropical forest and *6* tropical rainforest) (An et al. 1990)

Modelling has shown the significant influence of changing vegetation cover on the monsoon climate. For example, a numerical experiment employing a current regional climate model shows that the summer-monsoon circulation would be weakened and the summer rainfall would be reduced significantly if all the vegetation over the East Asia monsoon region were removed and replaced by bare soil (Wei and Fu 1998). The necessity of including changing land cover in the regional modelling of summer monsoon precipitation has been clearly demonstrated, especially when attempting to specify the detailed structure of the spatial rainfall distribution (Fu et al. 1993). Despite uncertainties in the models, the response of the summer monsoon to land cover changes illustrates clearly the feedback effects of changing ecosystems on the monsoon climate (Fu and Yuan 2001). The natural vegetation has been so altered in East Asia over millennia that its reconstruction other than by modelling is not possible. However, it is feasible to specify the equilibrium climax vegetation that might be expected at present based on the currently prevailing climate (Legates and Willmott 1990). Potential vegetation distribution over the region has been specified in this way using climate-biome matching approach based on a prescribed vegetation template. This surrogate is taken to be the natural land cover unaffected by human interference (Fig. 4.35). The actual current vegetation cover is derived from a land cover data set for the period of 1987–1988 created by NASA from satellite information (Meeson et al. 1995). Human-induced change is defined as the difference between potential and current conditions. More than 60% of the region has been affected by conversion of natural vegetation into farmland and grassland, accompanied in many areas by land degradation and desertification. The differences between the current and potential vegetation covers are significant, especially in the following respects:

- mixed forests in the east part of central China have been converted completely into crop land;
- a semi-desert area in northwest China has changed into desert; and
- an evergreen broadleaf forest area in southwest China has been reduced to various kinds of shrub vegetation.

As the pristine natural vegetation (the potential vegetation) has been converted by human activities into different types of land cover, consequential changes have occurred in those aspects of the cover that have direct feedbacks to the atmosphere. In order to understand and predict changes in the Asian monsoon, it is therefore necessary to couple the biological component with the physical monsoon climate system.

This interaction process is best described by means of a climate-vegetation interaction module inserted into

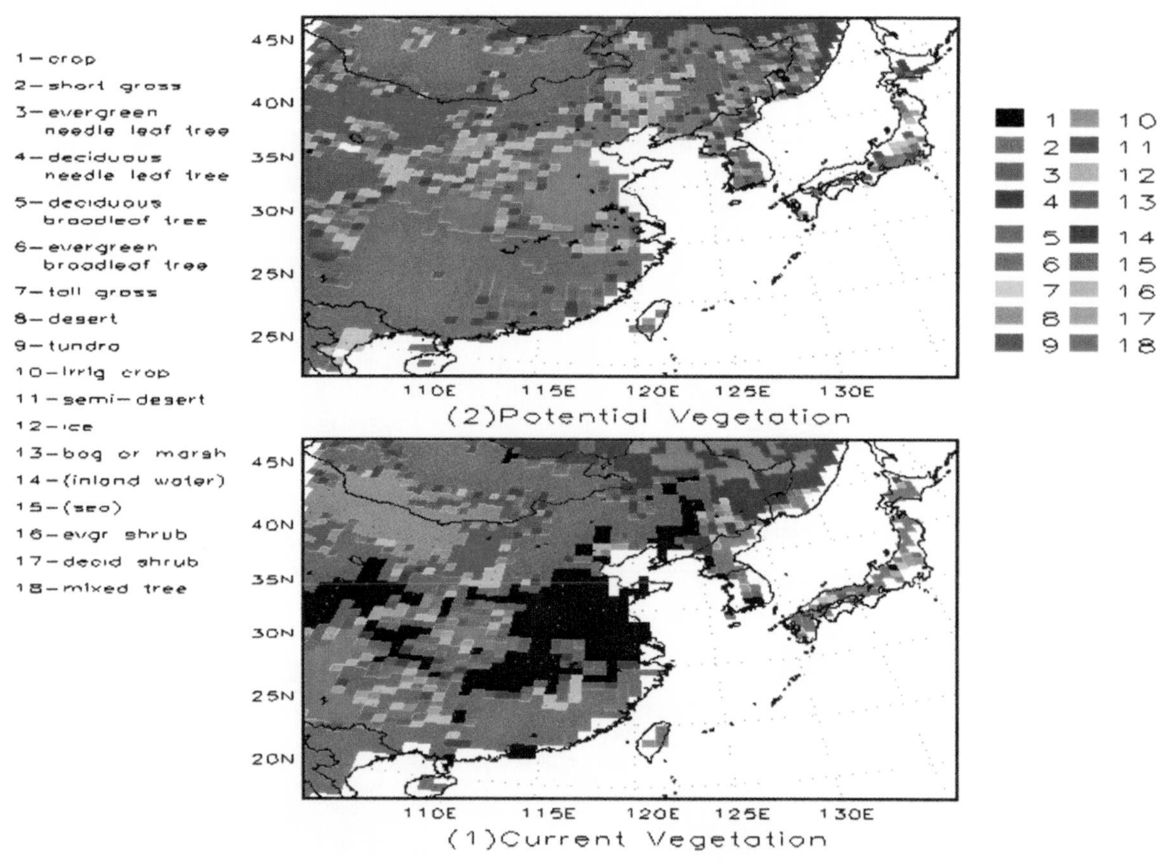

Fig. 4.35. Modelled summer potential and observed current land cover (Fu and Yuan 2001)

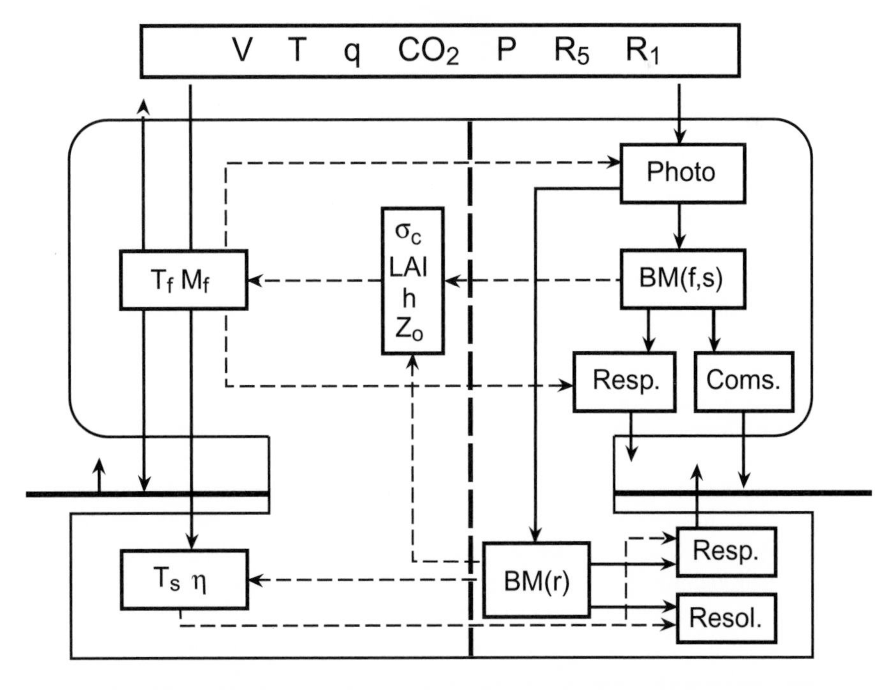

Fig. 4.36. A schematic diagram of the atmosphere-vegetation interaction model (Ji 1995)

the regional climate model that couples physical processes with the physiological processes of the vegetation, focusing on mature ecosystems on annual and interannual time scales (Fig. 4.36). The two-way interaction between the physical and physiological processes occurs mainly though changes of leaf area index which is a function of vegetation growing conditions, and at the same time changes the dynamic parameters of vegetation

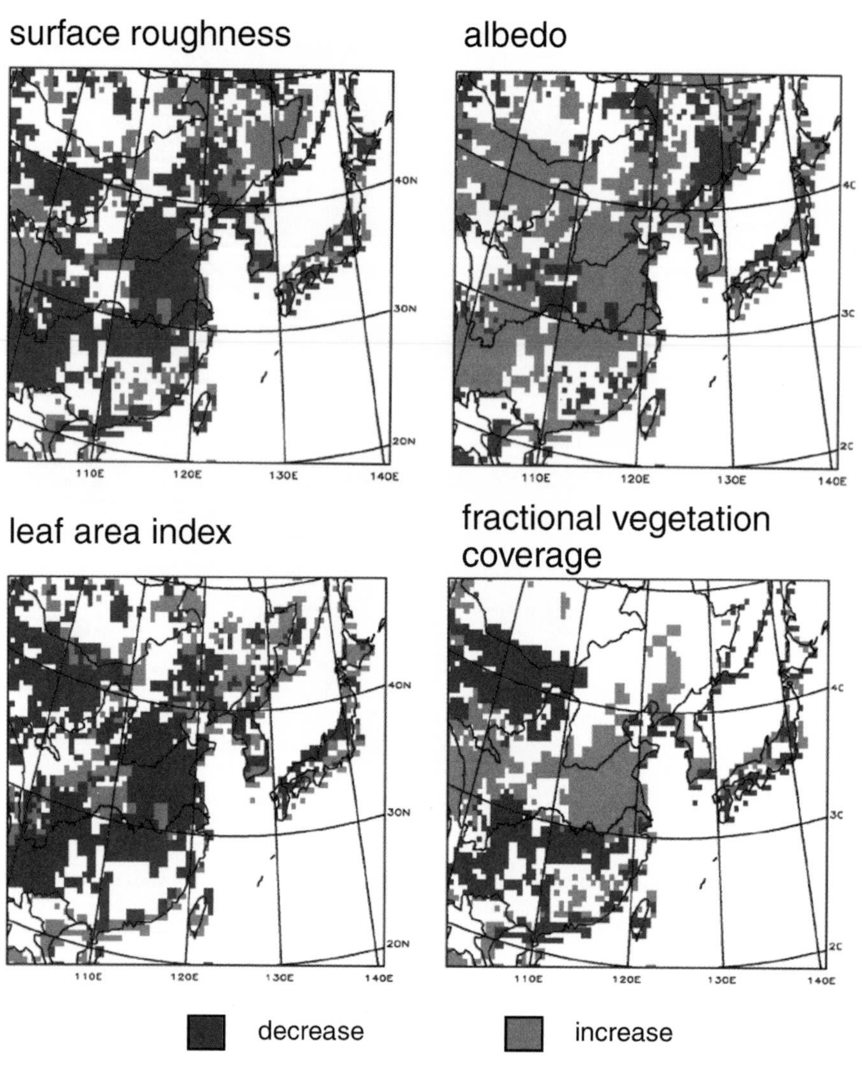

Fig. 4.37. Changes in surface roughness, surface albedo, leaf area index and fractional coverage as a consequence of the difference between potential and observed land cover (Fu and Yuan 2001)

cover such as surface albedo, surface roughness and zero level displacement. These parameters determine the energy and water exchange between the atmosphere and the vegetation canopy and therefore affect the climate. As climate changes, any alteration of parameters such as temperature, precipitation, radiation and wind speed will affect the physiological processes of photosynthesis, respiration, decomposition and vegetation growth in a feedback loop (Ji 1995).

Significant changes in surface roughness, surface albedo, leaf area index and vegetation fractional coverage have occurred as the land cover has been altered (Fig. 4.37). Changes in these dynamic parameters have had major effects on surface energy and water balances and consequently on the physical climate system.

Comparison of summer differences between currently observed and modelled surface air temperature and precipitation shows significant warming and drying of central China and an area to the northwest in response to changes of vegetation (Fig. 4.38). The response appears to be more significant in August than in June, suggesting a response time to changing vegetation of about two months.

The contrast between present and past conditions appears most clearly seen in the surface and lower-tropospheric (below 850 hpa) humidity field (Fig. 4.39). Atmospheric moisture appears to have been reduced by about 0.5 g kg^{-1} on average throughout most of east China. Changes in surface roughness, albedo and vegetation fractional cover (and hence evapotranspiration) translated directly into the weakening of the summer monsoon. The model shows anomalous northerly flow to dominate over most of the eastern China as a consequence of land cover changes, weakening of the monsoon low over the south and the development of an anomalous low-pressure system over the north. The changes of atmospheric circulation pattern lead to reduced atmospheric moisture transfer northward over eastern China (Fu and Yuan 2001).

The climate and ecosystems in the East Asia monsoon region appear to be coupled in clear two-way interac-

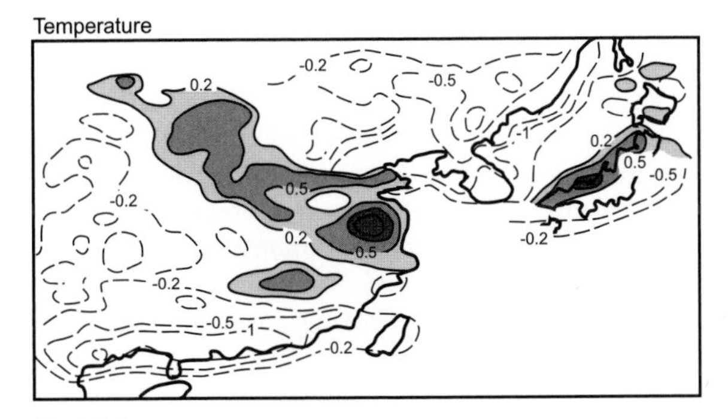

Fig. 4.38. Summer temperature (°C) and precipitation (mm d^{-1}) changes induced by land cover changes (Fu and Yuan 2001)

tions in which ecosystem variation of over East Asia is driven by variations of monsoon climate on various time scales. At the same time changes in vegetation cover, caused by both this natural variability and the changes induced by anthropogenic land use and land cover change, has significant feedback effects on the monsoon climate.

Changes from potential vegetation into current vegetation can be attributed mostly to human land use changes. Over time, intensive agriculture and industrial development have produced remarkable changes in land cover that have had significant impacts on the regional climate of East Asia and beyond. These impacts will continue long into the future.

During the past 280 years, the total decrease in the area of forests and grasslands in East Asia was 313 million ha, the largest change of any region in the world. The average annual rate of loss of tropical forests in East Asian countries was 0.87%, higher than in the Amazon (0.4%) (FAO 2000). Eighty five percent of western and southern Asia has become desertified; 70% in China and Mongolia – both higher than the global average of 62%. Population movements from rural areas to cities have accelerated the growth of cities. This growth has considerable implications for inducing future regional change. Vegetation cover in East China has been greatly modified by human activities over many millennia and, despite the many uncertainties associated with modelling the system, strong indications exist that these changes have led to a weakening of the East Asia monsoon system.

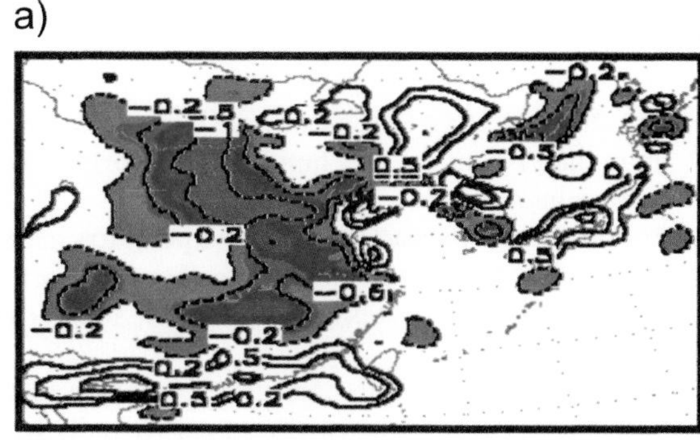

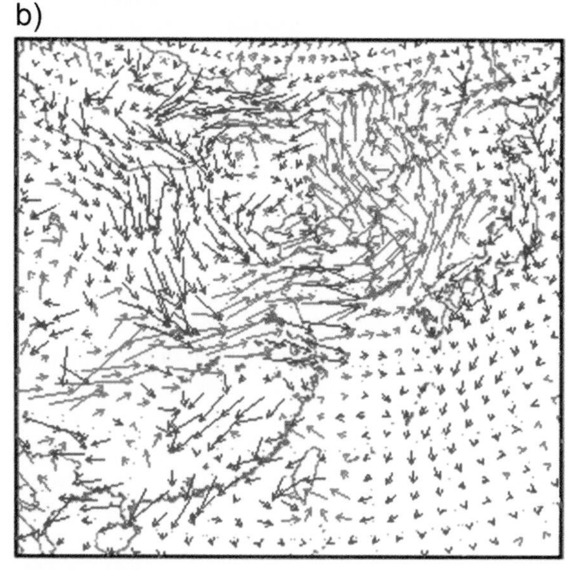

Fig. 4.39. Changes in humidity and circulation fields resulting from land cover changes. Humidity changes are given as g kg^{-1} (Fu and Yuan 2001)

4.6 Conclusions

The East Asia region comprises a major portion of the Earth's largest continent. It is bordered by the planet's highest mountains and largest ocean. Its climate and ecosystems are uniquely dominated by the East Asian monsoon. Moreover, East Asia is the homeland of some of the world's oldest, most populous, most advanced, and most rapidly evolving human civilisations. It is therefore not surprising that this region plays a major role in global processes, nor that human activities have powerfully influenced these processes on both regional and planetary scales. The overarching story of the region is of massive human-induced changes in every aspect of the environment and their consequences for natural resources, the global environment, and human welfare. The major conclusion is that the region's past, present, and future may be understood and predicted only in the context of a comprehensive integrated framework that includes the physical, chemical, biological and human components of the system. Specific findings resulting from the synthesis include:

- The climate of East Asia as a whole has warmed by more than 1 °C in the past hundred years, primarily in the winter season.
- Regionally, over the past 100 years, temperatures have increased by 2–4 °C in eastern and northeastern parts of the region and decreased by 1–2 °C in southern China.
- Negative forcing by anthropogenic aerosols over East Asia is of the same order as direct positive radiative forcing by CO_2, and can partially explain the cooling area observed in southwestern parts of China.
- The variable outflows of East Asian rivers, increasingly affected by human activities, greatly influence the hydrology, hydrochemistry and coastal dynamics of the marginal seas of the region.
- Sea surface temperature increases have led to rising sea level; with continued warming, tropical and subtropical marine species will shift to the north.
- Combined temperature increases and considerable inflow of nutrients from the land due to human activities is leading to eutrophication of coastal waters and potential decreases in marine biodiversity.
- Land use intensification, changes in consumption patterns, conversion of land from subsistence food production to cash crops and to urban use have greatly impacted environmental conditions throughout East Asia.
- Livestock numbers have greatly increased at the same time that grassland area has decreased and degraded; there is now a serious imbalance between the numbers of livestock and their sources of nutrition.
- Although farmland area has remained relatively constant in recent years, agricultural productivity has been markedly increased through irrigation, increased inputs of fertilisers, and genetic improvements. However, the ability of the region to maintain increases in crop production is questionable, and in many areas soil fertility and water availability have declined. The loss of nitrogen, phosphorus, and potassium through erosion in China exceeds the annual applied amount of chemical fertilisers.
- Over a quarter of China is threatened by desertification caused by over-grazing, deforestation, inappropriate farming practices and other human interventions.
- Land cover changes in East Asia have produced demonstrable changes to regional ecosystems and appear to have lead to significant atmospheric responses that have considerable ramifications for the monsoon circulation of the atmosphere over Asia as a whole, as well as implications for the globe as a whole.
- Despite the uncertainties associated with modelling, good evidence exists to suggest that the current East Asia monsoon system has been modified to a considerable extent by human activities, both by the change of vegetation cover from its natural condition (potential vegetation) to the current intensively exploited urban and agricultural landscapes, and by anthropogenic emission of greenhouse gases and aerosols due to the development of industry and agriculture. Albedo, surface roughness, leaf area index, and fraction of vegetation cover have all undergone significant change over large areas of the region. Simulations have shown how by altering the complex exchanges of water and energy from the surface to the atmosphere changes in land cover have brought about significant changes to the East Asian monsoon. These include weakening of the summer monsoon low-pressure system over the region and a commensurate increase in anomalous northerly flow. The consequent diminution of inland moisture transfer may be a significant factor in explaining observed decreases of precipitation and atmospheric humidity and the trend in aridification that has taken place in many parts of the region during the last century.
- The weakening of the summer monsoon due to the modification of surface conditions is probably one of the main reasons for the development of aridification and desertification trends in many parts of East Asia.
- The conceptual model of a general monsoon system that would include physical, chemical, biological, ecological and socio-economic processes provides a useful theoretical framework for assessing the regional aspects of global change in East Asia from the standpoint of earth system science.

Changes in the environment of East Asia have been considerable in comparison with many other regions of the world. These human-induced changes have had major impacts not only on regional conditions, but also on a vital component of the planet's general circulation of the atmosphere. Clear regional-global linkages in the Earth System are apparent in the case of East Asia.

Acknowledgement

The preparation of this chapter was conducted under the guidance of the Temperate East Asia START Scientific Steering Committee and with support from the International START Secretariat. The authors would like to thank members of TEACOM regional committee for their contributions to this chapter. Credit should also be given to all scientists from the region as well as outside the region that have provided the scientific literature for the synthesis study.

References

Aibulatov NA, Artukhin YV (1993) Geoecology of shelf and coasts of world ocean. Saint Petersburg, Hydrometizdat

An ZS, Wu XB, Lu YC, Zhang D, Sun XJ, Dong G (1990) A preliminary study on the palaeoenvironment changes of China during last 20 000 years. Quaternary Geology, Global Change. Beijing, China Science Press, 1–26

Bas C (1987) Role of oceans in carbon cycle. In: Carbon dioxide in atmosphere. Moscow, Mir Press. 43–67 (in Russian)

Belan TA (1997) Man-made changes in coastal ecosystems of the Okhotsk and Japan seas. In: Kasyanov VL (ed) Global change studies at the Russian far east. Vladivostok, Dalnauka Press, 7–8 (in Russian)

Bird ECF (1990) Changes of coastal line. The Global Review. Leningrad, Hydrometizdat (in Russian)

Budyko MI (1991) Carbon dioxide and climate. In: Man and element. Saint Petersburg, Hydrometizdat, 44–46 (in Russian)

Chen CTA, Tsunogai S (1998) Carbon and nutrients in the ocean. In: Galloway LN, Melillio JM (eds) Asian change in the context of global change. Cambridge, Cambridge University Press, 271–307

Chen J (1997) The impact of sea level rise on China's coastal areas and its disaster hazard evaluation. Journal of Coastal Research 13:925–930

Chen SP, Li GS (1995) Earth observation for global change study. In: Ye DZ, et al. (eds) China contribution to global change studies. Beijing Science Press, 189–192

Chen X, Zong Y (1998) Coastal erosion along Changjiang deltaic shoreline, China: History and perspective. Estuary and Coastal Shelf Science 46:733–742

China Statistics Bureau (1997) China statistics yearbook. China Statistics Press, Beijing, 1997

Cloern JE (1996) Phytoplankton bloom dynamics in coastal ecosystems: a review with some general lessons from sustained investigation of San Francisco Bay, California. Reviews in Geohpysics 34:127–168

Degterev AK, Eremeyeva LV, Ryabinin AI (1992) Ecology of warming. Moscow, Hydrometizdat (in Russian)

Deng JZ, et al. (1981) China agricultural geography. Beijing, Science Press (in Chinese)

Dulepova EP (1997) Some tendencies in inter-annual dynamics of plankton community of Eastern Kamchatka waters. Proceedings of Pacific Institute of Fisheries and Applied Oceanography 122:299–306

Ellison J, Stoddart DR (1991) Mangrove ecosystem collapse during predicted sea-level rise: Holocene analogues and implications. Journal of Coastal Research 7:151–165

FAO (1997) Review of the state of world fishery resources: Marine fisheries. FAO Circular No. 920 FRIM/C920, Northwest Pacific, FAO Statistical Area 61

FAO (2000) FAOSTAT agriculture data: FAO material for the animal numbers based on 1999 statistics. *http://www.apps.fao.org/page/collections?subet=agriculture*

Fu CB (1994a) An aridity trend in China in association with global warming. In: Zepp RG (ed) Climate biosphere interaction: Biogenic emissions and environmental effects of climate change. London, John Wiley, 1–17

Fu CB (1994b) Study on the abrupt climatic changes, Acta Atmospherica Sinica 18:373–374

Fu CB (1997) Towards the development of a regional "General Monsoon System" model for Asia. TEACOM Report: proceedings of the international workshop on regional modelling of the "General Monsoon System" in Asia. Beijing, 1–6

Fu CB, Fan HJ (1999) The impacts of El Niño on China: A historical review and current event analysis. In: Goebel K (ed) International seminar on nuclear war and planetary emergencies, 23rd Session. Erice, World Scientific Publishing, 69–76

Fu CB, Fletcher J (1988) Large signals of climate variation over the ocean in the Asia monsoon region. Advances in Atmospheric Sciences 5:389–404

Fu CB, Teng XL (1993) Relationship between summer climate in China and the El Niño/Southern Oscillation phenomenon. Frontiers in Atmospheric Sciences, Allerton Press, 166–178

Fu CB, Wang Q (1992) On the definition of abrupt climatic change and its detecting approaches. Acta. Atmospherica Sinica 16: 482–493

Fu CB, Wen G (1999) Variation of ecosystems over East Asia in association with seasonal, interannual and interdedacal monsoon variability. Climate Change 43:477–494

Fu CB, Xie L (1998) Global oceanic climate anomalies in 1980s. Advances in Atmospheric Science 15:167–178

Fu CB, Yuan HL (2001)A virtual numerical experiment to understand the impacts of recovering natural vegetation on summer climate and environmental conditions in East Asia. Chinese Science Bulletin 46:1199–1203

Fu CB, Zeng ZM (1988) Ten years forecasting experiment on long range variation of northwest Pacific high according to sea surface temperature anomalies in the tropical Pacific, in Proceedings of the First WMO Conference on long range forecasting: the practical problems and future prospects, WMO, TD No. 147:77–86

Fu CB, Zeng ZM (1998) Monsoon region: the highest rate of precipitation changes observed from global data. Chinese Science Bulletin, 43:662–666

Fu CB, Zheng WZ, Su BK (1993) Study on the sensitivity of mesoscale model in response to land cover classification over China. EOS, Transactions, American Geophysical Union, 74, No.43:172–74

Fu CB, Wei HL, Chen M, Su BK, Zhao M, Zheng WH (1998) Simulation of the evolution of summer monsoon rainbelts over East China from a regional climate model. Chinese Journal of Atmospheric Science 22:522–534

Fu CB, Diaz H, Dong DF, Fletcher J (1999) The changes of atmospheric circulation over Northern Hemispheric oceans association with the global rapid warming of the 1920s. International Journal of Climatology 19:581–606

Himiyama Y (1998) Land use/cover changes in Japan: from the past to the future. Hydrological Processes 12:1995–2001

Honda Y, Ono M, Sasaki A, Uchiyama I (1995) Relationship between daily maximum temperature and mortality in Kyusyu, Japan. Japanese Journal of Public Health, 42, 260–268 (in Japanese with English summary)

Hu D, Saito Y, Kempe S (1998) Sediment and nutrient transport to the coastal zone. Asia changes in the context of global climate change. In: Galloway J, Melillo JM (eds) IGBP Publication Series 3, Cambridge, Cambridge University Press, 245–270

Huang RH, Sun FY (1994) Impacts of the tropical western Pacific on the East Asian summer monsoon. Journal of the Meteorological Society of Japan 70:243–256

Ji JJ (1995) A climate-vegetation interaction model: simulating physical and biological processes at the surface. Journal of Biogeography 22:445–451

Jung HS, Lim GH, Oh JH (1999) Interpretation of the transient variations in the time series of precipitation amounts in Soul. Journal of the Korean Meteorological Society 35:354–371

Kaplin PA, Selivanov AO (1999) Development of coastal relief of Eastern China at Pleistocene-Holocene. Oceanologiya 39, 920–929 (in Russian)

Karl T (1998) Regional trends and variations of temperature and precipitation. In: Watson RT, et al. (eds) The regional impacts of climate change, an assessment of vulnerability. Cambridge, Cambridge University Press, 413–425

Kasyanov VL (1998a) Ecological aspects of the project of the Tumen River economic development area. In: Report of the 7th TEACOM Meeting and International Workshop. Vladivostok, Dalnauka Press, 111–126

Kasyanov VL (1998b) Sea level rise and colonization of new substrate by marine organisms. Herald of Far Eastern Branch, Russian Academy of Sciences FEB RAS. 1, 3–6. (in Russian)

Kasyanov VL, Fadeev VI, Fadeeva NP, Tarasov VG, Kamenev GM (2001) Distribution of benthic organisms in condition of chronic oil pollution in the Nakhodka Bay, Sea of Japan. Marine Pollution Bulletin, in press

Kim JW, Ha KJ (1987) Climatic changes and interannual fluctuations in the monthly amounts of precipitation at Seoul. Journal of the Korean Meteorological Society 23:54–69

Klige R.K. (1982) Changes of oceanic level in Earth history. In: Kaplin PA, et al. (eds) Oscillation of Sea Level for the Past 15 000 Years. Moscow, Nauka Press, 11–22 (in Russian)

Kondratyev KY (1993) News on the estimation of global changes. Proceedings of the Russian Geographical Society 4, 1–11 (in Russian)
Kondratyev KY, Grassl H (1993) Changes in global climate in the context of global ecodynamics. Saint Petersburg, PROPO. 195 p (in Russian)
Korotky AM (1994) Sea of Japan level fluctuations and coastal landscapes. Herald of Far Eastern Branch, Russian Academy of Sciences 3:29–42 (in Russian)
Korotky AM, Grebennikova TA, Pushkar VS, et al. (1997) Climatic change at the Far East in late Pleistocene-Holocene. Herald of Far Eastern Branch, Russian Academy of Sciences 3:121–143 (in Russian)
Kostina EE (1997) Global changes of climate and their possible consequences. A Review. Vladivostok, Dalnauka Press (in Russian)
Krause DC, Angel MV (1994) Marine biogeography, climate change and societal needs. Progress in Oceanography 34:221–235
Kussakin OG, Tsurpalo AP (1999) Long term changes in littoral macrobenthos of Krabovaya Bight, Shikotan Island, in conditions of different phases of organic pollution. Biologiya Morya. 25, 209–216 (in Russian)
Lasserre P (1992) Biodiversity in marine ecosystems. In: Solbrig OT (ed) Biodiversity and the global change. Paris: International Union of Biological Sciences Press, 105–130
Leatherman SP (1989) Impact of accelerated sea level rise on beaches and coastal wetlands. In: White JC,(ed) Global Climate Change Linkages, New York, Elsevier, 43–57
Legates DR, Willmott CJ (1990a) Mean seasonal and spatial variability in gauge-corrected global precipitation. International Journal of Climatolology 10:111–127
Legates DR, Willmott CJ (1990b) Mean seasonal and spatial variability in global surface air temperature. Theoretical and Applied Climatology 4l:11–21
Li SJ, Shi YF (1995) The climate implication of Quaternary glacier on the Tibetan Plateau. In: Qinghai-Tibetan Plateau and global variation. China Society of the Qinghai-Tibetan Plateau Research, Beijing, Meteorological Press, 30–40
Liu C (1997) The potential impact of climate change on hydrology and water resources in China. Advances in Water Science 8:21–23
Liu TS, An ZS, Chen MY, Sun DH (1996) A correlation between southern and northern hemispheres during the last 0.6 Ma. Science in China 39:113–120
Liu TS, Liu JQ, Guo ZT (1998) Climatic change in the historical time. Advances in Global Change Studies of China. Beijing China Ocean Press:17–21
Ma ZG and Fu CB (2000) Variation of humidity condition in China in last 50 years. Acta Meteorologica Sinica 58:278–287
Meeson BW, Gorprew FE, Mc Manus JMP, Myers DM, Class JW, Sun KJ, Sunday DJ, Sellers PJ (1995) Global data sets for land-atmosphere models, ISLSCP Initiative I, 1987–1988, vol.1–5
Mikishin YuA (1998) Prediction of development of lagoon coastal zone of Sakhalin Island at sea level rise. In: Geographical Studies of Marine Coasts. Vladivostok, Far Eastern State University Press, 66–74 (in Russian)
NASA (2001) *http://www.earthobservatory.nasa.gov/*
Nezlin NP, Musaeva EI, Dyakonov VY (1997) Evaluation of plankton stocks in the western part of Bering and Okhotsk Seas. Okeanologia 37:403–413
Niebauer HJ (1998) Variability in Bering Sea ice cover as affected by a regime shift in the North Pacific in the period 1947–1996. Journal of Geophysical Research 103:27717–27737
Nunn PD (1998) Sea level changes over the last 1000 years in the Pacific. Journal of Coastal Research 14:23–30
Oceanographic Encyclopaedia (1974) Leningrad. Hydrometizdat (in Russian)
Ojima DS, Parton WJ, Schimel DS, Scurlock JMO (1993) Modelling the effects of climatic and CO_2 changes on grassland storage of soil carbon. Water, Air and Soil Pollution 70:643–657
Orlova T, Selina M, Stonik I (2000) Harmful algal monitoring in Peter the Great Bay, the Sea of Japan. Abstracts of the 8th PICES Annual Meeting, Vladivostock
Pahlow M, Riebesell U (2000) Temporal trends in deep ocean Redfield ratio. Science 287:831–833
Pan YH (1982) Variation of Typhoon activities in western Pacific in association with El Niño. Acta Meteorologica Sinica 40:24–33
Petrenko VS (1998) Geoecological consequences of implementation of the "Tumangan" project for south-west of Primorye. In: Report of the 7th TEACOM Meeting and International Workshop. Vladivostok, Dalnauka, 127–134
Qian Y, Giorgi F (1999) Interactive coupling of regional climate and sulfate aerosol models over East Asia. Journal of Geophysical Research 104:6477–6499
Robin G (1986) Changing the sea level. In: Warrick B, Jaeger D (eds) Greenhouse effect. Climate change and ecosystems. SCOPE 29. New York, Wiley, 323–359
Sahagian DL, Schwartz FW, Jacobs DK (1994) Direct anthropogenic contributions to sea level rise in the twentieth century. Nature 367:54–57
Savelyev AV (1998) El Niño impact on Sea of Japan. Geographic Studies Marine Coasts. Vladivostok, Far Eastern State University Press, 54–70
Schellnbuber HJ (1999) "Earth system" analysis and the second Copernican revolution, Nature 402:19–23
Schindler DW (1999) The mysterious missing sink. Nature 398: 105–107
Shen HT (1995) Tendentious analysis of runoff discharge of the Yangtze River. In: Ye DZ, et al. (eds) China Contribution to Global Change Studies. Beijing, Science Press, 212
Shuntov VP, Dulepova EP, Radchenko VI, Lapko V.V. (1996) New data about communities of plankton and necton of Far Eastern Seas in connection to climatic oceanological re-organization. Fishery Oceanography 5:38–44
Shuntov VP, Radchenko VI, Dulepova EP, Temnykh OS (1997) Biological resources of Far Eastern Russian economic zone: structure of pelagic and bottom communities, present state, tendencies of multi-year dynamics. Proceedings of Pacific Institute of Fisheries and Applied Oceanography 122, 3–15 (in Russian)
Shushkina EA, Vinogradov ME, Shbertsov SV, Nezlin NP, Gagarin VI (1995) The characteristics of epipelagic ecosystems of the pacific Ocean on the base of satellite and field observations: the stock of plankton in the epipelagial. Okeanologia 35: 705–712
Stock JH (1992) Shifting ranges and biodiversity in animal ecosystems. Biodversity and Global Change. Paris, International Union of Biological Sciences Press 167–171
Taira K, Lutaenko KA (1993) Holocene palaeooceanographic changes in the Sea of Japan. Reports of Taisetsuzan Institute of Sciences 28:65–70
Tang MC (ed) (1989) The Introduction to Theoretical Climatology. Beijing, Meteorological Press, 25–41
UNEP (2000) Global environment outlook 2000 at *http://www.unep.org/Geo2000/english/0069.htm*
UNESCO at *http://www.unesco.org*
Vinogradov ME, Shushkina EA, Vedernikov VI (1996) Characteristics of the epipelagic ecosystems in the Pacific Ocean on the basis of the satellite and measured data: Primary production and its seasonal variations. Oceanologia 36:241–249
Wang B (1999) Sudden change of monsoon index. Presented in AMIP/EAC SSC Meeting, December 7–9, 1999, Honolulu, Hawaii
Wang PX (1999) Response of Western Pacific marginal seas to glacial cycle: palaeooceanographic and sedimentological features. Marine Geology 156:5–39
Wang SW, Zhao ZC (1981) Drought and flood in China, 1470–1979. In: Lamb H (ed) Climate and history. Cambridge, Cambridge University Press, 271–288
Webster P, Yang S (1992) Monsoon and ENSO: Selective interactive system. Quarterly Journal of the Royal Meteorological Society 118:877–926
Wei HL, Fu CB (1998) Study of the sensitivity of a regional model in land cover change over northern China. Hydrological Process 12:2249–2285
Wigley TMC, Raper SCB (1987) Thermal expansion of sea water associated with global warming. Nature 330:127–131
Wigley TMC, Raper SCB (1992) Implications for climate and sea level of revised IPCC emissions scenarios. Nature 357:293–300

Wu CJ, Guo HC (1994) Land use of China. Beijing. Science Press (in Chinese)
Xu ZY, Bai LH, Wei SL (1983) Cold summer in Northeast China in association with SST anomalies in North Pacific. Proceedings of Cold Summer Disaster Studies, Beijing, China Meteorological Press, 219–223
Yan ZW, Ji JJ, Ye DZ (1990) Northern hemisphere summer climate jump during the 1960s, Part I, Rainfall and temperature. Science in China, Series B 33, 1092–1101, Part II, Sea level pressure and 500 hpa height. Science in China, Series B 34:469–478
Yang ZC, Hu DX (1995) Impact of human activities and climate change on material discharge from Huanghe River to the seas, the Huanghe coastal environment and the delta socio-economic development. In: Ye DZ, et al. (eds) China contribution to global change studies. Beijing, Science Press, 213–217
Ye DZ, Yan ZW (1992) Climatic jumps in the history. Climate Variability. China Meteorlogical Press, 3–14
Zenkevich AA (1963) Biology of the seas of the USSR. Moscow, Academic Science Press (in Russian)
Zhao L, Chin YS (1985) Transgressions and sea level changes in the Eastern coastal region of China in the last 300 000 years. In: Liu DS (ed) Quarternary Geology and Environment of China. Beijing, Springer, 200–209
Zhao MS, Fu CB (2001) Study of the relationship between different ecosystems and climate in China using NOAA/AVHRR data. Acta Geographica Sinica 56:287–296
Zhirmunsky AV, Evseev GA, Latypov YY (1997) Possible changes of the shelf community composition at the rise of temperature and sea level. In: Kasyanov VL (ed) Global change studies at the Russian Far East. Vladivostok, Dalnauka Press

Chapter 5

Global Change and Development: a Synthesis for Southeast Asia

Louis Lebel[1]

5.1 The Global Change – Development Nexus

Global environmental change is important for human development. Development depends on ecosystem goods and services. There is already substantial evidence that some key resource systems are being depleted and that the capacity of some pollution sinks is being exceeded as a result of growing and cumulative impacts of human activities. Some of these changes, for example, losses in biodiversity, are irreversible. Consumption and population growth trends suggest that competition for these resources and alternatives will become more intense, challenging human ingenuity, technically and socially. Most resources and sinks systems are also sensitive to changes in climate, atmosphere and biogeochemical cycles. Consequently, global environmental changes will be a major constraint of future human development.

Development, on the other hand, is also important for global environmental change. The development pathways which societies follow may have profound impact on the way resources and sinks are used. Whereas, the nations that were first to industrialise are largely to blame for the elevated CO_2 levels now experienced, the contribution of emissions from rapidly industrialising regions such as in Asia as a whole, will increase substantially over the next couple of decades. Gobalisation of investment and trade is facilitating this rapid industrialisation. As much of the investments are still to come to agriculture-dominated regions like Southeast Asia, countries of the region (Fig. 5.1) still have opportunities to guide the way these are directed. Collectively these decisions about development trajectories will help determine the magnitude and rates of future global environmental changes.

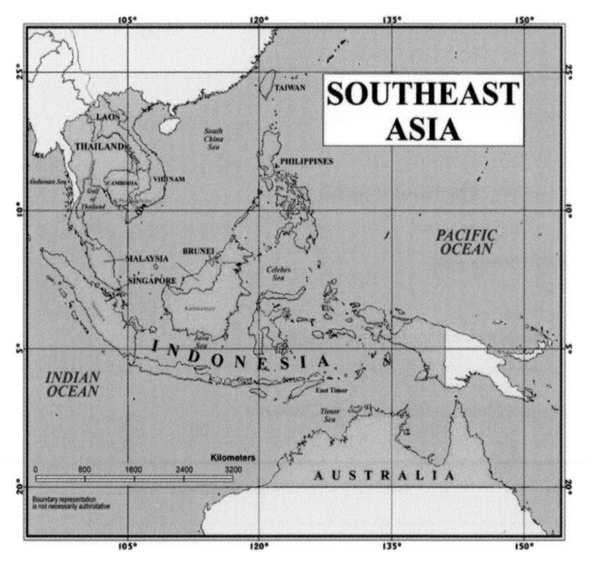

Fig. 5.1. The Southeast Asia region

Not all societies have contributed equally to current and future shared environmental changes. Likewise, not all societies stand to gain or lose the same amount from either development processes or global environmental changes. Nor do all societies have similar access to power to control or correct these inequities. For this reason, it is envisaged that those groups now disadvantaged for various, political, social or economic reasons, will be those most vulnerable to global environment changes. In other words, existing inequities will be enhanced as global environmental changes have disproportionate effects on the most vulnerable groups.

Environment-human interaction is an interactive, multi-scale loop. Human activities produce both local and wider environmental impacts, which in turn can result in changed ecosystem behaviour. Human beings respond to these by changing management practices, institutional arrangements and policies. As a result of these adaptations or coping strategies the behaviour of the factors or forces driving environmental changes in the first place are altered. Development thus is both a driver and a response to environmental changes. In this chapter the implications of development in Southeast Asia for regional and global environmental change and how global environmental change will affect development in Southeast Asia will be addressed. The synthesis will consider human-environment interactions taking place among three linked systems: terrestrial ecosystems, urban and industrial systems, and coastal and marine systems. These interact with each other physically through surface, water and atmospheric transfer processes, as well as through flows of information, money and institutions. Each of the systems also interacts with the climate system.

[1] *Contributing authors:* O. G. Ling, S. Tay, T. Moya, B. Malayang, D. Murdiyarso, A. Snidvongs, Y.K. Sheng.

5.2 Cities and Industrial Transformation

5.2.1 Industrialisation and Globalisation

The rate of growth of economic activity in eastern Asia has been faster than any part of the world for several decades (Fig. 5.2). Not surprisingly this has attracted a lot of attention of development experts, policy makers and economists, especially in the so-called newly industrialised economies of Taiwan, S. Korea, Singapore and Hong Kong. Most governments in the region have been focused on duplicating this upward mobility in the global system. The second tier of so-called tiger economies, including Malaysia, Thailand, Indonesia and the Philippines, sought and largely achieved high rates of economic expansion. Investment in industry by Japanese, US and transnational corporations is a crucial part of the story.

For decades the rapid economic growth of ASEAN economies has been led by industrial growth. Between 1970 and 1993 the contribution by industry to the ASEAN region Gross Domestic Product has increased from 25% to 40% and industrial output increased 25 times during the same period (ADB in ASEAN 1997). Manufacturing contributed more than two thirds of the Gross Domestic Product in 1994, having expanded at an annual rate of 19% since 1980. The transformation of the Thai economy illustrates the phenomenon (Fig. 5.3). Over the past four decades the contribution of the manufacturing sector to GDP has grown enormously, largely at the expense of agriculture. Over the same time period GDP increased at a constant price about 15 times. Cambodia, Laos PDR and Myanmar have hardly begun the process of transformation (Table 5.1).

Trade flows have grown rapidly. For example, during the period 1980 to 1990 ASEAN exports expanded at an annual rate of 10.9%, while at the same time imports grew at only 8.4% (ASEAN 1997). For almost two decades ASEAN markets have been the fastest growing for exporting countries and the region the fastest growing export region in the world. During this period the share of intra-ASEAN trade remained at less than 20%. Annual growth in intra-ASEAN exports, which had averaged almost 29% in the period 1993–1996, fell to 4.6% in 1997 as a result of the financial crisis of that year (Severino 1998). The most important trading partners for the region are the US, Japan and the European Union (Table 5.2). The most important trading nations in ASEAN are Malaysia, Singapore and Thailand with over 80% of the trade flows in 1996 (ASEAN 1997).

The economies of the Southeast Asian newly industrialised economies (Thailand, Malaysia and Indonesia) are more industrialized, diversified and integrated into the global economy than their counterparts elsewhere (Angel and Rock 2000: Knight 1998). The recent Asian economic crisis has been a very sharp reminder to the countries of Southeast Asia about how vulnerable their current development strategies are to the vagaries of investment flows. Just as huge inflows of foreign investment have driven economic expansion, their quick withdrawal resulted in a sharp contraction of the economy (Phongpaichit and Baker 1998; Jomo 1998) with many negative social consequences. The economic crises also provided insights and reminders about the workings of the globalised economic/production systems. For example, the virtual economy in which speculation in currency values and shares re-distributes huge sums of money with hardly a reference to commodities or services, can have a profound effect on the economy and government policy.

5.2.2 Urbanisation and Population Growth

Although cities only make up a small part of the landsurface they are an important driver of regional and global environmental change. At present, the impact of Southeast Asian cities on the global environment may

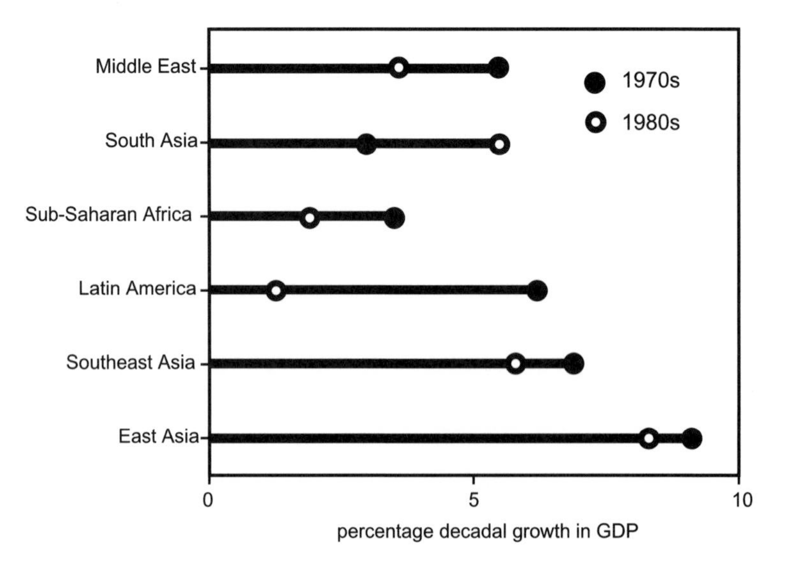

Fig. 5.2. Comparison of average annual regional GDP growth for six developing regions of the world (Stallings 1995)

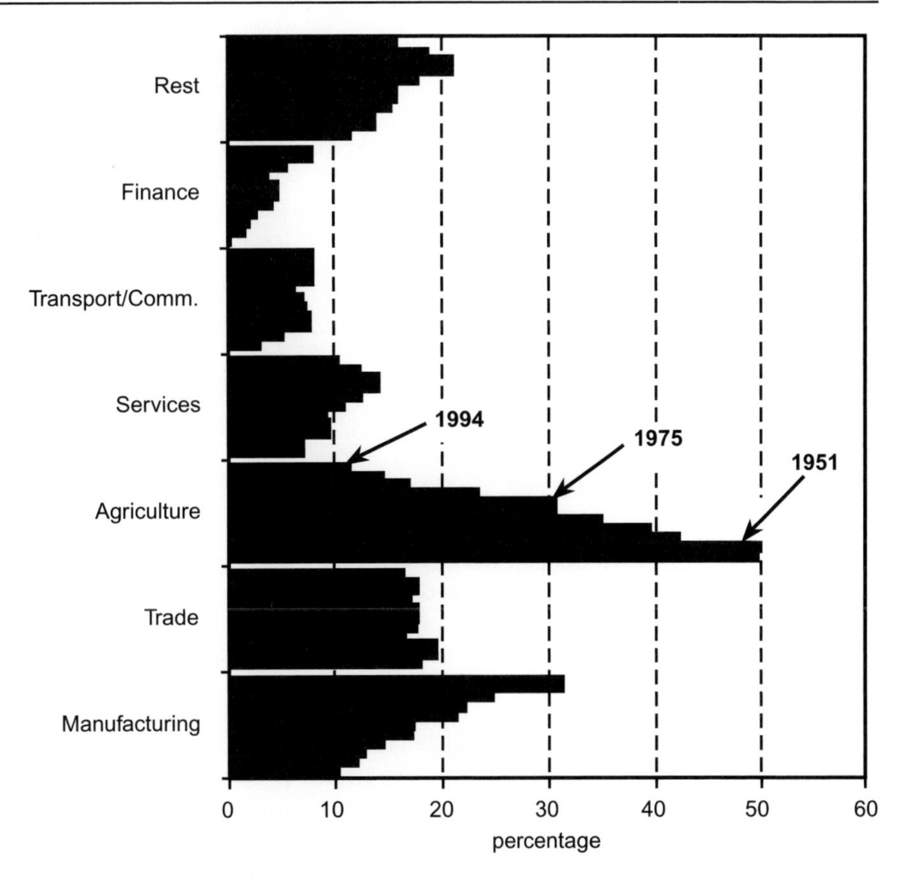

Fig. 5.3.
Percentage contribution of different sectors of the Thai economy to GDP at constant prices at 5-year intervals between 1951–1994. The time series within each sector starts from the lowest bar as 1951 up to the highest bar as 1994. The rest category consists of mining and quarrying, construction, electricity and water supply, ownership of dwellings, public administration and defence (Dixon 1999)

Table 5.1.
Structure of the economies by percentage of GDP (1998) (World Bank 1999)

Country	Structure of the economy (%)			
	Agriculture	Industry		Services
		Manufacturing	Other	
Cambodia	50.6	6.2	8.6	34.6
Indonesia	19.5	24.9	20.4	35.2
Lao PDR	55.6	15.1	5.7	23.6
Malaysia	9.4	27.9	15.6	47.1
Myanmar	57.9	7.7	3.4	31.0
Philippines	16.9	21.9	9.7	51.5
Singapore	0.1	23.3	11.9	64.6
Thailand	11.2	32.1	9.1	47.7
Vietnam	28.6	36.1		35.3

Table 5.2.
ASEAN trade with major trading partners in 1996 (ASEAN 1997)

Trading partner	Million US$		Percentage share	
	ASEAN's exports to	ASEAN's imports from	ASEAN's exports to	ASEAN's imports from
European Union	46926	57380	14	16
United States	59515	53011	18	15
Japan	43150	73310	13	20
China	18045	14574	5	4
Rest of World	80563	96705	24	27
Intra-ASEAN	82420	67139	25	19
TOTAL	330620	362120	*100*	*100*

Table 5.3. Population, growth rate and urban population and forecasts (World Bank 1999)

Country	Total population (million)	Population growth rate	Urban population as % of total population		Total population (million)
	1998	1992–1998	1992–1998	2025	2050
Brunei	0.31	2.4	71	–	0.53
Cambodia	10.7	2.6	22	–	20.7
Indonesia	204	1.6	38	60.7	312
Lao PDR	4.97	2.6	22	44.5	13.3
Malaysia	22.2	2.5	56	72.7	37.0
Myanmar	44.4	1.2	27	47.3	64.9
Philippines	75.1	2.3	57	74.3	131
Singapore	3.16	1.9	100	100.0	4.01
Thailand	61.1	1.1	21	39.1	74.2
Vietnam	77.6	1.7	24	39.0	127

not be as much as that of cities in the developed world, but as their economies grow so their impacts increase. Historically, Southeast Asia has been one of the least urbanised regions of the world. This is changing with high rates of urbanisation and population growth in most countries (Table 5.3).

Urbanisation in Southeast Asia is primarily a coastal phenomenon. The migration from upland rural areas to the major coastal cities has been very rapid, particularly around the biggest capitals like Manila, Bangkok and Jakarta. Infrastructure for water, electricity and roads has struggled to keep pace with demographic and consumption changes. These trends are expected to continue as on-going industrialisation attracts more labour, manufacturing and a large informal sector servicing the new migrants (Forbes 1996).

Much of the concern with urbanisation in Southeast Asia is has been with the growth of mega-cities like Bangkok, Manila, and Jakarta. As the core parts of cities expand and merge with surrounding satellite towns, administrative and planning boundaries are often redrawn. The Jakarta Metropolitan region, known as Jabotabek, for example, consists of seven administrative units, the total population of which was 17.1 million in 1990. Southeast Asian cities are largely the result of unregulated market forces, rather than of planning or consultation. Local governments have largely lost control of cities at a time when the problems and challenges they present are increasing rapidly (Yap and Mohit 1998). Uncontrolled planning of road infrastructure and waterways, for example, compounds flooding problems during the wet season in Jakarta and Bangkok. The contribution of domestic waste water to poor water quality is substantial, and by some measures is larger than that of industrial effluents, especially in areas around major urban centres. Thus, for example, in Bangkok, 75% of the biological oxygen demand load in the lower Chao Phraya River comes from urban activities (Sachasinh et al. 1992). The discharge of huge quantities of untreated sewerage into the sea near major population centres, like Manila Bay and Jakarta Bay, is leading to eutrophication of coastal areas, that in turn can lead to an increase in toxic algal blooms and red-tides (Soegiarto 1994).

The rapidly growing, often low-lying and coastal cities of Southeast Asia are vulnerable to global environmental changes, especially any increases in the frequency of droughts, heat waves, storm surges and floods. Inadequate planning and coordination of infrastructure projects, lack of effective regulations or their implementation, and poor standards and building codes are all factors that tend to increase vulnerability to extreme climatic events.

In the longer term, sea-level rise will be one of the most obvious impacts of global climate change in Southeast Asia. The coastal siting of cities and adjacent peri-urban areas makes them vulnerable to sea-level rise (Nicholls et al. 1999), the results of which are often exacerbated by subsidence caused by excessive and uncontrolled extraction of ground water. Globally, sea-level rise is estimated to rise by 15–95 cm by 2100 (IPCC 1995). However, local differences may be expected depending on regional changes in atmospheric pressure patterns, winds and tides, as well as with changes in ocean circulation. The delta regions of Myanmar, Vietnam and Thailand and the low-lying areas of Indonesia, the Philippines and Malaysia are particularly at risk (Nicholls et al. 1999). A 1-m rise in sea level could lead to land losses in Indonesia and Malaysia of 34 000 km^2 and 7 000 km^2 respectively. In Vietnam, 5 000 km^2 of land could be inundated in the Red River Delta and 15 000 to 20 000 km^2 of land could be threatened in the Mekong Delta (McLean et al. 1998).

The effect of heat waves, on mortality, morbidity, and energy use, are compounded by heat island effects around large cities. Thus, temperatures are already a few degrees warmer in Bangkok (Heitmann et al. 1999) and the Klang Valley around Kuala Lumpur (WWF Malaysia 2000) than in surrounding countryside. Climate warming could exacerbate such problems.

In general, increasing temperature can influence the incidence and emergence of infectious diseases in human populations through various pathways affecting the relationships between vectors, pathogens and hosts in both urban and rural areas (Hayes and Hussain 1995; Mayer 2000). Malaria, like many other important infectious diseases in the tropics is highly temperature sensitive (Martens et al. 1995). Already disease management in Southeast Asia is difficult because of the evolution of resistance, to pesticides in vectors and to anti-malarial drugs in the *Plasmodium* parasite. Unfortunately, few detailed studies of the interactions between global change and infectious diseases have been completed for Southeast Asia. What is certain, however, is that temperature changes will interact and co-evolve with other changes, such as those arising from land-use and cover changes, making it unlikely that responses will be linear (Woodward 1995; Janssen and Martens 1997).

From a demographic perspective the Southeast Asian region is still very dynamic. Population sizes continue to increase, albeit more slowly (Table 5.3). Fertility and mortality rates are falling as countries go through various versions of the demographic transition, with some countries facing relatively rapid changes in age structure over the coming decades. Migration, facilitated by improved infrastructure and stimulated by new economic opportunities, is high, but varies with economic conditions, regions and ethnic backgrounds. A nation wide migration survey in Thailand in 1992 found that almost 15% of the population had migrated for at least six months in the past five years (Chamratrithirong et al. 1995). Seasonal migration to the capital for employment in construction or manufacturing industries is especially common for farmers from the North and Northeast (Chamratrithirong et al. 1995). Muslims from the South, on the other hand, migrate much less (Guest and Uden 1995). Overall, the movements to and from urban and rural areas in Southeast Asia are linking these sub-systems much more closely than before (Rigg 1997). These population changes take on an added significance for energy and materials consumption when multiplied by the rapid rates of industrialization, adoption of urban life styles, and intensification of agricultural systems.

5.2.3 Energy Consumption

Patterns of growth in energy consumption, GDP and CO_2 emissions are usually closely associated. Like the economy in general, commercial energy consumption and production in Southeast Asia has grown rapidly. Average annual growth in energy consumption between 1990–1994 varied between 8–11% for most countries in Southeast Asian for which statistics are available. Electricity consumption has increased even faster (Fig. 5.4). Between 1986 and 1995 annual electricity consumption

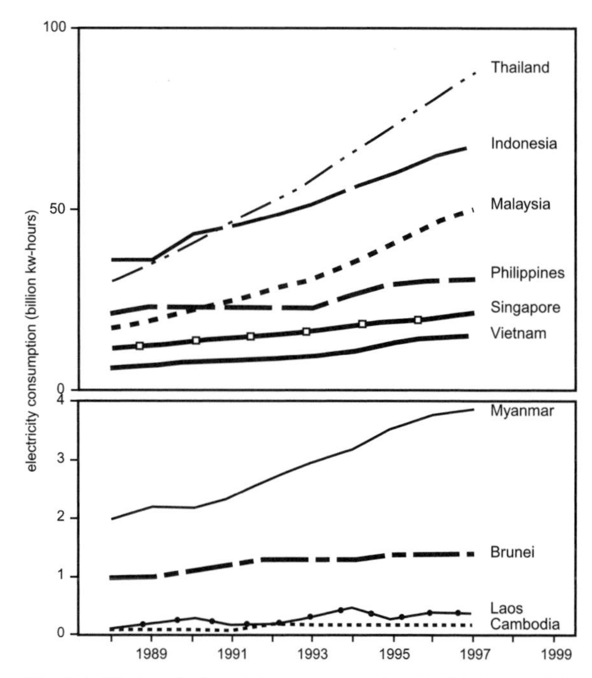

Fig. 5.4. National electricity consumption in Southeast Asia (1988–1997) in billions of kilowatt-hours. Note scale changes between upper and lower panels

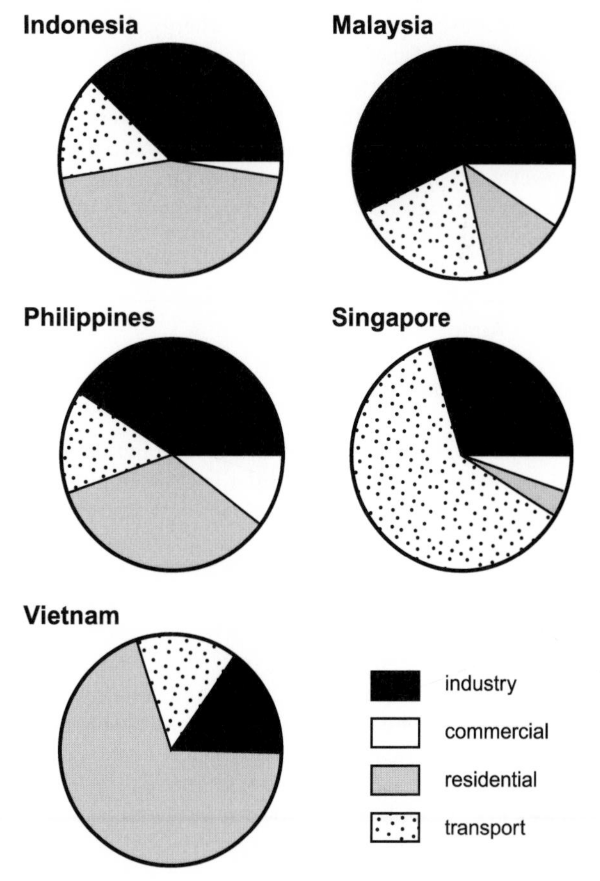

Fig. 5.5. Sectoral share of energy consumption for countries in Southeast Asia in 1997. Shade-coded sectors are labeled in the plot for Indonesia

Table 5.4. Share of sources of electricity (World Bank 1999). Figures do not always add to 100% as other sources of energy are not shown

	Hydropower		Coal		Oil		Gas		Nuclear	
	1980	1996	1980	1996	1980	1996	1980	1996	1980	1996
Indonesia	16.0	13.3	–	25.9	84.0	25.4	–	32.0	–	–
Malaysia	13.9	10.1	–	6.4	84.7	12.3	1.3	71.2	–	–
Myanmar	53.3	38.3	2.0	0.1	31.3	15.9	13.2	45.7	–	–
Philippines	19.6	19.3	1.0	13.2	67.9	49.6	–	0.1	–	–
Singapore	–	–	–	–	100.0	78.7	–	18.6	–	–
Thailand	8.8	8.4	9.8	20.0	81.4	29.3	9.9	42.0	–	–
Vietnam	41.8	0.0	39.9	14.0	18.3	8.6	0.4	77.3	–	–

increased 24% in Indonesia, 22% in Thailand and 19% in Malaysia, compared to only 3% per year for OECD countries (Forsyth 1998). The industry sector is the largest consumer of energy consumption in most countries, apart from Singapore where transport sector dominates, and the less developed nations, such as Vietnam, where residential uses still dominate (Fig. 5.5).

The countries of Southeast Asia are highly dependent on fossil fuels for electricity generation. Unlike China and India, however, most countries are more reliant on oil and natural gas (Table 5.4). In rural areas biomass is still the most important source of energy. Production of fuelwood and charcoals, largely from natural forests, increased by 27% between 1979 to 1992 in ASEAN countries (ASEAN 1997). In Thailand and Indonesia crop residues are also important sources of biomass energy. However, fossil fuels are likely to remain an important source of energy in the near future with biomass the major source of traditional fuels.

5.2.4 Atmospheric Pollution

5.2.4.1 *A Pollution Sink*

The atmosphere has been treated as a pollution sink. Pollutants in the atmosphere differ in their abundance, residence times, transport characteristics and the extent to which they affect climate. Important groups are the long-lived (CO_2, CH_4, N_2O) and short lived gases (CO, NO_x, SO_2), ozone and particulate carbon materials.

Atmospheric emissions have increased rapidly with industrialisation. For example, in Thailand during the 1980s, sulphur dioxide emissions from the industrial sector doubled, while nitrogen oxides almost trebled and suspended particulate matter increased almost fourfold (Sachasinh et al. 1992). Likewise, in Malaysia between 1988 and 1995 sulphur and nitrogen oxides almost doubled and emissions of particulates increased threefold (Tan and Kwong 1990).

Air pollution problems in Southeast Asia occur at a variety of scales. Photochemical smog, CO, N_2O, O_3 and lead are some of the unwelcome by-products of rapid industrial and urban transformation. Episodes of high levels of pollution in the mega-cities are commonplace (Table 5.5) and their effects on health are an issue of major concern to governments in the region (Table 5.6. ASEAN, for example, recently endorsed long-term environmental goals for ambient air quality of below 100 Pollutant Standards Index (PSI) by the year 2010 with priority on urban and industrialised areas (ASEAN 1997). Industrial pollution makes a significant contribution to these air quality problems. In Jakarta, for example, the industrial sector accounts for 15% of suspended particulates, 16% of N_2O and 63% of SO_2 loadings (World Bank 1994). In Thailand, 56% of emissions of suspended particulate matter and 22% of SO_2 can be attributed to industry (Sachasinh et al. 1992).

Table 5.5. Suspended particles ($\mu g\ m^{-3}$) in the urban environment (Asiaweek 1998; Asiaweek 1999; WRI 1999)

City	Suspended particles	
	1998	1999
Bandar SB	<20	28
Singapore	51	35
Cebu City	52	110
Chiang Mai	52	70
Davao City	55	98
Bandung	70	58
Yangon	75	75
Vientiane	75	150
Kuala Lumpur	120	122
Surabaya	89	102
Phnom Penh	100	217
Georgetown	110	115
Hanoi	150	300
Ho Chi Minh	180	n.a.
Manila	198	200
Bangkok	223	142
Jakarta	271	140

Table 5.6. Health impacts of air pollution (Hughes 1997)

City	Population	Premature deaths	Chronic bronchitis cases	Health benefits from better air quality as share of urban income
Bangkok	7300000	2800	28000	7
Jakarta	9700000	6300	47000	12
Kuala Lumpur	1500000	300	4000	4
Manila	9700000	3800	33000	7

5.2.4.2 *Carbon Dioxide Emissions*

The build-up of CO_2 in the global atmosphere as whole is mostly the result of industrial emissions from developed countries. In 1991, Southeast Asia, which is home to about 10% of the World's population, contributed about 5% of the total greenhouse gas emissions. Per capita emissions of CO_2 from most Southeast Asian countries are low to intermediate in comparisons with other parts of the world (Fig. 5.6). Singapore is an exception with relatively high per capita emissions. Rising emissions are strongly associated with increases in energy consumption associated with expansion and structural changes in economies, as well as population growth (Han and Chatterjee 1997; Engelman 1998). Cleaner production and consumption technologies, by promoting energy efficiency and changes to cleaner fuel types, can slow the growth rates of emissions. These kinds of projections, however, depend strongly on assumptions about industrialisation and the patterns of greenhouse gas and sulphur emissions.

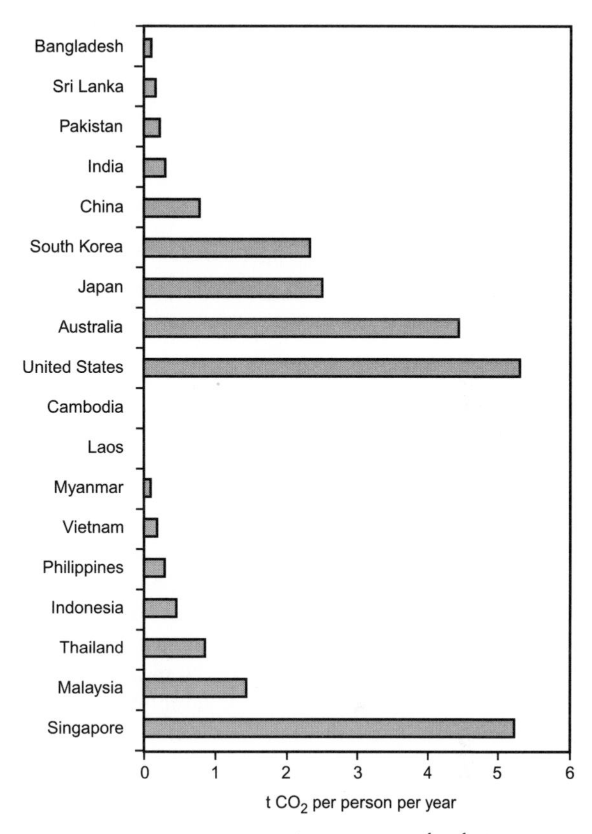

Fig. 5.6. Per capita emissions of CO_2 (t person^{-1} yr^{-1}) from fossil fuel combustion and cement production in 1995 for Southeast Asian and other countries (Engelmann 1998)

5.2.4.3 *Pollution Patterns with Development*

Pollution patterns accompanying industrial transformation in Southeast Asia follow a common pattern (Auty 1997). Pollution intensity of GDP, a measure of the amount of pollution per unit of economic activity, first intensifies then eases, whereas total emissions follow an S-shaped curve. The transition from agro-processing into capital and then into skill-intensive industries also implies changes in the composition of emissions. Typically, water-borne organic pollutants first dominate. Airborne pollution (from vehicle emissions and industrial activities) and solid wastes, especially in the newly forming urban centers, then increase. Growth in hazardous wastes appears later. Deviations from the general pattern occur among individual nations depending on differences in the availability of natural resources, the nature of the national industrial base and environment policies and institutions (Auty 1997).

In Thailand, major shifts have occurred in the last few decades in the sectoral composition of manufacturing and in the share of manufacturing in GDP. Pollution has increased rapidly and pollution intensities of GDP for all pollutants have risen over time rise over time (Angel and Rock 2000) (Fig. 5.7). All pollutants have increased faster than GDP; in some cases up to 10-fold increases have occurred. In Singapore CO, NO_2 and SO_2 are expected to rise 3- to 5-fold by 2025. Most other pollution levels are expected to increase much more slowly or remain stable. There is some suggestion from developed countries that the effects of affluence on CO_2 emissions reaches a maximum at about US$10 000 per capita GDP and declines thereafter, probably as a result of structural changes towards service-based economies and investments in energy efficiency (Dietz and Rosa 1997). Without special attention given to energy, material and pollution intensities of new urban and industrial investments (Angel et al. 2000) this transformation may not happen early or fast enough in Southeast Asia to avoid serious impacts on environmental health.

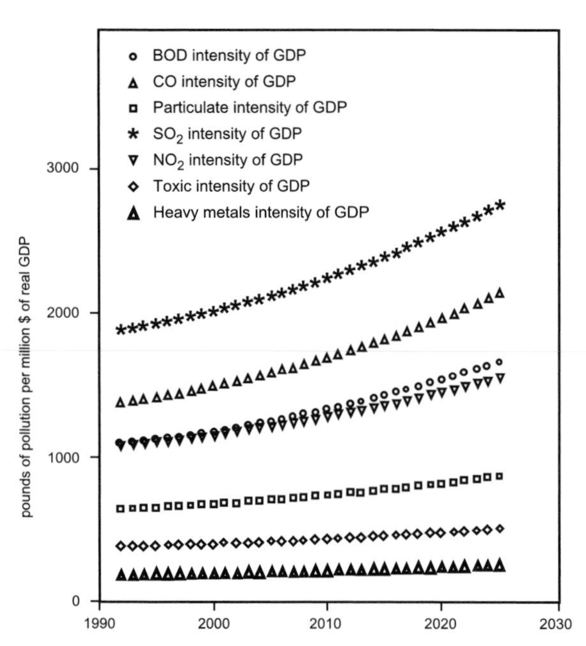

Fig. 5.7. Predicted pollution intensity of Thai GDP for 1992–2025 by pollutant (Angel and Rock 2000)

5.2.4.4 *Transboundary Acid Rain*

Transboundary atmospheric pollution is a significant phenomenon in the whole of Asia. On particular days the transport of aerosols and trace gases over the region in dust clouds may be striking (Fig. 5.8). The extent of transboundary pollution is exemplified by Vietnam where only 35% of the total annual sulphur deposition comes from sources within the country (Arndt et al. 1996). China contributes almost 39% and Thailand 19%. Similarly, Singapore accounts for 30% of Malaysia's deposition during the summer, but only 10% during the winter months (because of differences in precipitation patterns between seasons) (Arndt et al. 1996). Modelling studies suggest that 70% of regional ecosystems in Asia will be affected by transboundary advection of aerosols and trace gases by 2100 (Posch et al. 1996). The largest and most rapid impacts will be for those regions closest to northeast Asia, where projections suggest that in only two to three decades sulphur deposition levels will exceed those observed in Europe and North America in the 1970s and 1980s (Arndt and Carmichael 1995).

The key factor affecting deposition in southeastern and northeast Asia is the extent to which emission controls and energy efficiency improvements are made in China. Even with emission controls on major energy plants SO_2 emissions are projected to increase from 25 Mt in 1995 to 31 Mt in 2020; without such controls emissions could increase to 61 Mt (Streets and Waldhoff 2000). Emissions of NO_x, which are largely uncontrolled, are projected to more than double to 24 Mt by 2020. The extent of emission changes in the southeast provinces and Sichuan basin of China, will be of particular importance for the impact of transboundary pollution on mainland Southeast Asian countries.

Sulphate aerosols may affect climate both directly and indirectly. Early modelling studies suggest that the aerosols may result in a decline in summer monsoon rainfall over Southeast Asia (Mitchell et al. 1995). So far models have considered only the direct cooling effects. There is still substantial uncertainty about the indirect effects, for example on cloud formation, deep convection in the torpics and on Hadley Circulation, and therefore on the net impacts of increased sulphate aerosols in the atmosphere on climate (Lal et al. 1999). In any case these effects, which are relatively short term once emissions drop, are likely to be outweighed by the increases in greenhouse gases in the atmosphere (IPCC 1995).

5.2.5 Sustainable Cities and Industrial Transformation

Most of the investment in industrial transformation is still to come, a fact of tremendous significance for business, policy and future global environmental change. The opportunities for a cleaner, more environmentally benign transformation are in place. International pressures on business through evolving institutions such as the Kyoto Protocol and ISO certification should further encourage these trends. The cumulative impact of these development processes in the region on resources and pollution sinks, including greenhouse gas emissions, however, is likely to continue as some of the efficiency gains are wiped out by continuing expansion of economies within and outside the region (Schipper and Grubb 2000).

The current development strategy of pursuing integration in the global market economy links the activities of consumers and producers in Southeast Asia ever more closely with events elsewhere in the region and the globe. On the one hand, through trade and diversification of sources of information and resources, this can mean that food and resource security is enhanced, and hence overall vulnerabilities to global environmental challenges may be reduced. On the other hand, the same processes can result in dependencies, elimination of local diversity and capacities though competition, and susceptibility to fluctuations in global markets that are well beyond the control of individual governments, that would have the opposite effect. Moreover, the effect of these multiple exposures on particular nations, economic sectors, and segments of a population need not be the same. Finally, the environmental goods and services upon which all these economic activities depend is being altered by the scale of human activities on land and in the sea within the region itself. The prospects for sustainable cities and industrial transformation in Southeast Asia are thus dependent on the effective management of terrestrial and coastal and marine ecosystems.

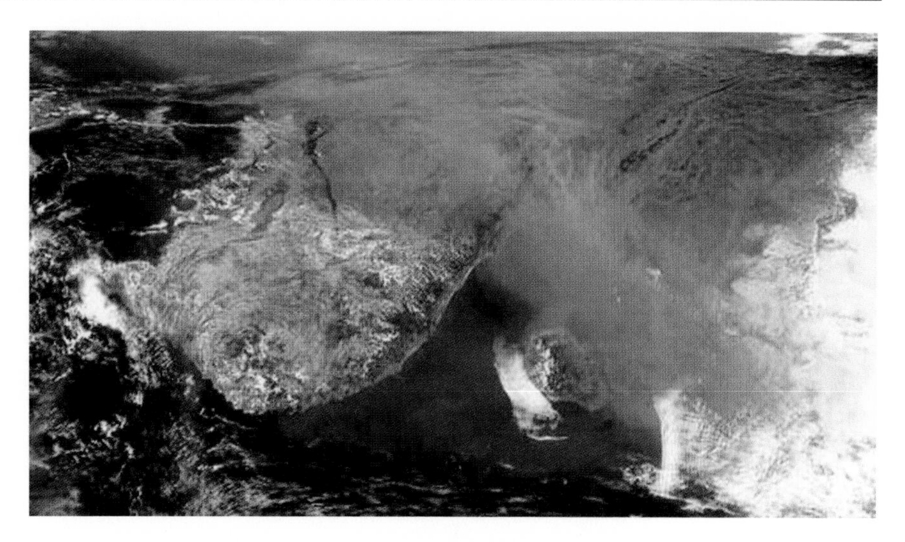

Fig. 5.8.
An example of the transport of aerosols and trace gases over and out of part of Southeast Asia (NASA 2001)

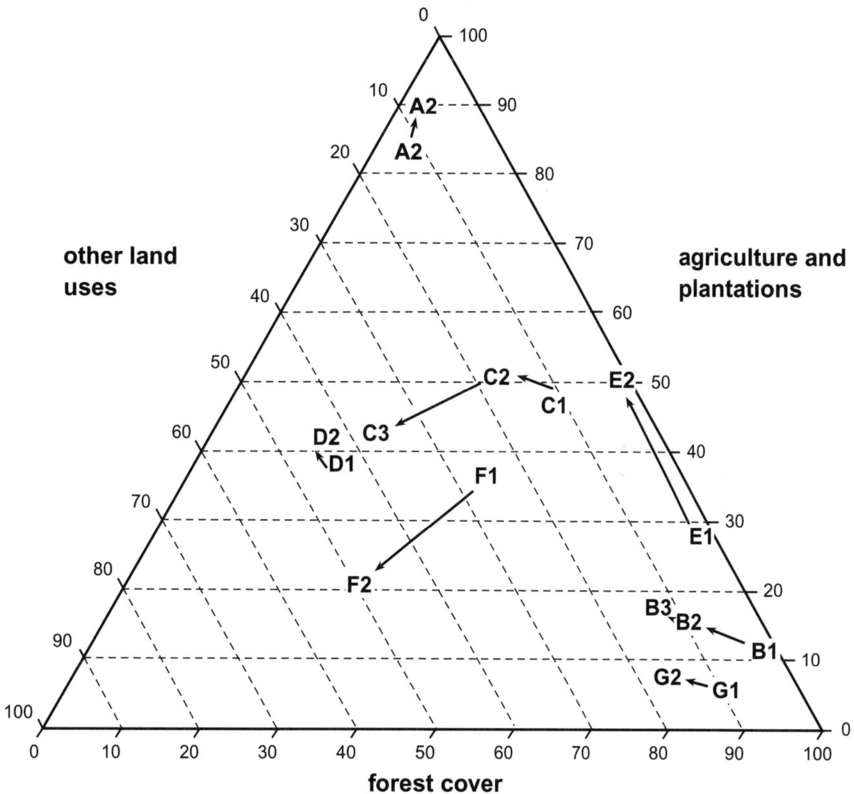

Fig. 5.9.
Some representative land-use and -cover changes (*1* denotes the earliest time, *2* and *3* subsequent times) for intermediate scale (ca. 2 500–50 000 km^2) sites in Southeast Asia: *A* Tapee coastal area of Surat Thani, Thailand, 1993–1998 (Wattayankorn et al. 1999), *B* Mae Rim and other watersheds north of Chiang Mai, Thailand, 1954–76–83 (Fox et al. 1995), *C* Klang-langat watershed, around Kuala Lumpur, Malaysia, 1974–89–99 (Moya 1997; Mastura et al. 1999), *D* Agno watershed, Luzon, Philippines, 1986–1993 (Gomez et al. 1999), *E* Jambi, Sumatra, Indonesia, 1986–1992 (Murdiyarso and Wasrin 1995), *F* Upper Citarum watershed, Java, Indonesia, 1984–1996 (Moya 1997) and *G* Mae Chaem, Chiang Mai, Thailand, 1985–1995 (Moya 1997)

5.3 Terrestrial Ecosystems

5.3.1 Land-Use and -Cover Change: Patterns and Processes

Land-use and -cover change is an important and on-going feature of development in the coastal plains and rural hinterlands of Southeast Asia (Fig. 5.9). A common pattern of transformation has been widespread deforestation, primarily as a result of government policies to obtain logging and timber export revenues, followed by both planned and spontaneous agricultural expansions (Kummer and Turner 1994; Brookfield and Byron 1990; Fox et al. 1994) by varying combinations of small-holders and large-scale plantation developers (Sutton and McMorrow 1998) (Fig. 5.10). This is likely to be followed by further changes in land use, including land abandonment where methods have been inappropriate, and intensification where infrastructure provides the support to market commodities.

Another common series of transitions is found in major cities that have grown rapidly as national economies have expanded. The Klang-langat basin around Kuala Lumpur in Malaysia illustrates these transformations. In the early 1960s, the main change was from for-

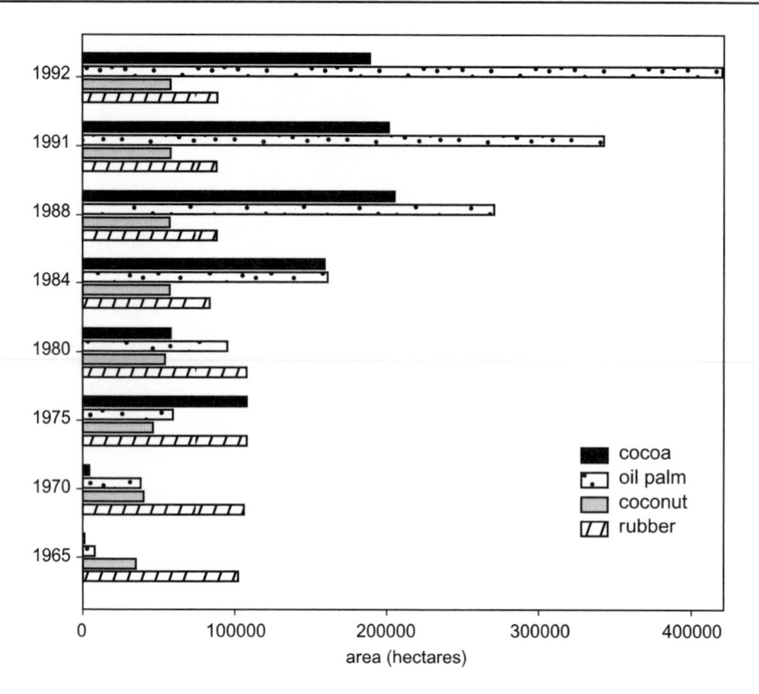

Fig. 5.10.
Land use change in Sabah, Malaysia 1965–1992 illustrating the rapid expansion of private commercial oil palm and cocoa plantation estates (Sutton and McMorrow 1998)

est to agriculture, especially plantation rubber (Mastura et al. 1999). During the 1970s much of the rubber was converted to oil palm. Since then the population has grown rapidly and oil palm is now being converted to urban development and infrastructure (Fig. 5.11). Land-use changes in Java, particularly around Jakarta, show similar dynamics, with prime paddy fields and agricultural lands giving way to urban developments and industrial estates (Verburget al. 1999). Land speculation around growing urban centres results in a time lag between the abandonment of agriculture and its replacement with urban settlements and other infrastructure. In Southeast Asia, land-use and -cover change, especially the conversion of forests to agricultural use and plantation forestry, are having significant impacts on above and below-ground biodiversity and biogeochemical cycles (Siebert 1987, 1990; van Noordwijk et al. 1995; Tomich et al. 1998). The initial impact occurs with clearing of forests and leads to immediate, but short-term alterations to biogeochemical cycles. Depending on the succeeding type and intensity of agricultural uses, the cycles change thereafter with longer-term effects.

Deforestation and conversion to grasslands, by increasing albedo, reducing surface roughness and decreasing soil porosity, have the potential to result in reductions in precipitation because of lower regional evaporation and regional moisture convergence. Understanding the consequence of land-use and -cover change for these interactions in the uplands of Southeast Asia is particularly difficult because of the dynamic and mosaic structure of many landscapes. Many areas are comprised of a mixture of various localities in different conditions of regrowth, tree plantation composition, orchards and valley crops (Giambelluca 1996).

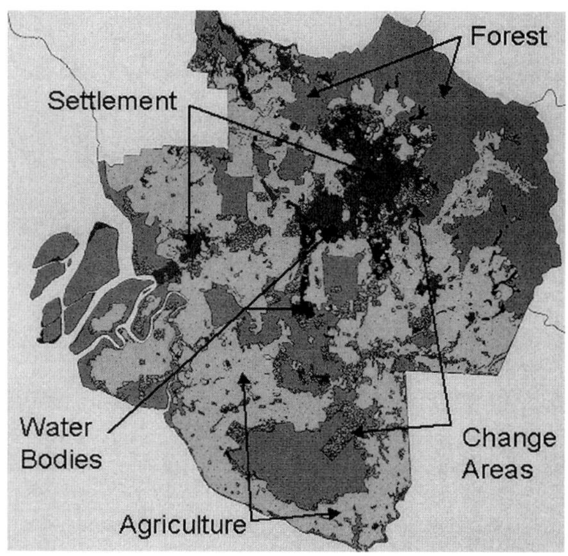

Fig. 5.11. Land-cover changes (*shaded*), 1974–1984, in the Klang-langat basin, Malaysia

5.3.2 Agricultural Expansion and Intensification

5.3.2.1 *Commercialisation*

Changes in land-use practices have occurred for at least two millennia in Southeast Asia. Those to facilitate cash cropping on a large scale are relatively recent and constituted a pivotal environmental development. Thereafter, the proportion of agricultural land used for cash crops increased rapidly in relation to that for subsistence. With cash cropping came the need for a transport infrastructure throughout the region and beyond to facilitate the export of cash crops and other commodities.

The expansion of agriculture, and its development into a major export industry from a subsistence base, was an important historical factor in the development of the major Chinese family business networks in Thailand, typified by the now giant conglomerate CP (Chairavanont Family) (Phongpaichit and Baker 1998). Much of the capital for industrialisation has come from the conversion of land-resources, in particular, the exploitation of tropical forests for timber (Brookfield and Byron 1990), and the expansion of export-oriented agriculture (Fig. 5.12).

The expansion of rubber plantations and land under rice cultivation illustrate the transformation. In peninsular Malaysia no rubber plantations existed before 1880; by 1970 they accounted for 65% of all cultivated land (Osborne 1997). Large areas of rubber plantations were also planted in Vietnam, Cambodia and Indonesia. Expansion of rice cultivation began in earnest in the mid-nineteenth century with the development of export-oriented production in the major deltas, including the Mekong (Vietnam), Chaophrya (Thailand) and Irrawaddy (Myanmar) (Osborne 1997). Large-scale changes in land use were also driven in part by political considerations, as states sought to expand their territories through development of forested lands by peasants (De Koninck 1996). Today, despite the overwhelming importance of urbanisation and manufacturing-oriented industrialisation in transforming the economies and societies of Southeast Asia, a high proportion of the population in all countries (apart from Singapore) is still dependent on or employed in agriculture and agriculture-related industries (Table 5.7).

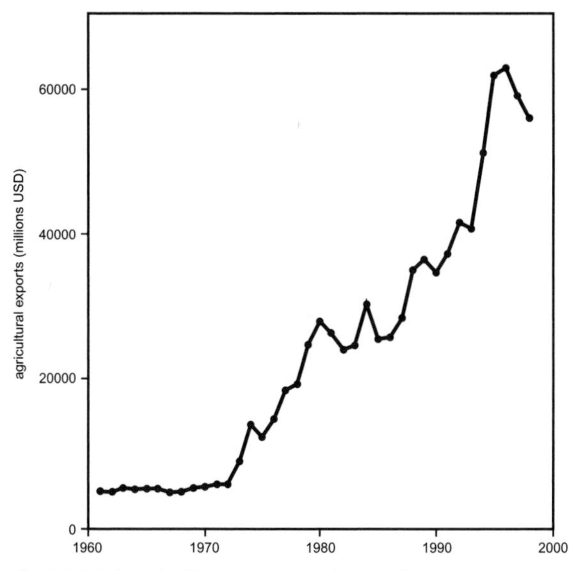

Fig. 5.12. Value of ASEAN country agricultural exports, 1960–1998 (FAO 2000)

Intensification and industrialisation of agriculture, although not yet complete, has already had a profound effect on ecological and social systems in Southeast Asia. Common features of the transformation include shifts away from subsistence to cash crops, increasing use of fertilizers, pesticides and herbicides and the expansion of irrigation. In the developed world, the intensification of agriculture has resulted in large increases in the application rates of fertilizer. Globally, the consequence is that the nitrogen cycle is now dominated by human activities (Vitousek et al. 1997). Intensification in Southeast Asia has also been rapid. Between 1961 and 1997 the area under permanent crops approximately doubled. The total consumption of fertilizers increased more than 19-fold, while the value of trade in pesticides, herbicides and fungicides increased more than 15 times (FAO statistical databases 2000). Intensification of agriculture in Thailand (1893–1994) illustrates these trends for the region (Fig. 5.13).

At the same time as agriculture has intensified, the sources of income for rural individuals, households and communities, have diversified through opportunities created by new markets for agricultural commodities as well as employment in the rapidly expanding manufacturing and construction sectors. As infrastructure outside the major capital cities improves, manufacturing industries are beginning to move into more rural settings. In some cases the exchange of ideas, labour and money between areas classified as rural and urban are so large that the distinction itself has begun to loose its usefulness for planning (Rigg 1997). The investments

Table 5.7. Agricultural land development relative to total land area and population growth in various Southeast Asian countries (Asian Development Bank 1996)

	Agriculture share of land area (%)		Cropped land per capita (ha)	
	1975	1993	1975	1993
Myanmar	15.2	15.3	0.33	0.23
Thailand	32.6	40.7	0.52	0.49
Vietnam	19.2	20.6	0.13	0.09
Laos	2.9	3.5	0.22	0.17
Cambodia	11.9	13.6	0.3	0.25
Indonesia	14.4	17.1	0.19	0.16
Malaysia	14.2	14.9	0.38	0.25
Philippines	24.6	30.8	0.18	0.14

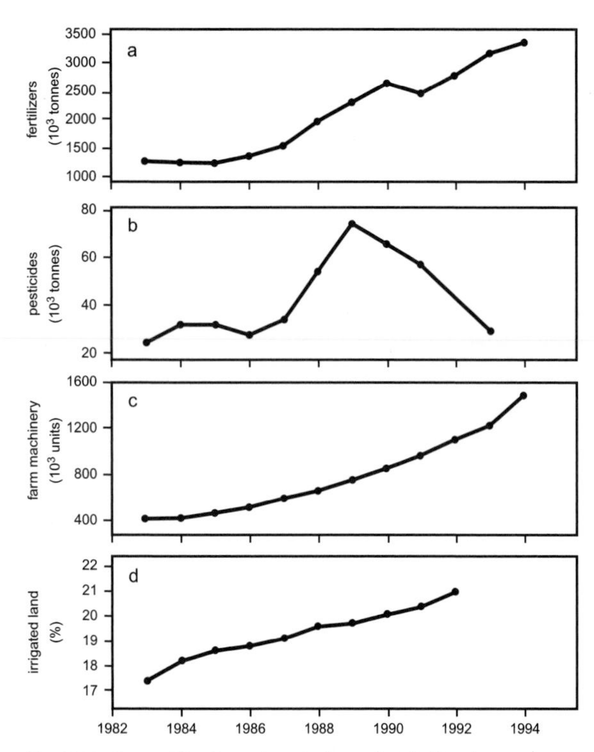

Fig. 5.13. Intensification of agriculture in Thailand, 1983–1994: **a** chemical fertilizers (thousands of t), **b** pesticides (thousands of t), **c** farm machine (thousands of units), **d** percentage of irrigated agricultural land (Office of Agricultural Economic, Thailand 2000)

back into agriculture from income and profits generated through industrialization have been important for the development and adoption of innovations that boost productivity and create new market crops. Overall, these linkages have improved living standards and probably reduced the vulnerability of rural communities to climatic variability and other environmental changes.

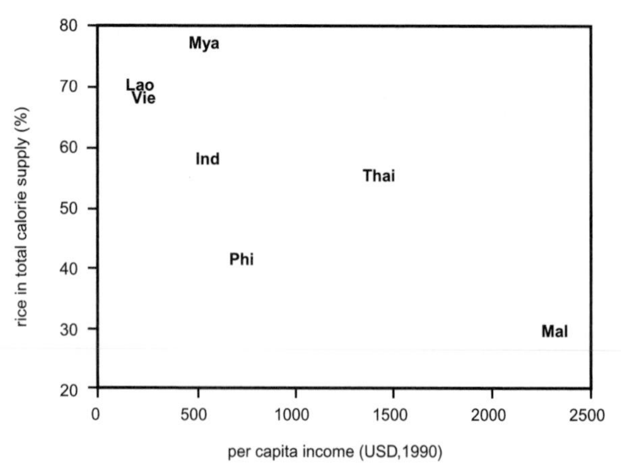

Fig. 5.14. Contribution of rice to total calorie supply in seven Southeast Asian countries. *Mya* denotes Mayanmar, *Lao* Laos, *Vie* Vietnam, *Ind* Indonesia, *Thai* Thailand and *Mal* Malaysia (Hossain and Fischer 1995)

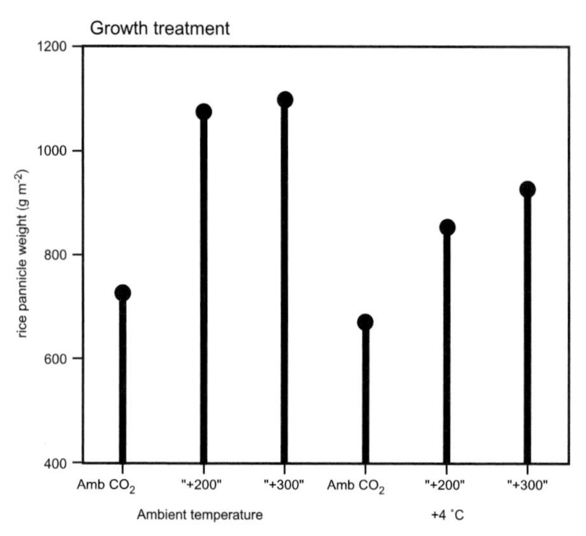

Fig. 5.15. Yield responses of rice (IR72) grown in open-top chambers to elevated temperatures and CO_2 at IRRI, The Philippines (Wassman et al. 1998)

5.3.2.2 *Rice*

Rice is a crucial commercial and subsistence crop for Southeast Asia (Fig. 5.14). It makes a major contribution to the economies of the poorest countries in the region. Traditional rice culture has been practiced in the same places for hundreds of years and is clearly a sustainable activity. The need for greater food production has resulted in widespread intensification of farming. Yields increased as a consequence, but over time could not be sustained. Longer periods of submergence and removal of rice straw resulted in nutrient depletion, reduced soil organic matter and altered soil properties. The emphasis in current practice on short-term gains, for instance through application of N-fertilizers, rather than sustaining long-term soil fertility and quality has frequently led to depletion of K, P, S and Zn contents of the soil. Improved inorganic and organic nutrient management and rotational cropping can help overcome some of these problems, but will also require advances in pest management (Reichardt et al. 1998). Increasing overall production of rice to meet growing demand will be a challenge as the already high level of inputs production is already approaching attainable yield levels (Gregory et al. 1999). Added to this, land available for conversion to rice production is becoming scarce in many countries and water for irrigation is becoming limited because of multiple and growing water demands.

With elevated CO_2 concentrations, rice produces more above- and below-ground biomass and gives higher yields at a given temperature (Fig. 5.15). As temperatures increase so does plant development, but as temperatures exceed 30 °C productivity begins to be curtailed as spikelet sterility increases with temperature (Nakagawa et al. 1998). As temperatures increases to 40 °C spikelet fertility decreases to near zero (Horie 1993). Comparison of the sensitivity of various rice models to temperature and CO_2 changes shows predicted yield increases of between 1.5 to 4.5% per °C increase (Mitchell 1996). Modelling suggests that doubling CO_2 in the atmosphere may increase yields by 10–40%, a result similar to that found in experiments (Peng et al. 1995).

A number of attempts have been made to assess the effect of future climate change on rice yields using GCM-based scenarios. One analysis suggests that yields may decreases in low-latitudes, while increasing at higher latitudes (Matthews et al. 1995). A review by Iglesias et al. (1996) concludes that in most areas rice production in Asia will not benefit from climate change. Impact modelling is sensitive to the scenarios used, and although general circulation models continue to be refined, uncertainties about changes in precipitation patterns remain very large. A further complication is that with time, the interaction of various components of the system is likely to change in ways that will influence yields.

An illustration of the complexity of environmental responses to global change is available from the study of Luo et al. (1995, 1998), who used a combined crop and disease model to assess the effects of changing temperature and enhanced UV-B radiation on rice leaf blast epidemics and rice yields. With higher temperatures epidemics are likely to increase in the cooler agricultural zones (Japan, North China), but increase in the warmer areas. Disease blasts were shown to be more severe and yield losses higher (typically 15–20%) as UV-B radiation increased, even without an increase in temperature.

Rice paddies are an important source of methane emissions and may contribute around 6–23% of the total anthropogenic sources of the greenhouse gas to the atmosphere (Neue and Sass 1994). Methane generated in aerobic layers of flooded rice soils comes mainly from soil organic matter, plant material and organic manure. In the absence of fertilization, emission rates tend to increase until the ripening stage. A variety of factors influence methane emissions paddy fields (Table 5.8). Fertilizer and water management practices, in particular, have a large impact on methane emissions, even in the absence of major differences in yield (Wassman et al. 1998). In Java it has been shown that water management regime and rice variety affect methane fluxes from tropical wetland rice (Husin et al. 1995). Continuously flooded fields emit more methane than those dried prior to harvest. The use of composted rice straw, rather than fresh staw, reduces seasonal emissions by 61% in Indonesia, the Philippines and Thailand (Buendia et al. 1998). Irrigated rice, because of its extent and relatively high rate of annual emissions, is therefore the most important overall source of CH_4 from rice ecosystems (Wassman et al. 1998). The total emission from just Indonesian wetland rice, for example, is estimated at around 4 Tg yr^{-1} (Husin et al. 1995).

Rice cultivation is not normally considered an important source of NO_x, since under anaerobic conditions it is reduced further to nitrogen gas. Intermittent inundation, however, for some rice varieties and fertilizer treatments, can result in substantially higher emissions of N_2O than continuous flooding. Flooding thus has opposite effects on CH_4 and N_2O emissions. Draining of paddies also results in reduction of CO_2 uptake from the atmosphere in the day and an increase in emissions at night mostly by removing the diffusion barrier caused by floodwater (Miyata et al. 2000).

Table 5.8. Some factors influencing methane emissions from irrigated rice (modified after Wassmann et al. 1998)

Component	Mechanism	Effect on CH_4 emissions
Soil	Texture with high porosity	++
	Indigineous methanogenic material	––
Water management	Long duration of flooding	+++
	Cont. flooding in early season	+++
	Strong leaching	–
Climate	High and even precipitation	++
	Low temperature	–
Organic manipulations	Removal of plant residues	–––
	High doses of manure	+++
	High organic inputs from floodwater	++
Nutrient and crop management	Sulfate fertilizers	–
	High N inputs	+
	Desne spacing	+
	Frequent soil disturbance	+
Rice cultivar	Strong root exudation	+++
	High diffusion resistance for methane transport	–––
	Short cropping period	–

5.3.3 Water Demand and Supply

Floods and droughts have long been part of the way of life of millions in Southeast Asia. Both affect food production and have considerable effect on national economies within the region. With rice as the staple food for the population throughout Asia, droughts hurt not only the food security of the region, but also exports from it.

This dependence is easily illustrated by the impact of ENSO dry events. Salafsky (1994), for example, showed how the 1991 dry event reduced annual income of a town in Kalimantan by 25–50% because of its impact on durian fruit harvest, coffee gardens and delayed rice crops. Even with advanced warning, and efforts made to coordinate government agencies beforehand, the impact of the 1997–1998 El Niño on the Philippines was still large (Jose 1999). The low rainfall greatly affected the Angat Dam the main source of potable water for Metro Manila, with typically delivery cuts lasting 4 hours each day. Across the country irrigation for agriculture was reduced to 38% of normal. In central Luzon, a key rice growing area, only 10% of the programmed area was irrigated. As a consequence of the drought, in the first six months of 1998 rice production fell 43%, corn 26%, and overall production 7%, the worst performance in two decades. The loss of agriculture production resulted in migration of farm labour to urban and industrial areas, where at the same time water shortages were leading to sanitation and hygiene problems.

Historically, most societies in Southeast Asia have developed strategies and institutions, sometimes quite sophisticated (Surarerks 1986), to cope with and exploit and fluctuations in water supply. For the most part these systems, however, were developed in a context where overall demand was low to moderate. The frequency of conflict and high political profile of water issues suggests that these systems and institutional arrangements are starting to struggle no that overall demand for water is much higher.

It is commonly asserted in Southeast Asia that deforestation of upland catchments is one of the primary causes of lowland flooding and droughts and high rates of downstream sedimentation. However, scientific evidence for the large claimed impacts is often weak and may even be in the opposite direction of conventional wisdom (Alford 1992, 1994). Many factors influence water use of forests at various scales (Calder 1996). Thus, re-forestation and upland orchard development may actually lead, especially when the young trees are growing quickly, to reduced and not increased streamflows. Water shortages are thus a function not only of drought, but also result from rapidly increasing demands for water for agriculture, industry and urban settlements (Forsyth 1996).

In future, water shortages will become increasingly difficult to manage as the already fierce competition for water between agriculture, industry, electricity generation and the domestic sector increases (Hirsch 1998) (Table 5.9). Improvements in efficiency and more fundamental changes in development strategies are needed.

There are opportunities for greatly improving efficiency of water use at farm-and irrigation system levels. It is estimated that each kilogram of irrigated rice needs about 5 000 l of water, although at the field level only about 25% of this amount is actually used (Bhuiyan et al. 1998). Wastages occur in land preparation and crop irrigation. High water depths help control weeds, but result in higher percolation losses. Problems are exacerbated by monsoon variability of rainfall. Most irrigation systems in Southeast Asia are designed for slow continuous water delivery, when they need to have the capacity to deliver on demand (World Bank 1994). Incentives for improved efficiency are often lacking, as water is provided free (i.e., subsidised by the state). The situation could be improved as new varieties with shorter growing seasons and direct wet and dry seeding are likely to result in less water demand (Bhuiyan et al. 1998).

A more fundamental development strategy, particularly for regions where water is seasonally scarce, but infrastructure is adequate, will be to reduce the dependence on water intensive agriculture and place more em-

Table 5.9. Water resources and sectoral withdrawal (WRI 1999)

Country	Annual internal renewable water sources (km^3)	Annual withdrawal (km^3)	% of water resources	Sectoral withdrawal (%)		
				Domestic	Industry	Agriculture
Cambodia	496.1	0.52	0	5	1	94
Indonesia	2530.0	16.59	1	13	11	76
Lao PDR	270.0	0.99	0	8	10	82
Malaysia	456.0	9.42	2	23	30	47
Myanmar	1082.0	3.96	0	7	3	90
Philippines	323.0	29.50	9	18	21	61
Singapore	0.6	0.19	32	45	51	4
Thailand				6	6	89

phasis on manufacturing industries. Trade policy because of its influence on the output structure of economies, has a role in determining future demands for freshwater resources (Rock 1998).

Large uncertainties about future precipitation trends (see section on climate) complicate assessments of the impacts of climate change on future water resources available for rain-fed and irrigated agriculture. Small changes in mean climatic conditions may be associated with large increases in extreme conditions, so affecting the frequencies of droughts and floods. Moreover, rainfall changes are amplified in run-off and streamflow.

Modelling studies are becoming important tools to explore possible consequences of increases and decreases in rainfall. Awadalla and Noor (1991), for example, model the responses of the Kelantan River in peninsular Malaysia to possible climate change. They found that a 15% increase in precipitation changed the flood peak from a 28- to a 35-year flood. The model is insensitive to temperature changes, with a 10% increase in evaporation having no detectable affect on river flow. Riebsame et al. (1995) preliminary analysis of the impacts of climate change on water resources suggested that there was not yet a need for structural adjustment or changes to current day-to-day operations.

However, given the large expected increases in demand from other sectors any reductions in water supply due to climate change would have important ramifications for agricultural development. In the short and medium term, greater use of seasonal and ENSO forecasting tools holds promise for improved control of water resources, preparation for floods and droughts in agriculture and disaster mitigation (Sivakumar 1999; Hammer et al. 2000). To what extent these skills may also help with long-term cumulative changes in climate and surprises remains to be seen.

5.3.3.1 *Water and Soil Pollution*

In most parts of Southeast Asia, apart from Singapore, resource-based industries account for the majority of water pollution emissions and have led to declining water quality in major rivers which discharge their waste to the sea. Recognition of these problems has led to new environmental laws and agencies to control and manage water quality.

In Malaysia, for example, concern over pollution of waterways from palm oil and rubber processing, led to the Environmental Quality Act of 1974 (Yaziz 1983). This did not prevent the Malaysian palm oil processing industry being responsible for 63% of Malaysia's total water pollution load nearly a decade later (Khalid and Braden 1993). From July 1978 through to 1984 the Department of Environment introduced progressively more restrictive standards on the palm oil effluent and required crude palm oil mills to apply for operating licenses every year (Vincent et al. 2000). License fees were adjusted according to quantities of wastes discharged. Eventually, as a result of the combination of mandatory standards and effluent charges substantial reductions in biological oxygen demand of wastes was achieved while at the same time as the industry continued to expand (Vincent at al. 2000).

Persistent organic pollutants resist natural decomposition processes and accumulate in living tissue. Not only do they persist, but they can be transported long distances as they often lack volatility. They include chemicals used in agriculture pest control, disease-vector control and industry. Severe restrictions on their use have been enforced in some countries of the region for more than 20 years; in other countries no restriction are in place. A serious problem is that there are substantial stockpiles of old chemicals (Watson et al. 1998). In many parts of Southeast Asia, pollutants derived from widespread application of pesticides, fungicides and herbicides in to crops and in plantations can be expected to affect surface and ground waters. Levels of organochlorine pesticide residues exceeding the Thai National Standard for surface water and unacceptable concentrations of heavy metals have been identified in the Mekong River near Vientianne, Laos PDR (Liwaruangrath et al. 1999). A variety of organochlorine pesticide residues are also present seasonally at higher than standard levels in the Ping River, one of the most important rivers in northern Thailand and the water source for the city of Chiang Mai (Rattanaphani 1999).

5.3.4 Upland Agriculture and Complex Agro-Ecosystems

A special feature of land use in the uplands of Southeast Asia is the prevalence of a wide range of intermediate mixed agro-forestry systems. These range from simultaneous tree-crop systems to agro-forests with only a short inter-cropping phase, together with various forms of shifting cultivation involving rotation of fallows and crops (Dove 1995; van Noordwijk et al. 1995; Tronkongsin 1987; Colfer et al. 1988; Tan-kim-yong 1997; Tomich et al. 1998). Studies of shifting cultivation and complex agro-ecosystems reveal a wide variety of practices having very different implications for biodiversity and sustainability (van Noordwijk et al. 1995; Bandy et al. 1993). Traditionally, many of these systems were practiced by subsistence small-holders. In recent times, however, resources in the uplands have become more seriously contested with the arrival of commercial agriculture, higher population densities and states interested in maintaining stronger control over forest resources (Chandraprasert 1997; Ganjanapan 1996; Rerkasem and Rerksasem 1995; Sikour and Troung 1998).

Land-use and management practices, especially on sloping lands, can have a profound effect on erosion and subsequent sedimentation in the high rainfall landscapes of Southeast Asia. Erosion rates of 40–80 t $ha^{-1} yr^{-1}$ have been recorded from vegetable plots in the Cameron Highlands, the main area for intensive commercial vegetation cultivation in Malaysia (Yusuf et al. 1998). The corresponding rates for tea plantations are ~0.5 t $ha^{-1} yr^{-1}$ and forests 10 t $ha^{-1} yr^{-1}$ on steeply sloping ground with high rainfall. Sedimentation rates in the highlands are high (2 000 $m^3 km^{-2} yr^{-1}$) and present major problems for downstream hydroelectric reservoirs. One method of control is re-forestation. The planting of trees and other productive soil conservation methods has major ecological benefits besides being economically beneficial. The overall impacts of deforestation on erosion and sedimentation, however, should always be evaluated against a variety of natural processes affecting erosion and sedimentation (Alford 1992; Schmidt-Vogt 1998; Forsyth 1999) and of course other activities, such as road construction.

Both traditional and recent experimental complex-agroecosystems provide plausible alternatives to standard agriculture/rural development models that depend on intensification through high inputs. For this reason they are likely to remain important for large numbers of marginalised, poor farmers with limited land. These systems are also attractive in connection with global change since they have the potential to maintain significant levels of biodiversity and an acceptable carbon balance, while at the same time providing opportunities for sustainable livelihoods. Shifting cultivation systems will have a near zero impact on net carbon balance, since the forest can return to its original biomass and soil status, in contrast to deforestation where large losses of carbon occur (Tinker et al. 1996).

Traditional and complex-agroecological systems have the added benefit in using an inherent diversity of products for subsistence use and income via traded commodities. This suggests they may be more resilient to global environmental change and shocks from the world economic system. The continuation of such systems, and the development of new ones, are under constant threat for a combination social, economic and institutional reasons associated with economic globalisation.

5.3.5 Tropical Forests and Plantations

5.3.5.1 *Political Economy of Forests*

The emphasis within the forestry sector in Southeast Asia has been on production of wood and wood products (including plywood and pulp and paper) rather than sustainable use of forest resources (Lohmann 1996; Thompson and Duggie 1996). Large-scale agri-business operations in Malaysia, Indonesia and Thailand have rapidly transformed vast areas of native forests, and sometimes traditional smallholder agricultural areas, into monoculture plantations of rubber, oil palm and tree plantations. Large-scale plantation development, although often driven to meet domestic and foreign market demand, may also be promoted to restore degraded lands as has happened in the case of *Imperata* grasslands. As a consequence, forest degradation and outright deforestation has been a prominent feature of development in Southeast Asia (Table 5.10). On the other hand, in some cases local communities have been able to recover previously degraded lands, even when under pressure of higher populations and production goals (e.g. Kummer et al. 1994; Batterbury and Forsyth 1999). Nevertheless, over the next several decades, the primary threat to terrestrial biodiversity in tropical Southeast Asia will remain habitat destruction from logging and conversion of forests for agriculture and infrastructure development.

Investment in Southeast Asia is politically constrained, as is the case in many other parts of the world.

Table 5.10. Changes in forest cover in Southeast Asia (WRI 1999)

Country	Forest cover in 1995 (× 1 000 ha)	% annual change in 1980–1995	Frontier forest cover as % of original forest (1996)	Present forest as % of original forest (1996)
Cambodia	9830	(2.71)	10.3	65.1
Indonesia	109791	(1.18)	28.5	64.6
Lao PDR	12435	(1.41)	2.1	30.0
Malaysia	15471	(2.83)	14.5	63.8
Myanmar	27151	(1.75)	0.0	40.6
Philippines	6766	(3.96)	0.0	6.0
Singapore	4	0.00	0.0	3.1
Thailand	11630	(3.58)	4.9	22.2
Vietnam	9117	(1.45)	1.9	17.2
Total for SE Asia	202195	(1.44)		

Membership of elite business networks is closely linked with political power and has often been a prerequisite for success in logging and wood product industries (Dauvergne 1997; Lohman 1996; Thompson and Duggie 1996; Vitug 1998). In Thailand, business has been quick to change strategies to maintain influence as the formal structures of power have evolved. Thus big firms sought privileged assistance from government, first through the general, later through the parliament and later still through technocrats (Phongpaichit and Baker 1997).

The distribution of benefits from the exploitation of natural resources (e.g. from forests, mangroves and coastal seas) is dependent on ownership and control of resources. Resource institutions are one of the ways of controlling the human use of the natural environment. They typically include structures of rights to resources, often known as property rights systems, rules under which those rights are exercised, and also carry additional responsibilities that guide resource use.

In Thailand, for example, the government has taken wide administrative control of property rights for forest lands. However, a lack of local input into land-use planning and competition among the government bureaucracy has meant that there has been little effective control of rural land-use (Vandergeest 1996). The ability to exclude the claims of others to a resource may be more important than the property rights themselves (Malayang 1991). Dove's (1995) study of cultivated rubber in Indonesia is a good illustration that relationships between the state and subsistence farmers can be much more important than purely economic or ecological factors in determining land-use. In other words, the key factor is who receives the benefits of use. In many places rainforest exploitation has benefitted the state and logging industry, whereas the environmental and social costs have been borne by local forest-dependent communities (Brookfield and Byron 1990; Dauvergne 1997; Pasong and Lebel 2000).

5.3.5.2 *Conservation of Biodiversity and Ecosystem Functions*

The main threat to biodiversity in Southeast Asia is habitat loss due to conversion of native forests to plantations, agriculture and infrastructure development, such as dams, roads and human settlements. Over the past two decades, average annual rates of loss of forest cover has varied been 1–4% among the countries in Southeast Asia with very little unfragmented or "frontier" forest remaining (Table 5.10). Degradation of remnant forests from non-sustainable logging practices is also an important process, hampering the regeneration of secondary vegetation, because of damaged soils, spread of weeds and changes to fire disturbance regimes. Hunting and poaching for the medicinal and wildlife trade also remains a serious threat to large mammals and birds in many areas. Round (1988), for example, assessed that 131 resident bird species in Thailand were at risk of extinction. Finally, encroachment by agriculture and spread of fires and weeds on forest margins, especially along roads, erode original habitats in fragments in many areas. The very high levels of endemic biodiversity in the tropical forests of Southeast Asia make these habitat losses and fragmentation of the original forest cover a global issue.

Changing biodiversity through hunting, harvesting and habitat losses can clearly have important repercussions for ecosystem functions, and ultimately the goods and services upon which humans depend (Chapin et al. 2000). Detailed knowledge about the effects of species reductions and removals on species interactions, resilience, and resistance to invasions in the tropical forest ecosystems in Southeast Asia, however, is still very limited. For example, relatively little is known about the impacts of local extinctions of top carnivore species on their herbivore prey and vegetation. It is, however, well recognized that logging, deforestation and other human activities, can lead to invasions of *Imperata* grasses and bamboo, which in turn alter fire disturbance regimes, making regeneration of many of the original species present difficult without intensive management.

Protected areas systems are growing in importance in the region. Some countries such as the Philippines, Cambodia and Laos still have only very limited areas officially protected, whereas in Indonesia and Thailand more than 10% of the country is now protected (Dinerstein and Wikramanayake 1993). In all countries, most of the protected areas are in harder to cultivate and develop uplands and mountainous regions. Lowland forests have already been largely cleared and are often poorly protected. These and other more restricted habitats, such as coastal wetlands, are among the most threatened.

Most protected areas in the region have serious management problems. In part, these management problems have resulted from the declaration of conservation areas by state agencies without taking into account existing settlements and property right arrangements of communities already living in those areas. The result has been chronic conflicts over hunting, collection of non-timber products and forest clearing for agriculture, seen as encroachment by the state, and as way of securing livelihoods by these communities. The interaction between states, local communities and civil society groups over the management of forested lands in protected areas and forest reserves has become more sophisticated and complex over time. One thing that emerges from these experiences in Southeast Asia is that some of the simplistic western models for management of protected areas that exclude consideration of people in planning and management are clearly not appropriate.

5.3.5.3 *Climate Change Impacts on Productivity and Composition*

In temperate and high latitude areas there is a growing consensus that climate change could have large impacts on tree growth and forests (Shugart et al. 1996). This understanding has come from the application of models, as well as from observations of the influence of climate variability on natural forests. Although in the tropics the expected magnitude of future warming is likely to be small, and precipitation changes are uncertain, there are still grounds for concern. In the first place, tropical trees are often highly specialised so even small changes in rainfall patterns and seasonality may have important consequences for plant reproduction and recruitment (Bawa and Markham 1995). Secondly, various land-use practices are likely to exacerbate the impacts of climate change on tropical forests. Thirdly, the environmental sensitivities and basic ecology of many tropical trees is poorly known.

Perhaps the most serious threat to tropical forests will come from drying trends (Condit et al. 1996). This would interact with existing process like fragmentation and harvesting, which all lead to an opening of the canopy. This, in turn, would increase the risks of fire and invasion by weeds. Condit et al. (1996), for example, reported that a 25-year-long drying trend in a tropical forest in Panama resulted in relative declines in species that specialise on moister slope habitats occurred, leading to the local extinction of 16 species. An analysis of the historical magnitude and frequency of droughts in different parts of Borneo suggests that, along with fire, these disturbance processes have important implications for forest composition and structure (Walsh 1996). For example, in areas with frequent drought, seasonal forest formations with many deciduous species are dominant, whereas in areas with episodic drought and fires, a range of forest formations is possible with age structures that are typically uneven.

Recent research suggests that, despite enhanced CO_2 forest uptake, the net effects on above-ground biomass, leaf area index, and below-ground biomass are likely to be much smaller than earlier thought (Mooney et al. 1999). In one of the few studies of tropical trees, Awang et al. (1998) found that elevated CO_2 increased short-term photosynthetic rates in seedlings in 7 out of 8 species of economically important tropical hardwoods, but that over a longer time the differences in rates disappeared.

In the absence of long-term experimental studies, models are the best tools available for examining longer-term impacts of climate and atmospheric change on tree growth and dynamics (Lebel and Jintana 1997). Booth et al. (1999) have analysed the sensitivity of two pulp-plantation species in Vietnam to climate change. For a range of plausible scenarios, the predicted effects on production in both northern and southern Vietnam are small.

5.3.5.4 *Disturbance, Fire and Community Dynamics*

Vegetation community dynamics in tropical forests are primarily influenced by land-use decisions and forest management practices, and secondarily by natural disturbances. Throughout Southeast Asia, even the remnant natural forests restricted to montane areas, are often used for products like fuelwood, timber, fodder and other non-timber products. Tree cutting in accessible primal forest areas not only leads to substantial declines in basal area, but also to changes in species composition (Smiet 1992). This is particularly clear in Java. Secondary forests recovering from logging or other human disturbance, now dominate tropical forest landscapes (Brown and Lugo 1990). Fire and the introduction of invasive species further alter patterns of forest community development as well as the characteristics of disturbance regimes themselves.

A striking example of these interactions is the replacement of moist, monsoonal forests by invasive grassy weeds, especially *Imperata cylindrica*. Garrity et al. (1997) estimate that the total area of *Imperata* grasslands in Asia now constitutes about 4% of the total land area, with the cover in the Philippines being 17%, that in Vietnam 9%, Laos PDR 4%, Indonesia 4%, Thailand 4%, Myanmar 3%, Cambodia 1% and Malaysia <1%. With increasing pressure to develop more land for oil palm, rubber and other agricultural crops, and to conserve biodiversity and forestry operations in remnant native forests, restoration and intensive use of *Imperata* grasslands is becoming a priority of a number of governments in the region. Developing successful policies and strategies for restoration, however, will require a strong research-based understanding of the many biophysical and socio-economic constraints and disincentives to conversion (Tomich et al. 1997; Foresta and Michon 1997; Potter 1997; Magcale-Macandog et al. 1998).

Fire in the humid and wet tropics of Southeast Asia biota has been occurring since the Pleistocene (Goldammer and Siebert 1989). Under normal current climatic conditions, natural fires from lightning are unlikely to gain access into a primary rain forest because of the high moisture conditions and high rates of decomposition. Over the past decades logging and the human use of fire for clearing tropical rain forests has made much more of the landscape susceptible to fire during periods of drought, especially during the dry phase of ENSO (Goldammer and Siebert 1990). In the large 1983 wildfires in Kalimantan heavily disturbed lowland forest was burned more than undisturbed forest (Table 5.11). Logged-over sites are also more likely to be burned at higher fire intensities than primary forest sites (Goldammer 1996).

Serious regional haze episodes and transboundary haze transport have drawn attention to biomass burn-

Table 5.11. Percentage areas of forest burning in the 1983 fires in the Mahakam Basin, Kalimantan, Indonesia (after Goldammer 1998)

Vegetation classification	Area (× 1 000 ha)	Burned (%)
Undisturbed lowland forest	410	11
Lightly disturbed lowland	1 096	58
Moderately disturbed lowland	984	84
Heavily disturbed lowland	727	88
Undisturbed swamp forest	181	17
Disturbed swamp forest	385	97

ing, most notably in ENSO drought years. Vegetation fires produce a range of gaseous and aerosol emissions that affect the atmosphere and climate. The fires arise from slash-and-burn agriculture, large-scale clearing of primary and secondary forests, regularly occurring fires in seasonally dry monsoon savanna forests, agricultural residue burning and use of wood fuel and arson (Tomich et al. 1998a; Goldammer 1996; Stolle and Tomich 1999).

The 1997/98 fires in Sumatra and Kalimantan, especially the smoldering fires in peat swamps, added millions of tons of emissions and smoke to the atmosphere. The emissions released from fires, while still small on a global scale compared to those emissions from industrial activities in developed nations, are nevertheless significant (Murdiyarso et al. 2000). Levine et al. (1999) estimated that these episodes produced 85–316 Mt of CO_2, 7–52 Mt of CO, 0.2–1.5 Mt of NO_x and 4–16 Mt of particulate matter, depending on assumptions about the amount of peat burnt. Many misconceptions exist about the source of the 1997/98 smoke and haze (Tomich et al. 1998a; Stolle and Tomich 1999). The first assumption was that the haze resulted from wildfires or from slash-and-burn agricultural activities of smallholders. Later it became apparent that many fires were deliberately lit to take advantage of the dry conditions either to clear land for agro-forestry development or to score points in land disputes.

Arising out of the debate on the causes of the haze generation is the notion that forestry development strategies in Indonesia are not sustainable (Brookfield and Byron 1990). The last Five-Year Development Plan (Pelita VI) designated 20–30 million ha of forest as Conversion Forests. The cumulative impacts of conversion are often compounded by poor site selection for large-scale agriculture and human settlement projects, the net result being vast areas of later abandoned and degraded land.

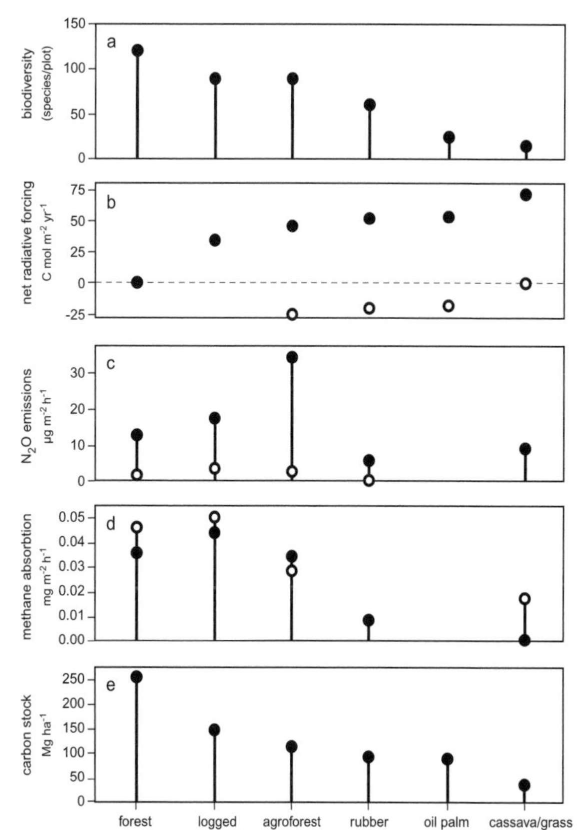

Fig. 5.16. Effects of current land-use and its change on **a** biodiversity, **b** net radiative forcing over 25 years (*black dots* denote conversion from forest, *circles* conversion from Imperata grassland), **c** N_2O emissions (*black dots* denote wet season, *circles* dry season), **d** methane absorption (*black dots* denote wet season, *circles* dry season), and **e** time-averaged carbon stock (Tomich et al. 1998c)

5.3.6 Sustainable Land Development Strategies

The dynamics of food, fibre, soil and biodiversity resources, and the terrestrial stocks and rates of sequestration of carbon, are closely linked through land-use and -cover change processes. Land-uses in the forest margins in Sumatra differ in significant ways with respect to profitability, carbon sequestration, biodiversity, agronomic sustainability and adoptability by smallholders (Tomich et al. 1998b). Along the land-use intensity gradient from unmanaged forests through agroforests, plantations and to croplands, biodiversity and time-averaged carbon stocks tend to decline together (Fig. 5.16). Nevertheless, for some traditional and modified land-use systems in Southeast Asia it is possible to conserve significant amounts of biodiversity and store substantial amounts of carbon in economically productive systems, though there are some trade-offs (Tomich et al. 1998b).

Overall soil carbon and biodiversity do not appear to be sensitive to land-use and -cover changes (Anas et al. 1999). Lighter fractions of soil organic matter (density <1.13 g cm^{-3}) are more sensitive to land-use changes than heavier fractions and decline with forest loss and land degradation (Hairiah et al. 1999). After fire, organic carbon content in soil goes up appreciably in the topsoil layer (0–5 cm), but with no effect on deeper layers of soil.

Table 5.12. Carbon loss and gain from land use change between 1986 and 1994 in Rantau Pandan, Sumatra, Indonesia (van Noordwijk et al. 1995; Murdiyarso and Wasrin 1995)

	Secondary forest		Plantation		Perennial crops		Annual crops		Imperata grasslands	
	Change in area (ha)	Change in carbon (t)	Change in area (ha)	Change in carbon (t)	Change in area (ha)	Change in carbon (t)	Change in area (ha)	Change in carbon (t)	Change in area (ha)	Change in carbon (t)
Logged-over	4073	−55	5616	−200	–	–	2359	−295	–	–
Secondary forest	–	–	17137	−145	7331	−195	27380	−240	–	–
Perennial crops	1852	195	101	50	–	–	17011	−45	432	−30
Annual crops	8786	240	3376	95	865	45	–	–	578	15
Imperata grasslands	–	–	–	80	3642	–	–	–	–	–

Table 5.13. Estimated total carbon balance of three sites in Sumatra, Indonesia over the period 1986–1994 (van Noordwijk et al. 1995)

Site	Area (ha)	Total C-loss (Mio. t)	Total C-gain (Mio. t)	Total C-change (Mio. t)	Rate C-change (t C ha^{-1} yr^{-1})
Rantau Pandan	63819	2.30	3.88	+1.58	3.1
Muaratebo	148571	11.95	3.93	−8.02	−6.8
North Lampung	141332	13.32	3.12	−10.19	−9.0

The effects of land use on microbial and soil macrofauna groups are generally small and counter-intuitive. *Imperata* grasslands, often considered prime examples of degraded land-cover, had the highest mycoorhizal spore diversity and abundance and highest densities of earthworms in comparison with forests, agroforests, reforested areas and cassava cultivation lands (Anas et al. 1999). By taking into account the average amounts of carbon in above-ground biomass (and sometimes also below ground), for different land-use types, and combining this information with observations of how land-use and -cover has changed (Table 5.12), it is possible to make estimates of carbon stock changes at sub-global scales (Table 5.13). The net emissions resulting from land-use changes over the period 1850 to 1980 in tropical and non-tropical regions were approximately the same, or about 50 Pg C each (Houghton and Hackler 1995). Emissions from temperate zones dominate global budgets prior to 1940, whereas the tropics have been dominant since. During the 1980s tropical Asia contributed almost half of the global emissions from land-use changes (Table 5.14).

Despite the growing amount of data on carbon stocks, and rates of sequestration and emission, there is still a substantial level of uncertainty in extrapolating results to estimate the net fluxes of CO_2 and other greenhouse gases resulting from land-use changes at regional, national and sub-national scales (e.g. Boonpragob 1998). On-going efforts to improve greenhouse gas inventory databases in the Asia-Pacific region (Magcale-Macandog 2000) should lead to improved budgets for the Southeast Asia as a whole.

Apart from impacts on biogeochemical cycles and in-situ biodiversity, land-use and land-cover changes also contribute to changes in coastal and marine systems.

Table 5.14. Net flux of carbon to the atmosphere from changes in land use in the tropics and in temperate and boreal zones (Houghton and Hackler 1995)

Region	Net flux (PgC yr^{-1})
Tropical America	0.6 ±0.3
Tropical Asia	0.7 ±0.3
Tropical Africa	0.3 ±0.2
Subtotal tropics	1.6 ±0.5
Temperate and boreal zones	0.0 ±0.5
Global total	1.6 ±0.7

5.4 Coastal and Marine Systems

5.4.1 Effects of Land-Based Activities on Coastal Biogeochemistry

In Southeast Asia, land-based activities, such as infrastructural development in coastal urban and industrial areas and the conversion of forested lands in the rural hinterland, are having a profound impact on the export of sediments and nutrients to many coastal zone ecosystems. These inputs to the coastal zone are modified in upland areas by the construction of hydro-electric and irrigation dams, and in lowland areas by the conversion of coastal wetlands and mangroves to paddy rice and aquaculture.

Sewerage effluent from urban, peri-urban and tourist centres can make a substantial contribution to pollution loads of coastal areas, for example, in Manila Bay and the upper Gulf of Thailand (Chongprasith and Srineth 1998). It is estimated that biological oxygen demand from domestic waste inputs of sixteen coastal provinces around the Gulf of Thailand was approximately 140 000 t yr^{-1} in 1994 and was increasing at a rate of about 7.5% per decade (Chongprasith and Srineth

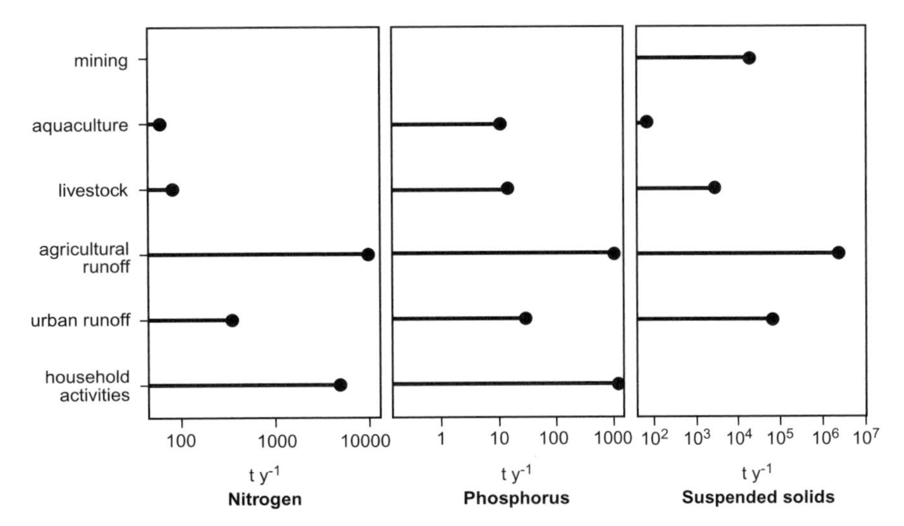

Fig. 5.17. Residuals (t yr^{-1}) entering coastal waters of the Lingayen Gulf, Philippines from various economic activities (Talaue-McManus et al. 1999; Gomez et al. 1999)

1998). Shrimp farms contribute approximately half this amount per year (FAO and NACA 1995). Considering human sewerage only, it is estimated that the seven countries around the South China Sea generate ~6 Mt yr^{-1} of organic matter, of which only ~11% is removed by effluent treatment (Talaue-McManus et al. 1999)

Average concentrations of nitrate in the Gulf of Thailand range from 5.2 µM in the upper gulf to less than 2.0 µM in the lower Gulf, and below 0.1 µM on the east coast of Peninsular Malaysia. Trace metals, petroleum hydrocarbons and most other pollutants tend to have higher concentrations in in-shore areas near major industrial centres, ports and cities (Chongprasith and Srineth 1998).

Marine organisms in the Gulf of Thailand form approximately 10 Mt yr^{-1} of calcite each year. Much of this carbon would be destined for subsequent release as CO_2 gas were it not for dissolution processes in the calcareous, magnesium-enriched sediments which help create a sink equivalent to about 4% of current fossil fuel emissions of Thailand (Snidvongs 1993).

The Lingayen Gulf on the northwest coast of Luzon, the Philippines, is fairly typical of some of the coastal zones areas in tropical Southeast Asia (Talaue-McManus 1999). Industrial development is still limited, with most of the growing population involved in agriculture. Agriculture and household activities make a significant contribution to loading of nutrients and suspended solids transported to the coastal zone (Fig. 5.17). Household activities account for approximately 32% of N and 52% of P that reach coastal waters (Talaue-McManus et al. 1999). Non-point agricultural runoff contributes 64% of the N and 45% of the P and 97% of suspended solids. Low overall annual changes in dissolved inorganic P flux and minimal differences between estimated production and respiration indicate that the gulf as a whole is metabolically approximately in balance or slightly heterotrophic (Talaue-McManus et al. 1999; Gomez et al. 1999). Dissolved organic P appears to be to be increasing at the rapid rate of 0.09 mol $m^{-2} yr^{-1}$. Based on different development scenarios, and considering the biogeochemical cycles involved, it appears that further coastal nutrient loading could lead to changes in recycling efficiency and greater metabolic imbalances. The already high contribution of human activities to the delivery of suspended solids to the coastal zone suggests that further uncontrolled economic development could have profound impacts.

Similar exercises using box models for biogeochemical budgets (Gordon et al. 1995) have been completed for sites in Malaysia, Vietnam and Thailand (Eong et al. 1999; Wattayakorn et al. 1999; Tri et al. 1999). These show that autotrophy-heterotrophy and nitrifying-denitrifying balances can change seasonally and from year to year. Most sites appear to be slowly accumulating N and P, with considerable seasonal and annual variability (Fig. 5.18).

Efforts are underway to upscale small-scale studies through developing a typology of coastal sites (Talaue-McManus 2000b) to assess cumulative regional impacts of land-based activities. At the same time coarse-resolution models of riverine transport to coastal zones for the whole of Southeast Asia (Richey and Snidvongs 1998) are being developed and will be validated against small-scale studies.

5.4.2 Mangroves and Shrimp Aquaculture

Mangroves and coastal wetlands are under pressure for a variety of reasons, including expansion of urban settlements, exploitation for fuel-wood, agricultural development and construction of fish and shrimp ponds. Between 1980 and 1994 most countries in Southeast Asia lost around half of their mangrove cover. The multiple-uses of, and ecosystem services provided by, mangroves makes economic valuation of their importance challenging. If for no other reason they are important for the large contribution they make to human well-being, as in the case of Bintuni Bay, Irian Jaya (Ruitenbeek 1994).

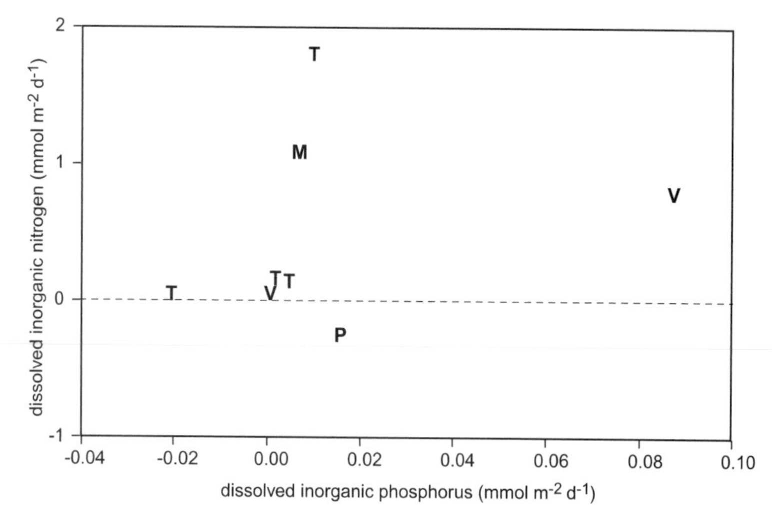

Fig. 5.18. Estimated changes in dissolved inorganic nitrogen and phosphorus for four coastal systems in Southeast Asia in different seasons and years. *V* denotes Red River Delta, Vietnam, *T* Ban Don Bay, Thailand, *L* Lingayen Gulf, Philippines and *M* Merbok Mangrove, Malaysia (Talaue-McManus et al. 1999; Tri et al. 1999; Wattayakorn et al. 1999)

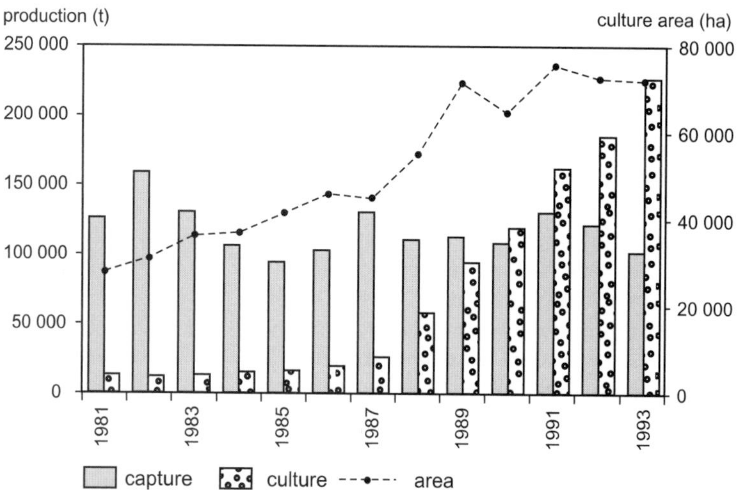

Fig. 5.19. Growth of the shrimp aquaculture industry by capture and culture (t) in Thailand (Menasveta 1998)

Here it has been shown that managed, conservative clearing, involving selective cutting of up to 25% of harvestable mangrove, is economically rewarding, while preserving an important environmental asset. In Vietnam the conversion of mangroves to shrimp farms, rice paddies and other uses increases the vulnerability of coastal communities to future sea-level rise and changes in the frequency or intensity of storms (Adger 1999; Tri et al. 1998). It has been concluded that rehabilitation of degraded mangrove forest could yield considerable economic benefits, particularly in mitigating storm damage (Tri et al. 1998).

The rapid expansion of shrimp farming, and secondary conversion of paddy rice farms, has placed mangrove swamps under pressure (Fig. 5.19). Over the past two decades, intensification of production systems, has greatly increased yields per unit area in Thailand. In other countries in Southeast Asia overall production yields have generally been much lower than in Thailand as many farms still use non-intensive systems. In 1994 there was approximately 300 000 ha of shrimp aqua-culture in Indonesia and 225 000 in Vietnam respectively compared to only 80 000 ha in Thailand. Apart from habitat losses, salinisation of freshwater, increasing nutrient loading and chemical pollution are having an effect on mangroves (Briggs and Funge-Smith 1994; Dierberg and Kiattisimkul 1996; Primavera et al. 1993). It is estimated that in 1992 alone, 40 000 ha of intensive shrimp ponds in Thailand produced N-waste equivalent to 3.1–3.6 million people and P equivalent to 4.6–7.3 mil-lion people (Briggs and Funge-Smith 1994). At the same time, 16.2 Mt yr^{-1} of sediments were deposited and biological oxygen demand reached 35 000 t yr^{-1}. The overall contribution to the waste loading in the Thai coastal environment was large.

As a short-term investment the shrimp industry is highly profitable. Shrimp products are now rank third in Thailand exports, with annual earnings exceeding US $1.6 billion. More than 200 000 people are employed in rural areas. Improved production technologies and water recycling systems have the potential to reduce pollution (Lin et al. 1995; Dierberg and Kitattismkul 1996; Kongkeo 1997). In the meantime, abandonment of ponds after a few years because of disease and declining productivity is common, as are periodic collapses in the industry (Lin 1989; Dierberg and Kitattismkul 1996).

5.4.3 Coastal and Marine Ecosystems

Coral reefs face a combination of local and global environmental threats, including sea-level rise, extreme temperatures, human damage, changing salinity and chemical pollution and storm damage (Pittock 1999). In addition, elevated CO_2 levels in the atmosphere may result in increasing acidification of the ocean surface, in turn may reduce calcification rates of corals.

Almost 60% of the world's reefs are threatened by human activities (Bryant et al. 1998). In Southeast Asia over 80% are at risk, primarily from coastal development and fishing. Since ~30% of the world's coral reefs occurs in the coastal zone of Southeast Asia, this is a source of considerable concern. Almost all the reefs of the Philippines and 83% in Indonesia are at risk. The key environmental threats include over-fishing, destructive fishing, sedimentation and pollution from coastal development (Bryant et al. 1998). It is likely that human activities will make coral reefs more prone to severe damage from the increasing frequencies of natural disturbances and bleaching that are expected to occur with continued global warming.

The annual marine fish catch in Southeast Asia is ~8 Mt and represents ~10% of the world total (ASEAN 1997). The Gulf of Thailand and continental shelf off Vietnam are particularly productive areas. Fish stocks are decreasing in many areas through over-fishing. This is illustrated by the decline that has been experienced in the Lingayen Gulf, where the production of 13 443 Mt in 1995 was just over half the peak extraction in 1987, despite an increase in the amount of fishing undertaken (Padilla and Morales 1997). Aquaculture has increased as open-sea fishing catches have declined; in 1993 aquaculture accounted for 60% of total Lingayen Gulf fish harvest (Gomez et al. 1999). Concern is growing about the impact of harvesting on fisheries resources and on biodiversity (Malakoff 1997; Watling and Norse 1998). The situation is likely to be exacerbated as global change continues.

The threats to marine biodiversity come from both land and sea activities of humans. In Southeast Asia, changes in coastal land-use are particular important. The loss of important nursery areas for many marine organisms in mangroves and coastal wetlands is a cause for considerable concern. Offshore transport of nutrients, sediments, and pollution is impacting not only coral reefs and mangrove swamps, but also sea-grass beds and other high diversity, limited-area ecosystems. In many continental shelf areas, as well as in the open seas, trawling and other forms of intense fishing are severely depleting stocks of target species, with unknown effects on non-targeted species.

The region has some of busiest oil and cargo shipping traffic in the world. Oil slicks and tar residues from spills and offshore oil and gas production are frequent, particularly in the Straits of Malacca and Johor and in the South China Sea (ASEAN 1997). The threat to marine ecosystems is serious and will unquestionably increase as shipping increases in the region.

5.5 Climate Variability and Change

5.5.1 Current Variability

The climate of the Southeast Asia is dominated by two monsoons. Current understanding of the Asian monsoon is based largely on observations collected in India and East Asia (e.g. Chang and Krishnamurti 1987). Statistical analysis of spatial correlations in rainfall indicates that areas in Southeast Asia are both in phase (Northern Thailand, Indonesia, Brunei, Borneo) and out of phase (surrounding the South China Seas) with Indian Monsoon rainfall (Kripalani and Kulkarni 1997). Rainfall patterns can be classified into three regimes across Southeast Asia. The high mountains of the Asian continental region contribute to an enhancement of the summer monsoon. The equatorial region consisting of Indonesia and Malaysia comes under the influence of the North Australian-Indonesian monsoon regime, and movements of the inter-tropical convergence zone bringing complex patterns of rainfall in most seasons. The Philippines region, in contrast, is much less influenced by continental land surface-ocean contrasts (Kripalani and Kulkarni 1998).

Overlying the seasonal Monsoon pattern, the El Niño-Southern Oscillation (ENSO) is the primary source of inter-annual variability in climate for Southeast Asia. The average effects on rainfall anomalies vary substantially between countries (Table 5.15).

5.5.2 History

Palaeoclimate records indicate that monsoon and ENSO variability have long histories in Southeast Asia. High-resolution sequences from China reveal that the East Asian monsoon may have commenced over 7 million years ago (An 1999). Millennial-scale variability is

Table 5.15. Impact of ENSO phases on mean standardized rainfall in El Niño and La Niña years in the Southeast Asia (Modified from Kripalani and Kulkarni 1997)

Region	Mean standardized rainfall	
	El Niño years	La Nina years
Thailand (monsoon)	−0.41	0.71
Malaysia	−0.12	0.26
Indonesia	−0.71	0.36

present in the East Asian monsoon with alternation between dominance by the dry-cold winter and warm-humid summer monsoons. Recent evidence from isotopic records in corals in the equatorial Pacific demonstrates the presence of ENSO-type variability over the past several centuries (Gagan et al. 2000). Other evidence from corals suggests that the tropical region has been sensitive to global forcing. Sea surface temperatures were depressed 4–6 °C during the Younger Dryas climatic event and rose episodically over the past 4 000 years (Gagan et al. 2000). Over the period of instrumental record, mean surface temperatures across the region have increased 0.3–0.8 °C, but little evidence exists for long-term trends in rainfall or the behaviour of cyclones.

5.5.3 The Future

Various global climate models have been used to generate regional scenarios of future climate changes. However, given that many of the models are not very successful at replicating present-day Asian climates, great care must be taken in interpreting scenarios of future change. Lal and Harasawa (2000a) compared seven atmosphere-ocean general circulation models with current climatologies for 1961–1990 using historical forcing for the whole of Asia and various sub-regions, including Southeast Asia. Four of the models, HadCM2, ECHAM4, CSIRO and CCSR preformed reasonably well in representing seasonally area averaged surface air temperatures and precipitation over Asia. Three others, the NCAR, CCCma and GFDL models, which generally have less vertical resolution in the atmosphere, performed poorly. For the Southeast Asia, the HadCM2, ECHAM4 and CSIRO models represented surface air temperatures reasonably. All three underestimated precipitation for most of the region (Lal and Harasawa 2000).

On the basis of the four best performing of the seven models, Lal and Harasawa (2001) constructed seasonal mean climate change scenarios for Asia (Table 5.16). They are similar to earlier simulations undertaken for Asian Development Bank (Asian Development Bank 1994). In general, the expected changes in mean temperature are small compared to other regions in Asia and other parts of the world, with the largest changes being for inland continental areas.

Future changes in precipitation are much more uncertain. Though different models give very different results, most suggest that more intense rainfall events are likely to occur. Owing to the archipelagic nature of most of the area, there is still nowhere near enough spatial detail in GCMs to capture important ocean and topographic influences on climate that are so crucial to understanding impacts on, for example, agricultural systems. For this reason attempts have been made to increase the resolution within the region using variable-mesh and nested models. The Australian CSIRO Division of Atmospheric Research Limited Area Model (DARLAM) reproduces the historical climate well (MacGregor et al. 1998) and predicts future small increases in precipitation over land, with larger increases over ocean areas and Sulawesi, Indonesia (Fig. 5.20). A mean temperature increase of about 1.5 °C is predicted for most of the region by the late twenty-first century.

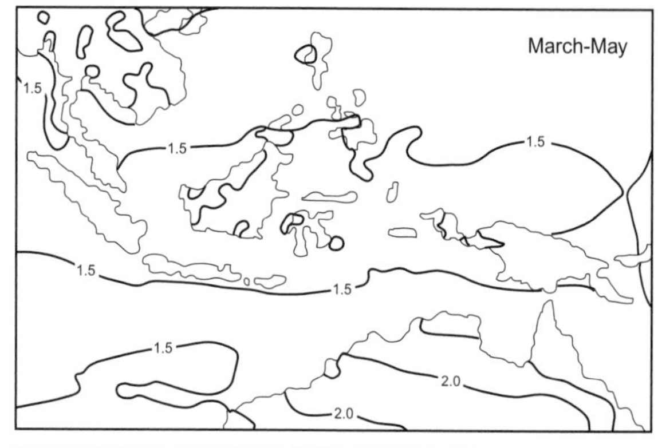

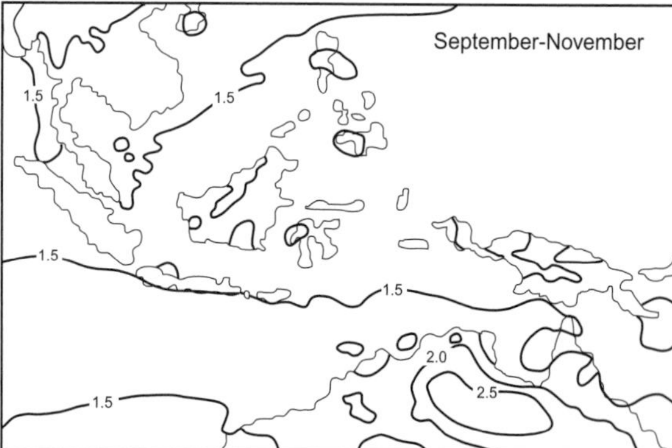

Fig. 5.20. Modelled March-May and September-November surface air temperature changes with doubled CO_2 based on the DARLAM model (McGregor et al. 1998)

Table 5.16. Mean seasonal climate change scenarios for Southeast Asia for the decade 2080s from four ocean-atmosphere coupled models with and without aerosols. Figures in parenthesis show the standard deviation between model projections (after Lal and Harasawa 2000b)

Climate variable	Season	GHG	GHG and aerosols
Seasonal mean temperature (°C)	Winter	3.2 (0.6)	2.5 (0.6)
	Summer	2.8 (0.6)	2.3 (0.6)
Diurnal temperature range (°C)	Winter	0.1 (1.0)	0.2 (1.1)
	Summer	–0.7 (3.6)	–1.1 (2.1)
Seasonal mean precipitation (%)	Winter	7.3 (3.4)	5.9 (2.5)
	Summer	6.1 (4.5)	4.9 (6.5)

The frequency, intensity and location of tropical cyclones are of great importance to the countries in Southeast Asia. Tropical cyclones are a major source of rainfall, damage from strong winds, floods and coastal storm surges. The behaviour of tropical cyclones is influenced by ENSO events. In general, tropical cyclones occur more frequently in the Australian-Asian region during La Niña years, but further east in the Pacific during El Niño years (Pittock 1995).

It is still fairly uncertain how tropical cyclones will be affected by climate warming. Early studies using climate models with relatively poor spatial resolution and representation of convective processes, when taken together with physically-based expectations and empirical observations, suggest at most a modest increase in heavy rainfall invents and possibly intensities of tropical cyclones (Fowler and Hennessy 1995; Walsh and Pittock 1998). More recently, Walsh and Ryan (2000) embedded tropical cyclone-like vortices into a regional climate model of Australia and lower part of Southeast Asia. They found weak evidence that storm intensities may increase under enhanced greenhouse conditions. In a related study, Walsh and Katzfey (2000) found stronger evidence for poleward movement of tropical cyclone-like vortices under a warmer climate.

These efforts notwithstanding, the key question of what the consequences of climate change will be for Southeast Asia remains largely unanswered. Perhaps all that it is appropriate to conclude at present is that as CO_2 levels are perturbed more and more surprises should be expected. The last decade of scientific research has underlined that climate is not just an external force affecting ecosystems, but that ecosystems themselves feedback and help determine weather and climate. A recent modelling study suggests that the onset of the Asian Monsoon, its evolution and intensity is sensitive to land surface characteristics at synoptic and regional scales (Xue et al. 1999).

Important though the effects of changing climate will be on Southeast Asia, the processes of urbanisation, industrialisation and globalisation are likely to have even greater consequences. Along with population growth and migration these processes include most of the major driving forces of land-use and -cover change, biodiversity loss, the exploitation of regional ecosystem goods and services, and other components of global change.

5.6 Regional Development

5.6.1 Human Development and Socio-Political Transformation

Most commonly used indicators of human development show positive trends over the past decades in most Southeast Asian countries. Thus, for example, infant mortality rates have fallen, life expectancies increased, and female literacy improved to fairly high levels in most countries (Table 5.17). Notwithstanding, wide variation in health, education and income levels continue to persist among the ten countries of Southeast Asia, and between rural and urban areas.

Table 5.17. Selected indices of human development in Southeast Asia and a selection of countries in the Asia and the Pacific for comparison (UNDP 1998). Per capita gross domestic product has been adjusted for purchasing power. The Human Development Index is a weighted sum of longevity, education measures and adjusted income

	Life expectancy at birth	Adult literacy rate (%) 1997	Enrolment ratio	Per capita GDP PPP$ 1997	Human development index 1997
Singapore	77.1	91.4	73	28460	0.888
Brunei	75.5	90.1	72	24350	0.878
Malaysia	72.0	85.7	65	8140	0.768
Thailand	68.8	94.7	59	6690	0.753
Philippines	68.3	94.6	82	3520	0.740
Indonesia	65.1	85.0	64	3490	0.681
Vietnam	67.4	91.9	62	1630	0.664
Myanmar	60.1	83.6	55	1199	0.580
Cambodia	53.4	66.0	61	1290	0.514
Laos	58.6	58.6	55	1300	0.491
Japan	80.0	99.0	85	24070	0.924
Australia	78.2	99.0	100	20210	0.922
Korea, Rep. of	72.4	97.2	90	13590	0.852
China	69.8	82.9	69	3130	0.701
India	62.6	53.5	55	1670	0.545

5.6.2 Common Pool Sinks

Many environmental problems, which in the past because of the small scale of cumulative impact, were either ignored or the concern of individual national governments only, are now becoming international issues because of the threat of transboundary impacts. International institutions to control the over-exploitation of resources and degradation of environments by transboundary pollution and waste are becoming more commonplace in Southeast Asia.

Understanding the global behaviour of carbon sinks is a challenging task, but crucial to answering questions about the role of the Southeast Asia region. Dynamic Global Vegetation Models (DGVMs) demonstrate the global-level importance of terrestrial ecosystem responses to elevated CO_2 and temperature. Recent comparison of six DGVMs suggests that net ecosystem productivity will continue to increase for several more decades, but that sometime within this century will level off or decline as the terrestrial carbon sink approach its limits (Cramer et al. 2001). If and when such limits are reached, atmospheric CO_2 concentrations will then begin to increase more rapidly than they are now. Land-use management practices in the tropical forests of Southeast Asia are of significance to the global carbon cycle. On the one hand, there is substantial capacity to sequester additional carbon. On the other hand, the history of the politics and economics of forest resource management in the region suggests that fundamental changes to governance will be needed before this can be effected.

It is still unclear what contribution the coastal zones and major ocean areas of Southeast Asia, such as the Andaman Sea, Gulf of Thailand and South China Sea, make to balancing global biogeochemical budgets. Is their behaviour as sources or sinks likely to change under the cumulative influence of land-derived inputs via the atmosphere or rivers?

Per capita emissions from industrial, domestic and transport sectors of economies in Southeast Asia are relatively small compared to the developed world and net emissions still small relative to the region's large neighbours, China and India. However, growth rates in emissions are high and when coupled with already important contributions from land-use and land-cover change processes it is clear that the region has a significant role in the carbon cycle.

5.6.3 Shared Resources and Human Security

Notwithstanding the substantial economic development that has occurred in recent decades, the people of Southeast Asia remain highly dependent on agriculture and coastal marine resources for their livelihoods and well being. The high diversity and productivity of the tropical ecosystems have traditionally provided a reliable resource base for society. Many modern rural and coastal development programmes are promoting greater economic specialisation, which could threaten the security of such resource-dependent systems. Moreover, long-term development plans for coastal and rural areas are often inconsistent with stated commitments to protection of biodiversity and sustainable development.

Trade can make an important contribution to food security, reducing vulnerability among countries and communities within countries. A number of global trade-production models have been used to explore the possible consequences of global change on agricultural production. Most predict net increases in high and mid-latitudes and decreases in the tropics (Parry et al. 1999). Net economic impacts of such changes can be reduced through trade, but negative impacts will remain high for the poorest countries (Takahashi et al. 1999; Parry et al. 1999).

Southeast Asia has some important characteristics that make development patterns different from recent western experience. Firstly, there are many indigenous land-use systems that are intermediate between crop agriculture and forestry. They involve long-term rotations of secondary forests and crops. Secondly, in the cities there is often a vibrant and large informal sector. Both these alternative systems have important implications for societies vulnerability to global change, but because they do not fit neatly into the standard categories of western development agencies, or have a large influence on GDP, they often tend to be ignored.

Another key transformation underway is the gradual opening up of what was originally very closed public policy process in many countries. This is reflected in the rise of civil society actors, as much as in the presence of formal democratic institutions. Continuing censorship of the media and military interference in some countries illustrates that the process of decentralization is far from complete. These transformations are likely to have important consequence for environmental governance (Smith 1994; Pasong and Lebel 2000), the weaknesses of which are underlined by the failure to manage the environmental consequences of rapid industrialisation.

Per capita levels of resource usage and emissions of major pollutants relevant to global change are still generally lower in Southeast Asia than in western developed countries. The ecological footprints are still comparatively small, but because of high population densities may still be exceeding national capacities (Table 5.18). These figures, however, oversimplify the situation. An important driver of degradation of resources and growth in emissions within Southeast Asia are in part the result of activities of consumers outside it consequent upon increasing trends in globalisation. Untangling interactions among various production-consumption systems

Table 5.18. Ecological footprints of nations (Wackernagel et al. 1997). All expressed in world average productivity 1993 data

Country	Population (1997)	Ecological footprint (ha cap^{-1})	Ecological capacity (ha cap^{-1})	Ecological deficit (ha cap^{-1})
Indonesia	203631000	1.6	0.9	-0.7
Malaysia	21018000	2.7	1.7	-1.0
Philippines	70375000	2.2	0.7	-1.5
Singapore	2899000	5.3	0.5	-4.8
Thailand	60046000	2.8	1.3	-1.5
Japan	125672000	6.3	1.7	-4.6
United States	26818900	8.4	6.2	-2.1
Australia	18550000	8.1	9.7	1.6

for food and manufactured products will be difficult, but will help better understand the implications of globalisation.

5.6.4 Regional and Global Cooperation and the Environment

In the period since the 1987 Brundtland Report and 1992 Rio Declaration, regional and international environmental organizations have grown in number and ambition. At the same time, nations have been working on international trade and investment agreements, which could have important repercussions for the environment. The countries of Southeast Asia countries are increasingly being co-opted into these agreements. That the nations in the region may be somewhat reluctant to participate, for example, in the reduction of atmospheric emissions is understandable given that they see the problem as not of their making (Smith 1994; Redclift and Sage 1998). Sustainability issues related to food security, human health and poverty, on the other hand, are central concerns of the region.

ASEAN provides a forum for discussing environmental and development issues of central concern to the region. So far, however, various plans and declarations have produced little in the way of demonstrable environmental protection, particularly in respect of the recurrent problem of transboundary haze transport from vegetation fires (e.g. ASEAN 1995). Strong barriers remain to prevent better regional cooperation on environmental matters. Not least of these are weak environmental protection institutions within individual countries, a problem exacerbated by the financial crisis of 1997/98.

The barriers are particularly strong in the case of marine resources problems (e.g. Talaue-McManus 2000a). Development of effective controls has been hampered by sovereignty and territorial disputes in the South China Seas. Conflicts over marine resources are frequent and have been accompanied by military action (Valencia 1990; Rosenberg 1999). UNEP's East Asian Seas Programme, although it has developed action plans, has no legal framework and fails to address fisheries management issues (Vanderzwaag and Johnston 1998). As elsewhere, international issues overlap with political opposition to environmental commitments. Lack of environmental law enforcement exacerbates the problem.

Illegal cross-border logging activities and large-scale development of hydroelectric power in the Mekong River Basin represent another set of problems requiring regional co-operation on environmental matters.

5.7 Conclusions

The global change-development nexus is particularly strong for Southeast Asia. Environmental changes within and outside the region are significantly affecting development patterns. Rapid, export-oriented economic growth and development are accelerating local change. Cumulative impacts are leading to regional change having global implications. The drivers of regional environmental change include key biophysical and socio-economic processes that originate to a large extend from outside the region. The strong interactions between human development processes and the environment are complex and involve multiple uncertainties. Notwithstanding, the forcing exerted by globalisation is clearly discernible.

Important findings from the Southeast Asia synthesis may be summarised as follows:

- Climate extremes, particularly those associated with ENSO, are likely to increase with global warming and affect the region accordingly. Floods and droughts will continue to have considerable effect on national economies within the region. Tropical cyclones are a major source of rainfall, damage from strong winds, floods and coastal storm surges. It is still uncertain how the frequency, intensity and location of tropical cyclones will be affected by climate warming.
- Most countries of the region are highly dependent on fossil fuels for electricity generation, but unlike China and India, are more reliant on oil and natural gas. Average annual growth in energy consumption has been around 10%. In rural areas fuel wood is still a very important source of energy.

- During most of the twentieth century the contribution from Southeast Asia to global greenhouse gas concentrations in the atmosphere was been small. At present, per capita CO_2 emissions from industrial, domestic and transport sectors of economies in Southeast Asia (with the exception of Singapore) are relatively small compared to the developed world and to the net emissions of the adjacent regions of South and East Asia. However, the prospect of rapid growth in future emissions is real.
- Transboundary atmospheric pollution from fossil fuel burning is significant in the whole of Asia. The key factor affecting deposition in Southeastern Asia in future is the extent to which emission controls and energy efficiency improvements are made in China.
- Land-use and land-cover changes in Southeast Asia have been and continue to be significant and have ramifications for the global carbon cycle. Despite the growing amount of data on carbon stocks, and rates of sequestration and emission, considerable uncertainties remain associated with extrapolating of net regional fluxes of CO_2 and other greenhouse gases resulting from land-use changes to the global scale.
- The environmental consequences of rapid, growth of air and water pollution in cities and contamination of soils and water in rural areas, are significant. The cumulative effects of these developments have resulted in formerly local problems being transformed into regional transboundary problems.
- The environmental consequences of the large-scale exploitation of terrestrial resources for economic development are significant. Land-use and -cover change is accelerated and remains an on-going feature of development in the coastal plains and rural hinterlands. Conversion of natural forests to plantation forestry and agricultural use is having significant impacts on soil resources, biogeochemical cycles and biodiversity.
- A special feature of land use in the uplands of Southeast Asia is the prevalence of a wide range of intermediate mixed agro-forestry systems. These systems have the potential to maintain significant levels of biodiversity and an acceptable carbon balance, while at the same time providing opportunities for sustainable livelihoods. Both traditional and complex agroecosystems provide plausible alternatives to standard agriculture/rural development models that depend on intensification through high inputs.
- Vegetation community dynamics in tropical forests are influenced primarily by land-use decisions and forest management practices, and secondarily by natural disturbances. Fire and the introduction of invasive species, especially *Imperata cylindrica* in moist, monsoonal forests, are further altering patterns of forest community development as well as the characteristics of disturbance regimes themselves.
- To date, the emphasis in forestry in Southeast Asia has been on production of wood and wood products rather than on the sustainable use of forest resources. Over the past decades, logging and the use of fire for clearing tropical rain forests has made much of the landscape susceptible to fire during periods of drought, particularly those associated with ENSO. With such droughts, the occurrence of vegetation fires are associated with massive inter-regional transport of aerosols and trace gases.
- At both national and regional scales in lowland areas increasing demands for food are likely to be met in future by further intensification and use of genetically modified crops. Increasing overall production of rice to meet growing local and export demands will not be easy as the already high level of inputs production is already approaching attainable yield levels. Added to this, land available for conversion to rice production is becoming scarce in many countries and water for irrigation is becoming limited because of multiple and growing water demands.
- The intensification and commercialisation of agriculture, although not yet complete, has already had a profound effect on ecological and social systems in Southeast Asia. Common features of the transformation include shifts away from subsistence to cash crops, increasing use of fertilizers, pesticides and herbicides and the expansion of irrigation. Pollutants derived from widespread application of pesticides, fungicides and herbicides are contaminating soils, surface water supplies and groundwater. The magnitude of river transport of fertilizer residues and sediments after crop production into coastal areas is a national problem in many cases and is becoming an important regional issue with implications for marine life and biodiversity.
- Although historically Southeast Asia has been one of the least urbanised regions of the world, this is quickly changing. The migration from upland rural areas to the major coastal cities has been very rapid, particularly around the big capitals like Manila, Bangkok and Jakarta. Untreated sewerage effluent from urban, peri-urban and tourist centres is making a substantial contribution to pollution loads of the coastal zone with significant consequences for marine life and biodiversity.
- In the coastal zones, mangroves and wetlands are under pressure for a variety of reasons, including expansion of urban settlements, exploitation for fuelwood, agricultural development and construction of fish and shrimp ponds. Between 1980 and 1994 Southeast Asia lost almost half of its mangroves. It is estimated that over 80% of coral reefs are at risk in the region, primarily from coastal development and fishing.

- It is not yet clear what the coastal zone ecological changes in the Andaman Sea, Gulf of Thailand and South China Sea of the Southeast Asian region, are making to global biogeochemical budgets.
- The Gulf of Thailand and continental shelf off Vietnam are particularly productive marine biological areas. The annual marine fish catch in Southeast Asia represents about 10% of the world catch. Fish stocks are decreasing in many areas through over-fishing and decreasing resources are becoming increasingly a regular source of conflict locally, nationally and regionally.
- Changing consumption and production patterns are key factors affecting the vulnerability of rural and fishing communities to natural variability and future global environmental changes. Despite the overwhelming importance of urbanisation and manufacturing-oriented industrialisation in transforming the economies and societies of Southeast Asia, a high proportion of the population in all countries (apart from Singapore) is still dependent on agriculture and fisheries. The high diversity and productivity of tropical ecosystems have traditionally provided a reliable and resource base for this section of society. Many modern developments place this security at considerable risk.
- Southeast Asia is a hot spot in both sustainable development and global change terms. The rates of economic growth have been faster than most parts of the world for several decades. As a result many parts of Southeast Asia are now more industrialized, diversified and integrated into the global economy than their counterparts elsewhere. Although there are still many disparities in the size and effectiveness of different national economies within the region, these are rapidly becoming more inter-dependent through commodity trade, investment flows and exchange and division of labour.
- Consumption of regionally produced goods is growing rapidly, both within and outside the region. Over the recent past, the most important socio-economic driving force for environmental change in the Southeast Asian region has been the shift from closed national and regional economies to integration into the global economy on a large scale. Untangling the domestic, regional and global interactions of various production-consumption systems for food and manufactured products is not easy, but it is essential to understand the positive and negative implications of gobalisation in the region.
- Large-scale agri-business operations in Malaysia, Indonesia and Thailand have transformed vast areas of native forests, and some traditional smallholder agricultural areas, into monoculture plantations of rubber, oil palm and tree plantations for the pulp and paper industry. Agricultural land-use has also been transformed across the region by export-lead boom and bust, production and abandonment, cycles of maize, cotton, tobacco, tea and cassava production.
- The gobalisation of trade and liberalisation of investment has had profound consequences for the environment and development in Southeast Asia. On the one hand, it has helped nations bring in foreign exchange through export-oriented agriculture, forestry and manufacturing. On the other hand, globalisation has progressively increased the distance between consumers and resource systems and pollution sinks and has strained existing local and national environmental regulatory mechanisms to breaking. International institutions to handle these problems are being developed in the region and globally, but are having little effect.
- In developed countries, pollution production per unit of economic activity (pollution intensity) has been found to intensify with development before easing, whereas total emissions follow an S-shaped curve. At current stages of development in Southeast Asia, only in Singapore is there any sign that the growth of pollution intensity may be reaching a plateau. In Thailand pollution intensities for all pollutants are increasing more rapidly than GDP, in some cases significantly so. Similar patterns can be expected for the other countries in the region for the next few decades as economic globalisation increases. Few indicators of change illustrate more graphically how accelerated development and globalisation (without adequate environmental protection) are leading to an unsustainable future.

Urbanisation, industrialisation, commercialisation and the transformation of coastal, lowland and upland rural landscape are linked processes. Associated changes in land-use and land-cover will affect the transport of aerosols and trace gases to the atmosphere and sediments, nutrients and pollutants to coastal zones. The condition and dynamics of terrestrial and marine ecosystem in turn constrain development processes. Understanding the many interactions between the atmosphere, climate, biogeochemistry, hydrology, land surface characteristics and biodiversity is a daunting challenge. But as this chapter shows, it is essential go further and try to understand how socio-economic changes at various scales affect global environmental change and its regional manifestations. Humans are an integral part of the ecosystems being changed. They have the capacity to learn, adapt and respond in a rich variety of ways. One of keys to attaining sustainability will be to cultivate these capacities.

Southeast Asia differs in important ways from other tropical regions of the world. It has achieved prolonged periods of rapid economic growth with relatively high population densities in a region dominated by oceans and coastal zones. The environmental consequences of these changes, in terms of degradation of shared natural resources and pollution sinks, are becoming important for local and region-wide development. Globalisation provides both opportunities and threats to the future of the development-environment interaction. The outcome of this interaction over the next couple of decades in Asia will have profound consequences for the people of Southeast Asia and for the global environment.

Acknowledgements

Thanks are extended to the lead authors of APN-START assessment book: Yap Kieo Sheng, Simon Tay, Ben Malayang, Tolentino Moya, Ooi Giok Ling, and Anond Snidvongs for sharing their knowledge about the policy implications of global change for the development process. Thanks also to Daniel Murdiyarso for his inputs about the effects of land-use and -cover changes on biogeochemical cycles.

References

Adger WN (1999) Social vulnerability to climate change and extremes in coastal Vietnam. World Development 27:249–269

Alford D (1992) Streaflow and sediment transport from mountain watersheds of the Chao Phraya Basin, Northern Thailand: a reconnaissance study. Mountain Research and Development 12:257–268

Alford D (1994) Water budgets and water regions: planning and managing water resources development in Thailand. Thailand Development Research Institute Quarterly 9 (4):14–23

An Z (2000) The history and variability of the East Asian palaeomonsoon climate. Quaternary Science Reviews 19:171–187

Anas I, Susilo FX, Hardiwinoto, Simanungkalit R, Setiadi Y, Djunaedy and van Noordwijk M (1999) Assessment of impacts of forest conversion on belowground biodiversity. In: Tomich TP, Thomas DE, Noordwijk M van (eds) Environmental services and land use change: Bridging the gap between policy and research in Southeast Asia. Methodology Workshop, 31 May to 2 June 1999, Chiang Mai, Thailand. ASB-Indonesia Report Number 19. Bogor, International Centre for Research in Agroforestry 59

Angel DP, Rock MT (eds) (2000) Asia's clean revolution industry, growth and the environment. Sheffield, Greenleaf

Arndt RL, Carmichael GR (1995) Long-range transport and deposition of sulfur in Asia. Water, Air and Soil Pollution 85(4): 2283–2288

Arndt RL, Carmichael GR, Roorda JM (1996) Seasonal source-receptor relationships in Asia. Atmospheric Environment 32(8):1397–1406

ASEAN (1995) ASEAN cooperation plan on transboundary pollution. Jakarta, ASEAN Secretariat

ASEAN (1997) First ASEAN state of the environment report. Jakarta, ASEAN Secretariat

Asian Development Bank (1994) Climate change in Asia: Thematic overview. Manila, Asian Development Bank

Asian Development Bank (1996) Key indicators of developing Asian and Pacific countries. Manila, Asian Development Bank, 27

Asiaweek (1998) The Asiaweek quality of life index: Special report best cities. Asiaweek, 11 December 1998, 46–51

Asiaweek (1999) The Asiaweek quality of life index, Asia's best cities. Asiaweek, 17 December 1999, 44–60

Auty RM (1997) Pollution patterns during the industrial transition. The Geographical Journal 163:206–213

Awadalla S, Noor IM (1991) Induced climate change on surface runoff in Kelantan Malaysia. Water Resources Development 7 (1):53–59

Awang MB, Abdullah AM, Johan S (1998) Effects of CO_2 enrichment on gas exchange of tropical hardwood forest species grown under different irradiance levels. In: The earth's changing land: GCTE-LUCC Open Science Conference on Global Change, Barcelona, Spain, 14–18 March 1998. International Geosphere Biosphere Programme and the International Human Dimensions Programme on Global Environmental Change

Bandy DE, Garrity DP, Sanchez PA (1993) The worldwide problem of slash-and-burn agriculture. Agroforestry Today July–September 1993, 2–6

Batterbury S, Forsyth T (1999) Fighting back: human adaptation in marginal environments. Environment 41 (6):6–11, 25–30

Bawa KS, Markham A (1995) Climate change and tropical forests. Trends in Ecology and Evolution 10:348–349

Bhuiyan SI, Toung TP, Wade LJ (1998) Management of water as a scarce resource: issues and options in rice culture. In: Dowling NG, Greenfield SM, Fischer KS (eds) Sustainability of rice in the global food system. Philippines, International Rice Research Institute, 175–192

Boonpragob K (1998) Estimating greenhouse gas emission and sequestration from land use change and forestry in Thailand. In: Moya T (ed) Proceedings of a synthesis workshop on greenhouse gas emissions, aerosols, land use and cover changes in Southeast Asia, 15–18 November 1997, Chung-Li, China-Taipei. Bangkok, Southeast Asia Regional Committeee for START, 18–25

Booth TH, Nghia NH, Kirschbaum MUF, Hackett C, Jovanovic T (1999) Assessing possible impacts of climate change on species important for forestry in Vietnam. Climatic Change 41:109–126

Briggs MRP, Funge-Smith SJ (1994) A nutrient budget of some intensive marine shrimp ponds in Thailand. Aquaculture and Fisheries Management 25789–25811

Brookfield H, Byron B (1990) Deforestation and timber extraction in Borneo and the Malay Peninsula. Global Environmental Change 1:52–56

Brown S, Lugo AE (1990) Tropical secondary forests. Journal of Tropical Ecology 6:1–32

Bryant D, Burke L, McManus JW, Spalding M (1998) Reefs at risk: A map-based indicator of potential threats to the world's coral reefs. World Resources Institute

Buendia LV, Wassman R, Lantin RS, Neue HU, Javellana AM, Lu W, Wang X, Makarim AK, Corton TM, Chreonsilp N, Nocon N (1998) Abstract 248 in The earth's changing land: GCTE-LUCC Open Science Conference on Global Change, Barcelona, Spain 14–18 March 1998. International Geosphere Biosphere Programme and the International Human Dimensions Programme on Global Environmental Change

Calder IR (1996) Water use by forests at the plot and catchment scale. Commonwealth Forestry Review 75 (1):19–30

Chamratrithirong A, Archavanitkul K, Richter K, Guest P, Thongthia V, Boonchalaksi W, Piriyathamwong N, Vong-ek P (1995) National migration survey of Thailand. Institute of Population and Social Research Publication No. 188. Bangkok, Mahidol University

Chandraprasert E (1997) The impact of development on the hilltribes of Thailand. In: McCaskill D, Kampe K (eds) Development or domestication? Indigenous peoples of Southeast Asia. Chiang Mai, Silkworm Books, 83–96

Chang CP, Krishnamurti TN (eds) (1987) Monsoon meteorology. Oxford Mongraphs on Geology and Geophysics No. 7. New York, Oxford University Press

Chapin FS, III, Zavaleta ES, Eviner VT, Naylor RL, Vitousek PM, Reynolds HL, Hooper DU, Lavorel S, Sala OE, Hobbie SE, Mack MC, Diaz S (2000) Consequences of changing biodiversity. Nature 405:234–242

Chongprasith P, Srineth V (1998) Marine water quality and pollution of the Gulf of Thailand. In: Johnston DM (ed) SEAPOL integrated studies of the Gulf of Thailand. Bangkok, SEAPOL, 137–204

Colfer CJP, Gill DW, Agus F (1988) An indigenous agricultural model from West Sumatra: a source of scientific insight. Agricultural Systems 26:191–209

Condit R, Hubbell SP, Foster RB (1996) Changes in tree species abundance in a neotropical forest: impact of climate change. Journal of Tropical Ecology 12:231–256

Cramer W, Bondeau A, Woodward FI, Prentice IC, Betts RA, Brovkin V, Cox PM, Fisher V, Foley J, Friend AD, Kucharik C, Lomas MR, Ramankutty N, Sitch S, Smith B, White A, Young-Molling C (2000) Global responses of terrestrial ecosystems structure and function to CO_2 and climate: results from six dynamic global vegetation models. Global Change Biology 7:357–373

Dauvergne P (1997) Shadows in the forest: Japan and the politics of timber in Southeast Asia. Massachusetts, MIT Press

De Koninck R (1996) The peasantry as the territorial spearhead of the state in Southeast Asia: the case of Vietnam. Sojourn 11:231–58

Dierberg FE, Kiattisimkul W (1996) Issues, impacts, implications of shrimp aquaculture in Thailand. Environmental Management 20 (5):649–666

Dietz T, Rosa EA (1997) Effects of population and affluence on CO_2 emissions. Proceedings of the National Academy of Science, 94:175–179

Dinerstein E, Wikramanayake ED (1993) Beyond "hotspots": how to prioritize investments to conserve biodiversity in the Indo-Pacific region. Conservation Biology 7 (1):53–65

Dixon JA (1990) Renewable resources, the environment and sustained growth: the next twenty-five years. ASEAN Economic Bulletin 7 (2):159–172

Dove MR (1995) Political versus techno-economic factors in the development of non-timber forest products: lessons from a comparison of natural and cultivated rubbers in Southeast Asia (and South America). Society and Natural Resources 8:193–208

Engelman R (1998) Profiles in carbon: An update on population, consumption and carbon dioxide emissions. Washington D.C., Population Action International

Eong OJ, Khhon GW, Rahim AB, Chiang CH (1999) Carbon and nutrient fluxes and socio-economic studies in the Merbok mangrove ecosystem. SARCS/WOTRO/LOICZ Malaysian Core Research Site. Synthesis Report 1996–1999. Universiti Sains Malaysia

FAO (2000) On-line statistical databases. Rome, Food and Agriculture Organization

FAO, NACA (1995) Report on a regional study and workshop on the environmental assessment and management of aquaculture development. NACA Environment and Aquaculture Development Series No. 1. Bangkok, Network of Aquaculture Centres in Asia-Pacific

Forbes D (1996) Asian metropolis: Urbanisation and the Southeast Asian city. Melbourne, Oxford University Press

Foresta HD, Michon G (1997) The agroforest alternative to *Imperata* grasslands: when smallholder agriculture and forestry reach sustainability. Agroforestry Systems 36:105–120

Forsyth T (1996) Science, myth and knowledge: testing Himalayan environmental degradation in Thailand. Geoforum 27:375–392

Forsyth T (1998) International investment in climate change: Energy technologies for developing countries. The Royal Institute for International Affairs. London, Earthscan

Forsyth T (1999) Questioning the impacts of shifting cultivation. Watershed 5:23–29

Fowler AM, Hennessy KJ (1995) Potential impacts of global warming on the frequency and magnitude of heavy precipitation. Natural Hazards 11:283–303

Fox J, Kanter R, Yarnasarn S, Ekasingh M, Jones R (1994) Farmer decision making and spatial variables in Northern Thailand. Environmental Management 18 (3):391–399

Gagan MK, Ayliffe LK, Beckb JW, Colec JE, Druffeld ERM, Dunbare RM, Schragf DP (2000) New views of tropical palaeoclimates from corals. Quaternary Science Reviews 19:45–64

Gangjanapan A (1996) The politics of environment in Northern Thailand: ethnicity and highland development programs. In: Hirsch P (ed) Seeing forests for trees: Environment and Environmentalism in Thailand. Chiang Mai, Silkworm Books, 202–222

Garrity DP, Soekardi M, Noordwijk M van, De La Cruz R, Pathak PS, Gunasean HPM, Van So N, Huijun G, Majid NM (1997) The *Imperata* grasslands of tropical Asia: area, distribution and typology. Agroforestry Systems 36:3–29

Giambelluca TW (1996) Tropical landcover change: characterizing the post-forest land surface. In: Giambelluca TW, Henderson-Sellers A (eds) Climate change: Developing southern hemisphere perspectives. Singapore, John Wiley and Sons, 293–317

Goldammer JG (1996) Overview of fire and smoke management issues and options in tropical vegetation. In: Proceedings of the AIFM Conference on Transboundary Pollution and the Sustainability of Tropical Forests: Towards Wise Forest Fire Management, 2–4 December 1996, Kuala Lumpur. Kuala Lumpur, ASEAN Institute for Forest Management, pp 189–217

Goldammer JG (1998) The role of fire in greenhouse gas and aerosol emissions and land use and cover change in Southeast Asia: ecological background and research needs. International Conference on Science and Technology for Assessment of Global environmental change and its impacts on the Indonesian maritime continent, 10–12 November 1997, Jakarta. Freiburg, University of Freiburg

Goldammer JG, Siebert B (1989) Natural rain forest fires in Eastern Borneo during the Pleistocene and Holocene. Naturwissenschaften 76:518–520

Goldammer JG, Siebert B (1990) The impact of droughts and forest fires on tropical lowland rainforest of East Kalimantan. In: Goldammer JG (ed) Fire in the tropical biota. Ecosystem Processes and Global Challenges. Ecological Studies 84. Berlin, Springer-Verlag, 11–31

Gomez ED, Talaue-McManus L, Lieuanana WY, McGlone DH, Siringan FP, San Diego-McGlone ML, Villanoy CL, Clemente RS (1999) Economic valuation and biogeochemical modelling of the Lingayen Gulf in support of management for sustainable use. SARCS/WOTRO/LOICZ Philippine Core Research Site. 1999 Annual Report. University of the Philippines, Marine Science Institute, November 1999

Gordon DC Jr, Boudreau PR, Mann HK, Eong JE, Silvert WL, Smith SV, Wattayakorn G, Wulff F, Yanagi T (1995) LOICZ biogeochemical modelling guidelines. LOICZ 95-5. Texel, LOICZ

Gregory PJ, Ingram JSI, Campbell B, Goudriaan J, Hunt IA, Landsberg JJ, Linder S, Stafford-Smith M, Sutherst RW, Valentin C (1999) Managed production systems. In: Walker B, Steffen W, Canadell J, Ingram J (eds) The terrestrial biosphere and global change: implications for natural and managed ecosystems. Cambridge, Cambridge University Press, 229–270

Guest P, Uden A (1995) Religion and migration in Southern Thailand: evidence from the 1970, 1980 and 1990 censuses. Institute for Population and Social Research Publication No. 175. Bangkok, Mahidol University

Hairiah K, Saptura AE, Noordwijk M van (1998) Carbon stocks in slash-and-burn agricultural systems in Indonesia. Abstract 470 in The Earth's Changing Land. GCTE-LUCC Open Science Conference on Global Change. Abstracts. Barcelona, Spain, 14–18 March 1998

Hammer GL, Nicholls N, Mitchell C (eds) (2000) Applications of seasonal climate forecasting in agricultural and natural ecosystems: the Australian experience. Atmospheric and Oceanographic Sciences Library Vol 21. Boston, Kluwer Academic Publishers

Han X, Chatterjee L (1997) Impacts of growth and structural change on CO_2 emissions in developing countries. World Development 25 (3):395–407

Hayes RL, Hussain ST (1995) Public health and force climate change: extreme temperature exposure and infectious disease. World Resource Review 7 (1):63–76

Heitmann HD, Surat L, Kiyoshi H, Lal S (1999) Turning down the heat in Bangkok. Geo Asia Pacific, December 1999/January 2000, 18–20

Hirsch P (1998) Dams, resources and the politics of environment in mainland Southeast Asia. In: Hirsch P, Warren C (eds) The politics of environment in Southeast Asia: Resources and resistance. London, Routledge, 55–70

Horie T (1993) Predicting the effects of climatic variation and effect of CO_2 on rice yield in Japan. Journal of Agricultural Meteorology 48:567–574

Hossain M, Fischer KS (1995) Rice research for food security and sustainable development in Asia: achievements and future challenges. Geojournal 35 (3):286–298

Houghton RA, Hackler J (1995) Continental scale estimates of the biotic carbon flux from land cover change: 1850–1980. ORNL/CDIAC-79, Carbon Dioxide Information Analysis Centre, Oak Ridge National Laboratory, Oak Ridge

Hughes G (1997) Can the Environment Wait? Priority Issues for East Asia. Washington DC, World Bank

Husin YA, Murdiyarso D, Khalil MAK, Rasmussen RA, Shearer MJ, Sabiham S, Sunar A, Adijuwana H (1995) Methane flux from Indonesian wetland rice: the effects of water management and rice variety. Chemosphere 31:3153–3180

Iglesias A, Erda L, Rosenzweig C (1996) Climate change in Asia: a review of the vulnerability and adaptation of crop production. Water, Air and Soil Pollution 92:13–27

Intergovernmental Panel on Climate Change (IPCC) (1995) Climate change, 1995. The IPCC Second Assessment Synthesis of Scientific-Technical Information Relevant to Interpreting Article 2 of the UN Framework Convention on Climate Change. IPCC

Janssen MA, Martens WJM (1997) Modelling malaria as a complex adaptive system. Artificial Life 3:213–236

Jomo KS (ed) (1998) Tigers in trouble: Financial governance, liberalisation and crises in East Asia. Bangkok, White Lotus

Jose AM (1999) Impact of the 1997–1998 El Niño event in the Philippines. Philippine Atmospheric, Geophysical and Astronomical Services Administration. September 1999

Khalid AR, Braden JB (1993) Welfare effects of environmental regulation in an open economy: the case of Malaysian palm oil. Journal of Agricultural Economics 44:25–37

Knight M (1998) Developing countries and the globalization of financial markets. World Development 26:1185–1200

Kongkeo H (1997) Comparison of intensive shrimp farming systems in Indonesia, Philippines, Taiwan and Thailand. Aquaculture Research 28:789–796

Kripalani RH, Kulkarni A (1997) Rainfall variability over South-east Asia – connections with Indian monsoon and ENSO extremes: new perspectives. International Journal of Climatology 17:1155–1168

Kripalani RH, Kulkarni A (1998) The relationship between some large-scale atmospheric parameters and rainfall over Southeast Asia: a comparison with features over India. Theoretical and Applied Climatology 59:1–11

Kummer DM, Concepcion R, Canizares B (1994) Environmental degradation in the uplands of Cebu. Geographical Review 84 (3):266–276

Kummer DM, Turner BJ II (1994) The human causes of deforestation in Southeast Asia. BioScience 44 (5):323–28

Lal M, Meehl GA, Marblaster JM (1999) Simulation of Indian summer monsoon rainfall and its interseasonal variability. Regional Environmental Change (November)

Lal M, Harasawa H (2000) Comparison of the present-day climate simulation over Asia in coupled atmosphere-ocean global climate models. Journal of the Meteorological Society of Japan 78:871–878

Lal M, Harasawa H (2001) Climate change scenarios for Asia. Journal of the Meteorological Society of Japan 79:219–227

Lebel L, Jintana V (1997) Modelling the impacts of global change on tropical forests. In: Khenmark C, Thaiuts B, Puangchit L, Thammincha S (eds) Proceedings of the FORTROP'96: Tropical Forestry in the 21st century, Volume 2. Global Changes in the Tropical Contexts, 25–28 November 1996. Bangkok, Kasetsart University, 1–26

Levine J, Bobbe T, Ray N, Singh A, Witt RG (1999) Wildland fires and the environment: A global synthesis. Division of Environmental Information, Assessment and Early Warning, United Nations Environment Programme

Liawaruangrath S, Simonkhoun P, Rattanaphani S, Suttheewasmnout S (1999) Monitoring of some heavy metals and organochlorine pesticide residues in Mekong River. In: Asnachinda P, Lerthusnee S (eds) Proceedings of International Conference on Water Resource Management in Intermontane Basins, 2–6 February 1999, Chiang Mai University, Thailand. Chiang Mai University, Water Research Centre, 469–483

Lin CK, Browdy CE, Hopkins JS (1995) Progression of intensive marine shrimp culture in Thailand. In: Swimming Through Troubled Water, Proceedings of the Special Session on Shrimp Farming, San Diego, California, USA, 1–4 February 1995. World Aquaculture Society:13–23

Lin CK (1989) Prawn culture in Taiwan. What went wrong. World Aquaculture 20:19–20

Lohmann L (1996) Freedom to plant: Indonesia and Thailand in a globalizing pulp and paper industry. In: Parnwell MJG, Bryant RL (eds) Environmental change in South-East Asia: People, politics and sustainable development. London, Routledge, 23–48

Luo Y, TeBeest DO, Teng PS, Fabellar NG (1995) Simulation studies of risk analysis of rice leaf blast epidemics associated with global climate change in several Asian countries. Journal of Biogeography 22:673–678

Luo Y, Tenga PS, Fabellara NG, TeBees DO (1998) The effects of global temperature change on rice leaf blast epidemics: a simulation study in three agroecological zones. Agriculture, Ecosystems and Environment 68:187–196

Magcale-Macandog DB, Predo CD, Rocamora PM (1998) Environmental and economic impacts of land-use change in tropical *Imperata* areas. Abstract 472 in The Earth's Changing Land. GCTE-LUCC Open Science Conference on Global Change. Abstracts. Barcelona, Spain, 14–18 March 1998

Malakoff D (1997) Extinction on the high seas. Science 277:486–88

Malayang BS III (1991) Tenure rights and exclusion in the Philippines. Nature and Resources 27 (4):18–23

Martens WJM, Jetten TH, Rotmans J, Niessen LW (1995) Climate change and vector-borne diseases. Global Environmental Change 5:195–209

Mastura S, Zakaria Z, Abdulla M (eds) (1999) Environmental imperatives of land use and land cover change in the Klang-langat watershed. Univerisiti Kebangsaan, Earth Observation Centre

Matthews RB, Kropff MJ, Bachelett D, Laer HH van (1995) Modelling the impacts of climate change on rice production in Asia. Oxford, CAB International

Mayer JD (2000) Geography, ecology and emerging infectious diseases. Social Science and Medicine 50:937–952

McGregor JL, Katzfey JJ, Nguyen KC (1998) Fine resolution simulations of climate change for Southeast Asia. Final report for a research project commissioned by Southeast Asian Committee for START (SARCS), December 1998

McLean RF, Sinha SK, Mirza MQ, Lal M (1998) Tropical Asia. In: Intergovernmental Panel on Climate Change (ed) Regional impacts of climate change: an assessment of vulnerability. Geneva, 383–407

Menasveta P (1998) Mangrove destruction and shrimp culture systems. World Aquaculture 28:36–42

Mitchell PL (1996) Comparison of five models of rice yield showing the effects of change in temperature and in carbon dioxide concentration. In: Ingram JSI (ed) Report of the GCTE Rice Network Experimentation Planning Workshop. GCTE Working Document No. 19. Canberra, GCTE IPO

Mitchell JFB, John TC, Gregory JM, Tett SFB (1995) Climate response to increasing levels of greenhouse gases and sulphate aerosols. Nature 376:501–504

Miyata A, Leuning R, Denmead OT, Kim J, Harazono Y (2000) Carbon dioxide and methane fluxes from an intermittently flooded paddy field. Agricultural and Forest Meteorology 102:287–303

Mooney HA, Canadell J, Chapin FS, III, Ehleringer JR, Corner C, McMurtrie RE, Parton WJ, Pitelka LF, Schulze E-D (1999) Ecosystem physiology responses to global change. In: Walker B, Steffen W, Canadell J, Ingram J (eds) The terrestrial biosphere and global change: Implications for natural and managed ecosystems. Cambridge, Cambridge University Press, 141–189

Moya TB (ed) (1997) Proceedings of a Synthesis Workshop on Greenhouse Gas Emissions, Aerosols, Land Use and Cover Changes in Southeast Asia. National Central University, China-Taipie, 15–18 November 1997. Bangkok, Southeast Asian Regional Committee of START

Murdiyarso D, Lebel L, Gintings AN, Tampubolon SMH, Heil A, Wasson M (2000) Policy responses to complex environmental problems: Insights from a science-policy activity on transboundary haze from vegetation fires in Southeast Asia. Journal of Agriculture, Ecosystems, Environment

Murdiyarso D, Wasrin UR (1995) Estimating land use change and carbon release from tropical forest conversion using remote sensing technique. Journal of Biogeography 22:715–721

Nakagawa H, Horie T, Kim HY, Homma K (1998) The interactive effects of elevated CO_2 concentration and temperature on rice growth and yield. Abstract 210 in The Earth's Changing Land: GCTE-LUCC Open Science Conference on Global Change, Barcelona, Spain, 14–18 March 1998. International Geosphere Biosphere Programme and the International Human Dimensions Programme on Global Environmental Change

NASA (2001) *http://www.earthobservatory.nasa.gov/*
Neue H-U, Sass RL (1994) Trace gas emissions from rice fields. In: Prinn RG (ed) Global atmospheric-biospheric chemistry. New York, Plenum
Nicholls RJ, Hoozemans FMJ, Marchaud M (1999) Increasing flood risk and wetland losses due to global sea-level rise: regional and global analyses. Global Environmental Change 9:69–87
Osborne M (1997) Southeast Asia: An introductory history. Sydney, Allen and Unwin
Padilla J, Morales A (1997) Evaluation of fisheries management alternatives for Lingayen Gulf: An options paper. Manila, DENR/USAID
Parry M, Rosenzweig C, Iglesiasc A, Fischer G, Livermore M (1999) Climate change and world food security: a new assessment. Global Environmental Change 99:51–67
Pasong S, Lebel L (2000) Political transformation and the environment in Southeast Asia. Environment 42 (5):8–19
Peng S, Ingram KT, Neue, H-U, Zisika LH (eds) (1995) Climate change and rice. Los Banos, International Rice Research Institute
Phongpaichit P, Baker C (1998) Thailand's boom and bust. Chiang Mai, Silkworm Books
Pittock BA (1995) Climate change scenarios for the Asia-Pacific region. Proceedings of the Asia-Pacific Leaders' Conference on Climate Change, Manila, 17–19 February 1995
Pittock BA (1999) Coral reefs and environmental change: adaptation to what? American Zoologist 39:10–29
Posch M, Hettelingh, J-P, Alcamo J, Krol M (1996) Integrated scenarios of acidification and climate change in Asia and Europe. Global Environmental Change 6 (4):375–394
Potter IM (1997) The dynamics of *Imperata*: historical overview and current farmer perspectives, with special reference to South Kalimantan, Indonesia. Agroforestry Systems 36: 31–51
Primavera JH, Lavilla-Pitogo CR, Ladja JM, Pena MR (1993) A survey of chemical and biological products used in intensive prawn farms in the Philippines. Marine Pollution Bulletin 26:35–40
Rattanaphani S (1999) Monitoring of some organochlorine pesticide residues in Ping River. In: Asnachinda P, Lerthusnee S (eds) Proceedings of International Conference on Water Resource Management in Intermontane Basins, 2–6 February 1999, Chiang Mai University, Thailand. Chiang Mai University, Water Research Centre 455–467
Redclift M, Sage C (1998) Global environmental change and global inequality. International Sociology 13:499–516
Reichardt W, Dobermann A, George T (1998) Intensification of rice production systems: opportunities and limits. In: Dowling NG, Greenfield SM, Fischer KS (eds) Sustainability of rice in the global food system. Los Banos: International Rice Research Institute, 127–144
Rerkasem K, Rerkasem B (1995) Montane mainland South-east Asia: agroecosystems in transition. Global Environmental Change 5 (4):313–322
Richey JE, Snidvongs A (1998) Proceedings of the Initial Orientation Workshop: Southeast Asia integrated regional model: river basin inputs to the coastal zones (SEA/BASINS). Ubon Rachathani, Thailand, 14–17 July 1998. Bangkok, Southeast Asia START Regional Centre
Riebsame WE, Strzepek KM, Wescoat JL Jr., Perritt R, Gaile GL, Jacobs J, Leichenko R, Magadza C, Phien H, Urbiztondo BJ, Restrepo P, Rose WR, Saleh M, Ti LH, Tucci C, Yates D (1995) Complex river basins. In: Strzepek KM, Smith JB (eds) As climate changes: international impacts and implications. New York, Cambridge University Press, 57–91
Rigg J (1997) Southeast Asia: The human landscape of modernization and development. London, Routledge
Rock MT (1998) Freshwater use, freshwater scarcity, socio-economic development. Journal of Environment and Development 7 (3):278–301
Rosenberg D (1999) Environmental pollution around the South China Sea: Developing a regional response to a regional problem. Resource Management in Asia-Pacific Working Paper No. 20, Division of Pacific and Asian History, Research School for Pacific and Asian Studies, The Australian National University, Canberra
Round PD (1988) Resident forest birds in Thailand: their status and conservation. ICBP Monograph No. 2. Cambridge, International Council for Bird Preservation
Ruitenbeek HJ (1994) Modelling economy-ecology linkages in mangroves: economic evidence for promoting conservation in Bintuni Bay, Indonesia. Ecological Economics 10:233–247
Sachasinh R, Phantumvanit D, Tridech S (1992) Thailand: Challenges and responses in environmental management. Paper presetned to the Workshop on Environmental Management in East Asia: Challenges and Responses. OECD Development Centre, Paris, 6–7 August 1992
Salafsky N (1994) Drought in the rain forest: effects of the 1991 El Niño-Southern Oscillaton event on rural economy in West Kalimantan, Indonesia. Climatic Change 27:373–396
Schipper L, Grubb M (2000) On the rebound? Feedback between energy intensities and energy uses in IEA countries. Energy Policy 28:367–388
Schmidt-Vogt D (1998) Defining degradation: The impacts of swidden on forests in Northern Thailand. Mountain Research and Development 18:135–149
Severino RC (1998) The impact of the crisis on ASEAN trade: a preliminary assessment. Paper presented by the ASEAN Secretary General at the 1998 East Asia Economic Summit of the World Economic Forum. URL: *http://www.asean.or.id/secgen/cris_trd.htm*
Shugart HH, Emanuel WR, Shao G (1996) Models of forest structure for conditions of climatic change. Commonwealth Forestry Review 75:51–64
Siebert SF (1987) Land use intensification in tropical uplands: effects on vegetation, soil fertility and erosion. Forest Ecology and Management 21:37–56
Siebert SF (1990) Hillside farming, soil erosion, forest conversion in two southeast Asian national parks. Mountain Research and Development 10:64–72
Sikor T, Troung DM (1998) Sticky rice, collective fields: community-based development among the black Thai. Centre for Natural Resources and Environmental Studies, Vietnam National University, Hanoi, Vietnam
Sivakumar MVK (ed) (1999) Climate prediction and agricuture: Proceedings of the START/WMO International Workshop held in Geneva, Switzerland, 27–29 September, 1999. Washington DC, International START Secretariat
Smiet AC (1992) Forest ecology on Java: human impact and vegetation of montane forest. Journal of Tropical Ecology 8:129–52
Smith TB (1994) Global climate change in Asia: the politics of public policy-making and science agenda setting. Science and Public Policy 21 (4):249–259
Snidvongs A (1993) Sedimentary calcium carbonate dissolution in the Gulf of Thailand and its role as a minor carbon dioxide sink. Chemosphere 27:1083–1095
Soegiarto A (1994) Sustainable fisheries, environment and the prospects of regional cooperation in Southeast Asia. East West Centre – Nautilus Institute – Monterey Institute of International Studies Workshop on Trade and Environment in Asia-Pacific: prospects for regional cooperation
Stallings B (ed) (1995) Global change, regional response: The new international context of development. Cambridge, Cambridge University Press
Streets DG, Waldhoff ST (2000) Present and future emissions of air pollutants in China: SO_2, NO_x, CO. Atmospheric Environment 34:363–374
Stolle F, Tomich TP (1999) The 1997–1998 fire event in Indonesia. Nature and Resources 35:22–30
Surarerks V (1986) Historical development and management of irrigation system in Northern Thailand. Chiang Mai, Department of Geography, Chiang Mai University
Sutton K, McMorrow J (1998) Land use change in Eastern Sabah. In: King VT (ed) Environmental Challenges in South-East Asia. Surrey, Curzon Press, 259–281
Takahashi K, Harasawa H, Matuoka Y (1999) Impacts of environmental change on food security in the Asian region. Abstract 59 in 1999 Open Meeting of the Human Dimensions of Global Environmental Change Research Community, Shonan Village, Japan, 24–26 June 1999

Talaue-McManus L, McGlone D, Sand Diego-McGlone ML, Siringan F, Villanoy C, Licuanan W (1999) The impact of economic activities on biogeochemical cycling in Lingayen Gulf, northern Philippines: a preliminary synthesis. Land-Ocean Interactions in the Coastal Zone Newsletter, 10, March 1999
Talaue-McManus L (2000a) Transboundary diagnostic analysis for the South China Sea. EAS/RCU Technical Report Series No. 14. Bangkok, United Nations Environment Program
Talaue-McManus L (2000b) A preliminary typology of watersheds of the South China Sea. LOICZ/GEF Studies and Reports Series on Estuarine Systems of the South China Sea 14:131–136
Tan CY, Kwong LY (1990) Industrial and the natural environment policies on pollution. Vol 8. of UNIDO, Policy Aassessment of the Malaysian Industrial Policy Studies and the Industrial Master Plan, Vienna, UNIDO
Tan-kim-yong U (1997) The Karen culture: a co-existence of two forest conservation systems. Pages 219–236 in McCaskill D, Kampe K (Eds) Development or domestication? Indigenous peoples of Southeast Asia. Silkworm Books, Chiang Mai, Thailand
Thompson H, Duggie J (1996) Political economy of the forestry industry in Indonesia. Journal of Contemporary Asia 26: 352–365
Tinker PB, Ingram JSI, Struwe S (1996) Effects of slash-and-burn agriculture and deforestation on climate change. Agriculture, Ecosystems and Environment 58:13–22
Tomich TP, Kuusipalo J, Menz K, Byron N (1997) *Imperata* economics and policy. Agroforestry Systems 36:233–261
Tomich TP, Fagi AM, Foresta H de, Michon G, Murdiyarso D, Stolle F, Noordwijk M van (1998a) Indonesia's fires: smoke as problem, smoke as a symptom. Agroforestry Today, January–March 1998, 4–7
Tomich TP, Noordwijk M van, Vosti SA, Witcover J (1998b) Agricultural development with rainforest conservation: methods for seeking best bet alternatives to slash-and-burn, with applications to Brazil and Indonesia. Agricultural Economics 19:159–174
Tomich TP, Noordwijk M van, Budidarsono S, Gillison A, Kusumanto T, Murdiyarso D, Stolle F, Fagi AM (eds) (1998c) Alternatives to slash-and-burn in Indonesia. Summary Report and Synthesis of Phase II. ASB-Indonesia Report No. 8. ASB-Indonesia, Bogor
Tri NH, Adger WN, Kelly PM (1998) Natural resource management in mitigating climate impacts: the example of mangrove restoration in Vietnam. Global Environmental Change 8: 49–61
Tri NH, Ninh NH, Lien TV, Trnh B, Chinh NT, Trong TD, Secretario FT (1999) Economic-environmental modelling of coastal zones in the Red River Delta, Vietnam. SARCS/WOTRO/LOICZ Vietnam Core Research Site. Final Report. Hanoi, University of Vietnam
Trongkongsin K (1987) Land use and agroforestry in Thailand with special reference to the Southern Region. Kasetsart University Journal of Social Science 8:189–202
United Nations (1998) World urbanization prospects. The 1996 Revision. Estimates and projections of urban and rural populations and of urban agglomerations. New York, Department of Economic and Social Affairs, United Nations Population Division
United Nations Development Programme (1998) Human development report 1999. New York, UNDP
Valencia MJ (1990) International conflict over marine resources in South-east Asia: trends in politicization and militarization. In: Ghee LT, Valencia MJ (eds) Conflicts over natural resources in Southeast Asia and the Pacific. Singapore, United Nations University Press, 94–144
Van Noordwijk M, Tomich TP, Winahyu R, Murdiyarso D, Suyanto, Partoharjono S, Fagi AM (1995) Alternatives to slash-and-burn in Indonesia. Summary Report of Phase 1. ICRAF Southeast Asia., Bogor, Indonesia
Vandergeest P (1996) Mapping nature: territorialization of forest rights in Thailand. Society and Natural Resources 9:159–75
Vanderzwaag D, Johnston DM (1998) Toward the management of the Gulf of Thailand: charting the course of cooperation. In: Johnston DM (ed) SEAPOL integrated studies of the Gulf of Thailand. Volume 1. Bangkok, Southeast Asian Programme in Ocean Law, Policy and Management
Verburg P, Veldkamp T, Bouma J (1999) Land-use change under conditions of high population pressure: the case of Java. Global Environmental Change 9:303–312
Vincent JR, Rozali MA, Khalid AR (2000) Water pollution abatement in Malaysia. In: Angel DP, Rock MT (eds) Asia's clean revolution: Industry, growth and the environment. Sheffield, Greenleaf Publishing, 173–193
Vitousek PM, Aber JD, Howarth RW, Likens GE, Matson PA, Schindler DW, Schlesinger WH, Tilman DG (1997) Human alterations of the global nitrogen cycle: sources and consequences. Ecological Applications 7:737–50
Vitug MD (1998) The politics of logging in the Philippines. In: Hirsch P, Warren C (eds) The politics of environment in Southeast Asia. Resources and Resistance. London, Routledge, 122–136
Wackernagel M, Onisto L, Linares AC, Falfán ISL, García JM, Guerrero AIS, Guerrero MGS (1997) Ecological footprints of nations: how much nature do they use? – how much nature do they have? Prepared for the Rio+5 Forum held in Rio de Janeiro, 13–19 March, 1997
Walsh RPD (1996) Drought frequency changes in Sabah and adjacent parts of northern Borneo since the late nineteenth century and possible implications for tropical rain forest dynamics. Journal of Tropical Ecology 12:385–407
Walsh KJE, Katzfey JJ (2000) The impact of climate change on the poleward movement of tropical cyclone-like vortices in a regional climate model. J of Climate 13:1116–1132
Walsh K, Pittock AB (1998) Potential changes in tropical storms, hurricanes, extreme rainfall events as a result of climate change. Climatic Change 39:199–213
Walsh KJE, Ryan BF (2000) Tropical cyclone intensity increase near Australia as a result of climate change. J of Climate 13:3029–3036
Wassmann R, Moya TB, Lantin RS (1998) Rice and the global environment. In: Dowling NG, Greenfield SM, Fischer KS (eds) Sustainability of Rice in the Global Food System. Los Banos, International Rice Research Institute, 205–224
Watling L, Norse EA (1998) Disturbance of the seabed by mobile fishing gear: a comparison to forest clearcutting. Conservation Biology 12:1180–1197
Watson RT, Dixon JA, Hamburg SP, Janetos AC, Moss RH (1998) Protecting our planet, securing our future: linkages among global environmental issues and human needs. United Nations Environment Programme, US National Aeronautics and Space Administration and The World Bank
Wattayakorn G, Tingsabadh C, Piumsomboon A, Paphavasit N, Aksornkoae S, Sathirathai S, Prapong P, Thananuparppaisan S (1999) Economic valuation and biogeochemical modelling of Bandon Bay, Suratthani, Thailand. SARCS/WOTRO/LOICZ Thailand Core Research Site. 1999 Annual Report. Chulalongkorn University, Thailand, December 1999
Woodward A (1995) Doctoring the planet: health effects of global change. Australian New Zealand Journal of Medicine 25:46–53
World Bank (1994) A review of World Bank experience in irrigation. Report No. 13676. Washington DC, World Bank
World Bank (1999) World development index database. Washington DC, World Bank (URL: *www.worldbank.org*)
World Resources Institute, United Nations Environmental Programme, United Nations Development Programme and the World Bank (1998) World Resources 1998–1999: A Guide to the Global Environment. New York, Oxford University Press
WRI (1999) World resources 1998–1999: A Guide to the Global Environment, New York, Oxford University Press
Xue Y, Juang HH, Kanamitsu M, Hansen M (1999) Asian monsoon and vegetation interactions. GEWEX News 9:8–9
Yap KS, Mohit RS (1998) Reinventing local government for sustainable cities in Asia. Regional Development Dialogue 19 (1):87–94
Yaziz MI (1983) Control and management of water quality in Malaysia. Pertanika 6:69–87

Chapter 6

Summary and Conclusions

Regional syntheses have been carried out for four regions of the developing world: Southern Africa, South Asia, East Asia and Southeast Asia. Understanding the Earth System requires that the two-way linkages between regions and the global system be well understood and predictable. In each of the regions considered, distinctive manifestations of global change are evident; each is contributing in individual ways to global change.

Common findings are emerging in all the regions examined. In all regions major ecosystem adjustments have resulted from global change; in all instances future food security is a matter of concern. In all drainage basins hydrology is tending to amplify rainfall extremes; water amount and quality is a common concern; everywhere damage to coastal zones is increasing. Spatial gradients of change, particularly latitudinal and altitudinal, are important in all cases. In all the regions clear global-regional linkages are evident. Likewise, the inter-linkage of natural and anthropogenic driving forces of change is ubiquitous, with a clear tendency for the human drivers to be assuming an ever-greater importance as levels of economic development increase. All the regions are vulnerable to systematic global change and contribute to such change in varying degrees depending on levels of economic development. Just as the synthesis has revealed common features of regional change, so clear differences have emerged that illustrate distinctive regional differences in regional-global linkage patterns and processes.

The Southern Africa region, despite its importance in the global system in respect of aerosol and trace gas transports, ocean circulation and extensive savanna ecosystems, is unlikely to be a significant global player in the sense that the South, Southeast and East Asian regions are rapidly becoming. Southern Africa, except possibly in the case of southernmost Africa, is unlikely to undergo economic development in the foreseeable future on the scale and at the rate seen in Asia in recent times. Consequently its emissions of greenhouse gases and aerosols, and land use changes, are likely to change only slowly in comparison to the Asian regions. While southern Africa is unlikely to contribute greatly to global change in the near future, the manner in which regional processes operate in the subcontinent offers important insights into global regional linkages in the Earth System that have cogency far beyond the region.

In Asia generally, rates of change are exceptionally high, both biophysically and socio-economically. Land use and cover changes have occurred extensively, affecting the carbon cycle, the atmosphere and its circulation and horizontal nutrient and sediment fluxes to the coastal zone. Mega-cities are expanding rapidly and massive inter-linked conurbations are certain to develop to an ever-greater extent in future. Accelerating change is the rate at which regional populations are being drawn into the global economic system. The capacity of biophysical sinks and human institutions of governance to absorb the changes is under severe strain, without immediate prospects of improvement.

The nexus between changes in the Asian monsoon system and the global general circulation of the atmosphere is widely accepted. The link between human-induced land use and cover changes and weakening of the Asian monsoon have implications that transcend the region. The rate at which regional greenhouse gas and aerosol emissions are likely to increase in future is high in all realistic scenarios. The potential for regional changes in Asia to have major global implications is considerable. It appears that economic globalisation without stringent environmental controls may enhance this potential. The consequences of global market demands on regional resources and biodiversity in Southeast Asia is a test case.

In order to illustrate similarities and differences in global-regional linkages in the system in detail, it is useful to summarise the main findings of the synthesis, as they were presented for each region.

6.1 Southern Africa

Perhaps the most striking results of the Southern African synthesis relate to the atmospheric transport of aerosols and trace gases and the impacts of this transport on regional ecosystems. For example, small changes in the anticyclonicity of airflow and subsidence over the region produce large effects that modulate environmental change. Anticylonic recirculation of aerosols and trace gases, trapped beneath stable layers in the atmosphere, occurs on a subcontinental scale to an extent hitherto unforeseen. Over 40% of all aerosol-laden air recirculates over southern Africa at least once.

Atmospheric transport and deposition of N and P nutrients appear to play a significant role in sustaining the wetland ecosystem of the Okavango Delta and, by extension, may play equally important roles in other terrestrial and marine ecosystems. On a larger scale, the plume of aerosols and trace gases moving out of South Africa to the Indian Ocean at ~30° S is a large, persistent and a major circulation feature with hemispheric implications. In comparison, the plume to the Atlantic Ocean in tropical and subtropical latitudes is small and infrequent. The resulting aeolian transport of aerosols from South Africa, and atmospheric iron fertilisation of marine biota, supports enhanced biological productivity in the central South Indian Ocean between South Africa and Australia and provides a significant carbon sink. Sulphate aerosols are transported as oxide adhering to small dust nuclei; those originating from the industrial heartland of South Africa may be transported over long distances across the Indian Ocean or, less frequently, by inter-hemispheric transport into the northern hemisphere over Kenya.

Out-of-phase climatic teleconnections between the equatorial/tropical region and the subtropics, similar to those resulting from present-day ENSO variability, have occurred for at least a thousand years on centennial time scales and clearly transcend the ENSO phenomenon. The resultant changing gradients of climatic change have had important consequences in the region. The evolutionary response to environmental change may be dramatic. In the case of Lake Victoria, a whole universe of cichlid fishes evolved in the last ~12 000 years in response to drying and reflooding of the lake.

Production of aerosols and trace gases by biomass burning is only significant in the tropics; south of 20° S biomass burning products constitute an insignificant fraction of the total aerosol loading in the haze layer that blankets the entire region and extends to the ~500 hPa level in the atmosphere. Industrially derived sulphate aerosols contribute significantly to the haze-layer loading, especially in summer. In the savannas of southern Africa, fluxes of trace gases from soil appear to be significantly greater than in savannas elsewhere in the world and approximately equal trace gases resulting from vegetation fires. Biogenic fluxes of NO exceed pyrogenic fluxes for a period of a few weeks following the onset of the first rains.

It is possible that global change could induce changes in the Agulhas and Benguela Currents off southern Africa that may be of significance, not only regionally, but perhaps globally. Over the past 150 000 years the carbon budget of the Benguela upwelling system has varied markedly; first estimates suggest that at present the system may be a net carbon sink.

As elsewhere changes in land use have profound effects on runoff and water quality. Moreover, hydrological regimes in southern Africa amplify inter-annual variability in rainfall. Net primary productivity appears unlikely to change much with increasing global change, however, this inference may change as trends in future precipitation become clearer. Significant loss of biodiversity is likely in the savannas, woodlands and forests and in the Cape Floral Kingdom with future changes in temperature and rainfall regimes. The effect of human depredation is certain to increase the loss unless strict conservation is practiced. Much of the savannas, woodlands and forests are likely to be converted to cropland and pasture during the 21st century. The main driver of this change is likely to be population growth coupled with unrelieved poverty, rather than climatic change. Poor land management, expansion of agricultural land into marginal areas, overstocking, accelerating urbanisation and increasing industrialisation will all contribute negatively to the environment in the region, unless timely ameliorative action is taken. Commercial agriculture and forestry will have to adapt, often at considerable cost, to the changing environmental conditions. However, not all changes will be negative and new opportunities will be presented for the enterprising to exploit.

6.2 South Asia

High-resolution Himalayan ice-core records show that the South Asian monsoon has failed repeatedly during the past millennium, catastrophically so from 1790–1796. They also show conclusively that during the twentieth century the increase in anthropogenic activity in South Asia, together with global change, is recorded by a doubling of chloride concentrations and a fourfold increase in dust at altitudes exceeding 7 000 m. The natural variability of monsoonal rainfall is marked and extremes are of particular concern to economic life in the region. Water stress, already great and growing because of population increase, may be exacerbated by climate change.

Mean annual surface air temperature is increasing at 0.4 °C per century over the region. The warming is mainly due to an increase in maximum temperatures, in contrast to many other parts of the globe where increases in minimum temperatures are responsible for much of the warming. Over the region twentieth-century surface warming appears to be amplified at higher elevations.

Although more than 20% of the global population lives in South Asia, greenhouse gas emissions at present are low (3%) as a proportion of global emissions. However, projections indicate substantial increases (six-fold by 2020) in such emissions from the region. Enhanced emissions will significantly impact the biogeochemistry and the regional radiative balance. Emission products from the region are transported over substantial distances and may have hemispheric and even global consequences.

Aerosol emissions are considerable in the region and as much as 50% arise from anthropogenic activities such as biomass burning and changing land use and land cover.

The coastal areas are experiencing severe biogeo-chemical modifications due to major human interference, including increased runoff of nutrients from the land. Eutrophication of surface water and hypoxia in bottom waters have been increasing leading to large depletion of marine life in the affected region One of the largest low-oxygen zones in the ocean develops naturally over the western Indian Continental Shelf during the late summer and autumn, and is associated with the highest accumulation of hydrogen sulphide (H_2S) and nitrous oxide (N_2O) yet observed. A global expansion of hypoxic zones may lead to an increase in marine production and emission of N_2O, which, as a potent greenhouse gas could contribute significantly to the accumulation of radiatively active trace gases in the atmosphere.

Meeting the food demands of its increasing population in the face of climate, atmospheric and land use change is a major challenge facing South Asia. Future variability of the monsoon, on which agriculture is highly dependent, is of great concern. While a rise in atmospheric CO_2 levels will raise grain yields, this may be offset by temperature increases of >2 °C. Judging by findings elsewhere, aerosol-reduced sunlight may substantially reduce agricultural yields. Changes in UV-B levels may inhibit growth in many crop species. Of particular concern is future agriculture production in the Indo-Gangetic plain, the breadbasket of South Asia where yields have begun to level off or decline at constant input levels, possibly because of changes in soil chemistry and physics. The vulnerability of the region to multiple stresses of global change is therefore a matter of profound concern.

6.3 East Asia

The overarching story of the region is one of massive human-induced change in every aspect of the environment and their consequences for the regional and global environment and human welfare.

During the past 3 centuries East Asia has had the greatest decreases of forests and grasslands of any major region in the world. Over one quarter of China is threatened by desertification caused by overgrazing, deforestation, inappropriate farming practices and other human interventions.

Changes in land use and cover have significantly altered albedo, surface roughness, leaf area index, and fractional vegetation cover over large areas of the region. The vertical and horizontal fluxes of water and energy from the surface to the atmosphere have consequently been altered to the point where they have brought about significant changes to the East Asian monsoon. The summer monsoon low-pressure system over the region has been weakened to bring about a commensurate increase in anomalous northerly flow and diminution of inland moisture transport leading to acceleration of acidification and desertification in many parts of the region. Land cover changes in East Asia have also produced demonstrable changes to regional ecosystems with significant atmospheric responses that have considerable ramifications, not only for the Asian monsoon circulation as a whole, but also for the general circulation of the atmosphere.

Although farmland area has remained relatively constant in recent years, agricultural productivity has been markedly increased through irrigation, increased inputs of fertilizers, and genetic improvements. However, the ability of the region to maintain increases in crop production is questionable, and in many areas soil fertility and water availability have declined. The loss of nitrogen, phosphorus, and potassium through erosion in China exceeds the annual amounts applied as chemical fertilizers.

The variable outflows of the great Asian rivers, increasingly affected by human activities, substantially influence the hydrology, hydrochemistry and coastal dynamics of the marginal seas of East Asia. Sea surface temperature increases have led to rising sea level; with continued warming, tropical and subtropical marine species will shift to the north. Combined temperature increases and considerable inflow of nutrients from the land due to human activities is leading to eutrophication of coastal waters and potential decreases in marine biodiversity.

6.4 Southeast Asia

Climate extremes such as floods and droughts in this region, particularly those associated with ENSO, are likely to become more frequent and more intense with global warming. For example, tropical cyclones provide a major source of rainfall, but also produce floods and coastal storm surges, and changes in their intensity, frequency, or paths of tropical cyclones would have major effects.

Most countries of the region are highly dependent on fossil fuels for electricity generation, although in rural areas fuel wood is still a very important energy source. Energy consumption grows at about ten percent per year, but the contribution of Southeast Asia to global greenhouse gas emissions has as yet been small. However, rapid growth in future emissions is likely.

Rapid growth in regional populations and economies, coupled with changing settlement patterns, may be expected to produce major environmental changes. Transboundary atmospheric pollution from fossil fuel burning, and land use/land cover changes continue to be significant. The cumulative environmental consequences of rapid growth of air and water pollution in cities and contamination of soils and water in rural areas have transformed formerly local problems into regional problems.

The environmental consequences of the large-scale exploitation of terrestrial resources in Southeast Asia for economic development are significant. Conversion of natural forests to plantation forestry and agricultural use is having significant impacts on soil resources, biogeochemical cycles, and biodiversity. However, both traditional and complex agro-economic systems provide plausible alternatives to standard high-intensity/high-input agriculture and rural development.

At both national and regional scales in lowland areas increasing demands for food are likely to be met in future by further intensification of agriculture and use of genetically modified crops. However, land available for conversion to rice production is becoming scarce in many countries, and water for irrigation is becoming limited because of multiple and growing water demands.

The intensification and commercialisation of agriculture has already had a profound effect on ecological and social systems in Southeast Asia through shifts from subsistence to cash crops; increasing use of fertilizers, pesticides and herbicides; and expansion of irrigation. Resulting pollutants are contaminating soils, surface water supplies and groundwater, and increasing river transport of fertilizer residues and sediments into coastal areas, with implications for marine life and biodiversity.

Southeast Asia is rapidly urbanising. Untreated sewerage effluent from urban centres is increasing pollution loads of the coastal zone with significant consequences for marine life and biodiversity. In the coastal zones, mangroves and wetlands are under pressure for a variety of reasons, and over 80% of coral reefs are at risk from coastal development and fishing. Fish stocks are decreasing in many areas through over exploitation.

Southeast Asia is a hot spot for both sustainable development and global change, because its economic growth has been faster than most parts of the world for several decades. Despite the overwhelming importance of urbanisation and industrialisation in transforming the economies and societies of Southeast Asia, a high proportion of the population in most countries is still dependent on agriculture and fisheries.

Many parts of Southeast Asia are now more industrialised, diversified and integrated into the global economy than their counterparts elsewhere. Large-scale agro-business operations have transformed vast areas of native forests, and some traditional smallholder agricultural areas, into monoculture plantations.

Although there are still many disparities in the size and effectiveness of different national economies within the region, these are rapidly becoming more inter-dependent through commodity trade, investment flows and exchange and division of labour. In recent years, the most important socio-economic driving force for environmental change in Southeast Asia has been the shift from closed national and regional economies to large-scale integration into the global economy. The gobalisation of trade and liberalisation of investment has helped nations bring in foreign exchange through export-oriented agriculture, forestry and manufacturing, but globalisation has progressively increased the distance between consumers and resource systems and pollution sinks and has strained existing local and national environmental regulatory mechanisms.

Untangling the domestic, regional and global interactions of various production-consumption systems for food and manufactured products is essential to understand the positive and negative implications of globalisation in the region. International institutions to handle these problems are being developed in the region and globally, but are having little effect. In any event, this synthesis has clearly shown that accelerated development and globalisation (without adequate environmental protection) are leading to an unsustainable future.

6.5 Some Final Thoughts

The decision not to follow a common methodology in the regional syntheses appears to have been justified. Varied explanations of global regional linkages in the Earth System have emerged. Two unstated contrasting underpinnings have been used in the syntheses. On the one hand, the tacit assumption has been that change results from natural variability in the system and is significantly modulated by anthropogenic activities. On the other, explanations are underpinned by the notion that change is brought about by human forcing modulated by natural variability. Both approaches, and those between, have provided fascinating insights into global-regional linkages and change. They illustrate that no one approach is necessarily correct for regional syntheses. Methodologies may differ legitimately depending on the nature of the regions themselves, their stages of development and the contributions different drivers of change are making in each case.

In looking to the future, it is clear that sustainability means different things in different regions and what is sustainable in one region is not necessarily so in another. As both regional and global change continues, surprises must be expected. Multiple stresses in the system add considerably to the difficulties of handling uncertainty and many uncertainties remain in the estimation of possible future conditions. New knowledge is required of climate variability at different time scales and of the probabilities of extreme events in future. A greater understanding of ecosystem responses to change is needed; more appreciation of the interaction in managed systems of crop production and pests and diseases with global change is urgently required. It is also clear that without a much greater understanding of social systems and institutions, and how they influence change and are impacted by it, it will never be possible to understand and predict regional-global linkages in the Earth System.

Unraveling the tangled skein of cause and effect in global-regional linkages of the Earth System is a non-trivial task that will take time. All regions are vulnerable to the global changes taking place at present. By the same token, the global system is sensitive to changes taking place in the regions. No region may be considered an island unto itself.

Regional studies add substantially to the understanding of the nature and functioning of the global-regional linkages and overall behavior of the Earth System during the Anthropocene. Integrated regional analysis is a powerful tool in global environmental change science and needs to be undertaken more frequently in future. A start has been made for four developing regions.

Acknowledgements

The research on which this book is based has been supported by many organizations. START wishes to acknowledge the following with gratitude: the Danish Ministry of Foreign Affairs, European Commission, German Federal Ministry for Education and Research, Japan Ministry of Education, Science, Sports, and Culture, the International Geosphere-Biosphere Programme, Netherlands Ministry of Foreign Affairs, Norwegian Agency for Development Co-operation, Swiss Agency for Development and Co-operation, United Nations Development Programme/Global Environment Facility, United Nations Environment Programme/Global Environment Facility, United States Global Change Research Program, United States National Aeronautics and Space Administration, United States National Science Foundation, Asia-Pacific Network for Global Change Research, European Network for Research in Global Change, John D. and Catherine T. MacArthur Foundation, United States National Oceanic and Atmospheric Administration and the Chinese Academy of Sciences.

The authors and publisher would like to thank those who provided illustrative material and are grateful to the following for permission to reproduce copyright material (for which acknowledgement and citation are made in figure captions and lists of references): **Academic Press** for Fig. 2.3c, 2.4, 2.40, 2.11b and 2.13; **ACTA Geographica Sinica** for 4.31, 4.32 and 4.33; **ACTA Meteorologica Sinica** for 4.12; **American Geophysical Union** for 2.16a, 2.21a, 2.31c, 2.21b, 2.27d, 2.28, 2.22b, 2.24, 2.27b, 2.27c, 2.29d, 3.10 and 3.16; **American Society of Limnology & Oceanography** for 3.23; **Arnold Publishers** for 2.7, 2.12, 2.11a, 2.14, 2.17d, 2.21d, 2.31a and 2.64; **Beijing Science Press** for 4.18; **Blackwell Publishers** for 2.63 and 4.36; **Cambridge University Press** for 2.16a from *The Science of Climate Change: Contribution of Working Group I to the Second Assessment Report of the IPCC* by Houghton JT et al. (1996), for 4.10 and 4.11 from *The Regional Impacts of Climate Change – An Assessment of Vulnerability: Regional Trends and Variations of Temperature and Precipitation. A Special Report of IPCC Working Group II* by Watson RT et al. (1997), for 4.19 from *IGBP Publication Series 3* by Galloway J and Melillo JM (eds) (1998) and for 5.2 from *Global Change, Regional Response: the New International Context of Development* by Stallings B (1995); **China Ocean Press** for 4.6 and 4.7 from *Climatic Change in the Historical Time, Advances in Global Change Studies of China* by Liu TS et al. (1998); **Chinese Science Bulletin** for 4.35, 4.37, 4.38 and 4.39; **China Science Press** for 4.34 *from Quaternary Geology, Global Change: A Preliminary Study on the Paleoenvironment Changes of China During last 2 000 years* by An ZS et al. (1990); **China Statistics Bureau** for 4.27; **CSIRO Publishing** for 2.29c; **Curzon Press** for 5.10 from *Land use changes in Eastern Sabah* by Sutton K and McMorrow J (1998); **Elsevier Science** for 2.5, 2.32a, 2.32b, 2.36b, 2.65, 3.21, 3.26, 4.20 and for 2.51 from *Global Change Scenarios of the 21st Century, Results from the IMAGE 2.1 Model* by Alcamo J et al. (1998); **FAO** for 4.24, 4.26, 4.28 and 4.29; the **Finnish Environment Institute** for 2.19; **Gordon and Beach** for 2.8c from *The Limnology, Climatology and Palaeoclimatology of the East African Lakes* by Johnson TC and Odada E (eds) (1996); **Institute for Southeast Asian Studies** for 5.3; **International Rice Research Institute** for 5.15; the **IUGG** for 2.31b; **John Wiley & Sons** for 2.17c, 3.11, 3.12, 3.14 and 3.15 and for 4.8 from *Climate Biosphere Interaction: Biogenic emissions and environmental effects of climate change* by Zepp, RG (1994); **Kluwer Academic Publishers** for 2.6b, 2.15, 3.13, 4.30 and 5.14 and for 2.10b from *Environmental Change and Response in East African* Lakes by Lehman JT (ed) (1998) and 2.42 from *Ocean Circulation Models: Combining Dynamics and Data* by Anderson DLT and Wildebrand J (eds) (1994); **Manohar Publishers** for 3.3a and 3.3b from *India's Engergy: Future Energy Trends and Carbon Mitigation Strategies for India* by Shukla PR and Grare F (eds) (2000); **Japan Journal of Public Health** for 4.17; **Journal of Marine Research** for 2.39; **Maik Nauka Interperiodica Publishing** for 4.21a, 4.21b, 4.21d and 4.21c; **Mausam Journal** for 3.18 and 3.20; **National Academy of Sciences, U.S.A.** for 3.4; **MIT Press** for 2.17a (with modification) from *The General Circulation of the Tropical Atmosphere and Interactions with Extratropical Latitudes* by Newell RE et al. (1972); **Narosa Publishing House** for 3.24 and 3.25; **NASA** for 2.30, 3.7, 4.14, and 5.8; **Nature** for 2.3a from *Climate and atmospheric history of the past 420 000 years from the Vostok ice core, Antarctic* by Petit JR et al. *Nature* 399: 429–436 (1999), for 2.8b from *Rainfall and drought in equatorial east Africa* *during the past 1,100 years* by Verschuren D et al. *Nature* 403:410–414 (2000), for 3.22 from *Budgetary and biogeochemical implications of N_2O isotope signatures in the Arabian Sea* by Naqvi SWA et al. *Nature* 394: 462–464 (1998), Macmillan Magazines Ltd.; **The Oceanography Society** for 2.37; **Oxford University Press** for 2.18 from *The Cenozoic of Southern Africa* by Partridge TC and Maud RR (eds) (2000) and for 5.21 from *Human Development Report 1999* by United Nations Development Programme (1999); **Population Action International** for 5.6 from *Countries' Annual Per Capita Carbon Dioxide Emissions from Fossil Fuel Combustion and Cement Production, 1950–1995* first appeared in *Profiles in Carbon: an Update on Population, Consumption and Carbon Dioxide Emissions* by Engelman R (1998); **Routledge Publishing** for 2.16b from *Drought, Volume 1, A Global Assessment* by Whilhite DA (ed) (2000); **Royal Society of New Zealand** for 2.29b; **Royal Swedish Academy of Sciences** for 2.26 and 2.43; **Science** for 2.8a, 2.9 and 2.10a from *Late Pleistocene desiccation of Lake Victoria and rapid evolution of cichlid fishes* by Johnson TC et al. (1996), for 3.6, 3.9a and 3.9b from *The Indian Ocean Experiment: widespread air pollution from South and Southeast Asia* by Lelieveld J et al. (2001), for 3.17 from *On the weakening relationship between the Indian monsoon and ENSO* by Krishna-Kumar K et al. (1999) copyright American Association for the Advancement of Science; **South African Journal of Marine Science** for 2.41; **South African Journal of Science** for 2.20a, 2.20b, 2.27a, 2.32c, 2.33, 2.34a, 2.34b, 2.34c, and 2.52; **Southeast Asia Regional Committee of START** for 5.9 and 5.20; **Swets & Zeitlinger Publishers** for 2.6a; **Tellus** for 2.22a; **UNESCO** for 4.23; **UNEP** for 4.13; **World Aquaculture** for 5.19; **World Scientific Publishing Company** for 4.4 and 4.5; **World Meteorological Organization** for 2.2.

Every effort has been made to trace and acknowledge copyright holders. Should any infringements have occurred, apologies are tended and omissions will be rectified in the event of a reprint of the book.

Index

C

D

E

N

O

P

X

Y

Z